BIOLOGY OF THE ARTERIAL WALL

BASIC SCIENCE FOR THE CARDIOLOGIST

1. B. Levy, A. Tedgui (eds.): *Biology of the Arterial Wall*
   1999 ISBN 0-7923-8458-X

KLUWER ACADEMIC PUBLISHERS - DORDRECHT/BOSTON/LONDON

Biology of the Arterial Wall

edited by

Bernard I. Levy

and

Alain Tedgui

INSERM U141
Lariboisière hospital
Paris, France

Distributors for North, Central and South America:
Kluwer Academic Publishers
101 Philip Drive
Assinippi Park
Norwell, Massachusetts 02061 USA

Distributors for all other countries:
Kluwer Academic Publishers Group
Distribution Centre
Post Office Box 322
3300 AH Dordrecht, THE NETHERLANDS

Library of Congress Cataloging-in-Publication Data

Biology of the arterial wall / edited by Bernard I. Levy and Alain Tedgui.
p. cm. -- (Basic science for the cardiologist ; 1)
Includes index.
ISBN 0-7923-8458-X
1. Arteries--Physiology. 2. Arteries--Pathophysiology. I. Lévy, Bernard I. II. Tedgui, Alain. III. Series.
[DNLM: 1. Arteries--physiology. 2. Arteries--pathology. WG 510 B615 1999]
QP106.2.B56 1999
612. 1'33--dc21
DNLM/DLC 98-55059
for Library of Congress CIP

Printed on acid-free paper.

Printed in the United States of America

Dedicated to Jacques Maclouf

Contents

List of Contributors

Joël BELMIN, M.D.
Department of Gerontology
Hôpital René-Bigotini
Sevran, France
Vascular Aging

Catherine BERNARD, M.D., Ph.D.
Department of Anesthesiology
Hôpital Lariboisière
Paris, France
Vessel and Inflammation
Atherosclerosis

Chantal BOULANGER, Ph.D.
INSERM U141
Hôpital Lariboisière
Paris, France
Endothelial Function and Dysfunction

Cedric J. GAULTIER, M.D.
INSERM U460 Remodelage cardiovasculaire
Faculté de Médecine X. Bichat
Paris, France
Angiogenic Growth Factors

Laurent J. FELDMAN, M.D., Ph.D.
Cardiology Department
Hôpital Bichat
Paris, France
Arterial Gene Tansfer

Stéphane LAURENT, M.D., Ph.D.
Department of Pharmacology
Hôpital Broussais
Paris, France
Neurohumoral Control of the Vascular System

Stéphanie LEHOUX, Ph.D.
INSERM U141
Hôpital Lariboisière
Paris, France
Mechanical Factors and Vascular Biology

Bernard I. LEVY, M.D., Ph.D.
INSERM U141
Hôpital Lariboisière
Paris, France
Morphologic Aspects of the Large Artery Vascular Wall
Mechanics of the Large Artery Vascular Wall
Mechanical Factors and Vascular Biology

Ferdinand A.C. le NOBLE, Ph.D.
Department of Pharmacology, Cardiovascular Research Institute
Maastricht (CARIM)
Maastricht, The Netherlands
Angiogenesis

Ziad MALLAT, M.D.
INSERM U141
Hôpital Lariboisière
Paris, France
Apoptosis in the Vascular System
Atherosclerosis

Jean-Baptiste MICHEL, M.D., Ph.D.
INSERM U460 Remodelage cardiovasculaire
Faculté de Médecine X. Bichat
Paris, France
Angiogenic Growth Factors

Michel SAFAR, M.D.
Médecine Interne 1
Hôpital Broussais
Paris, France
Stiffness of Wall Material in Human Hypertension

Frank R.M. STASSEN, Ph.D.
Department of Pharmacology, Cardiovascular Research Institute
Maastricht (CARIM)
Maastricht, The Netherlands
Angiogenesis

Ph. Gabriel STEG, M.D., Ph.D.
Cardiology Department
Hôpital Bichat
Paris, France
Arterial Gene Tansfer

Harry STRUIJKER-BOUDIER, Ph.D.
Department of Pharmacology, Cardiovascular Research Institute
Maastricht (CARIM)
Maastricht, The Netherlands
Angiogenesis

Alain TEDGUI, Ph.D.
INSERM U141
Hôpital Lariboisière
Paris, France
Morphologic Aspects of the Large Artery Vascular Wall
Mechanical Factors and Vascular Biology
Apoptosis in normal and pathological vessels Atherosclerosis

Paul VANHOUTTE, M.D., Ph.D.
IRIS
Courbevoie , France
Endothelial Function and Dysfunction

Preface

Adequate function of the arterial tree depends on the maintenance of lumen patencies and blood pressure levels which assure continuous blood flow to the organs and tissues of the body. In advanced societies, disease of the blood vessel wall is responsible for a great deal of mortality and morbidity. Each organ functions in relation to a set of hemodynamic variables and regulatory mechanisms peculiar to its role in the integration of the entire organism. With access to the same genome, artery cells of each individual collaborate to differentiate into a set of vessels of precisely the appropriate length, diameter, structure and composition, organ development, to immediate, short-term, reversible metabolic and physical variations and to long-term persistent changes in these factors. Metabolic or physical stresses which may result in cell dysfunction and apoptosis/necrosis or in tissue disruption and disorganization nevertheless induce adaptive or healing processes which tend to preserve adequate function or at least to postpone total loss of integrity. Besides modifications in tissue response and structure which have been relegated to the non-specific waste-basket called aging, reactions are also modulated by structural changes which can be related to specific metabolic and physical stresses. Thus, changes in mural structure and composition which at one stage seems to be adaptive may in the long-term cause modifications which inhibit access to the cells of conditioning stimuli and enabling metabolites. For arteries, as for other organs and tissues, our knowledge of both physiology and pathology is increased by exploration of the boundaries among homeostatic interactions between genome and micro- environment, reactions which induce abnormal but stable configurations and those which result in irreversible, destructive processes.

This book is therefore intended as a general reference text concerned with the *biology of the vascular cells and the blood vessel wall under physiological and pathological conditions*. The early chapters describe ***the structure of the vessels and their mechanical properties***. One the major functions of the arteries is to maintain a continuous blood flow to the organs whatever the pressure conditions thanks to the vasomotor tone of the smooth muscle cells. A specific chapter deals with the *neuro-humoral control of the vascular system*. Great advances have been made for the last decade in the understanding of the endothelial cells as integrators and transducers of signals originating from the the blood stream. The pluripotent control functions of the endothelial cells in vessel wall is now well recognized. *A review of endothelial functions and dysfunctions follows.*

Structural adaptive responses of the artery wall are normally elicited by two principle mechanical factors, wall shear stress and tensile stress. Wall shear is a frictional deforming force at the blood-endothelium interface which depends on the gradient of near-wall blood flow velocities. The magnitude of wall shear stress is related directly to flow rate and blood viscosity and inversely to the third power of the vessel radius. Changes in flow engender changes in radius which stabilize when baseline wall shear stress is restored. Wall tension, the force tending to stretch the wall, is determined by distending pressure and also by the radius. Thus, a change in radius in response to a change in wall shear stress results in a change in wall tension. Corresponding modifications in thickness and/or composition assure stability of the wall. Changes in distending pressure or of effective radius at geometric transitions such as branches, bifurcations and bends result in local redistributions of tensile stresses and also elicit changes in wall thickness and composition. These modifications obviously require responses by the component cells of the artery wall. The endothelial cells lining the lumen are directly exposed to the wall shear stress. Communication between the endothelial cells and the smooth muscle cells of the media is essential if wall shear stress is to result in vessel enlargement, immediate or gradual, and the restoration of baseline levels. Furthermore, the response to altered tensile stress, whether due to enlargement in relation to wall shear or to altered pressure or geometry, requires that the cells of the media enlarge and/or proliferate and/or adjust wall matrix composition and structure in appropriate proportions to assure stability. The terms modeling and remodeling summarize the results of these putative compensatory processes at the tissue level. A review of *the role of mechanical factors in vascular remodeling*, including discussion of mechanotransduction in the vascular cells is proposed.

Cell biology and molecular genetic studies have now identified an array of molecules, elaborated by endothelial cells and vascular smooth muscle cells and by the blood born elements which interact with artery cells, defending the artery against injury and modulating evolving abnormal processes. These include leukocytes and platelets as well as cells differentiated into macrophages. Molecules which induce or inhibit smooth muscle cell proliferation and those which mediate macrophage adhesion and ingress are currently under great scrutiny. Developments in this field have been explosive. A chapter on *growth factors and vascular biology, including differentiation and phenotype modulation of smooth muscle cells* is discussed and followed by a chapter on *vasculogenesis*, essential in the normal development of the arterial tree, and *angiogenesis* which plays a major role in tumor growth but may be also beneficial as a healing process in muscle ischemia.

It is increasingly evident that the adjustments of the blood vessel wall are made in the presence of deforming disease processes such as hypertension and arteriosclerosis and in the presence of pathogenic agents presented to the endothelium from the bloodstream. The second part of this book is concerned with the blood vessel wall in diseased conditions. Several chapters review the role of *the vessel and vascular cells in inflammation, the vascular remodeling during arterial hypertension* and *aging*. A specific chapter is concerned with atherogenesis, atheroma and plaque *unstability, followed with the pathophysiology of post-angioplasty restonosis*, which is a crucial issue in modern interventional cardiology.

The panoply of possible and probable interaction is staggering. We are only at the threshold of assigning to each its proper place in the network of control and communication systems which regulate artery wall structure and function and its relationships to disease.

This book is aimed to provide clinicians and researchers involved in the field of vascular biology with a concise review of the principal factors which could participate in the induction and modulation of several common arterial disease states.

Bernard I. LEVY and Alain TEDGUI

I. The Vascular Wall under Physiological Conditions

I. MORPHOLOGIC ASPECTS OF THE LARGE ARTERY VASCULAR WALL

Bernard I. Levy, Alain Tedgui
INSERM U 141, Lariboisière Hospital, Paris, France

All arteries show a common pattern of organization and are made up of similar materials, though the proportions vary in different parts of the circulation.

The arterial walls are well-organized connective tissue structures composed of cells and matrix fibers arranged in three tunicae: – the intima, the media and the adventitia.

Intima

The innermost subluminal layer consists of endothelium and a variable quantity of underlying cells and matrix elements, constituting, from the lumen to the outer part of the arterial wall : the basement membrane, the sub-endothelial layer and the internal elastic lamina (Figure 1).

A continuous monolayer of polygonal flat cells (0.3 to 0.5 µm), the endothelium, lines the luminal surface of all arteries. The endothelium layer extends as a continuous lining right through the circulation, covering all the surfaces which come in contact with the blood-arteries, capillaries, veins, heart valves, and endocardial surfaces.

Endothelium

The endothelium is continuously submitted to shearing forces related to blood flow and exposed to circulating cells and plasma components. In relation to shear stress, endothelial cells tends to be elongated (± 100 µm) in the direction of blood flow, particularly where the latter is rapid, laminar and unidirectional (Figure 2). Where flow is slow, complex, turbulent, or nearly stagnant, the endothelial cells are less distinctly elongated or oriented. In vitro experiments confirmed that cultured endothelial cells are oriented and elongated in the direction of the shear stresses [1]. The luminal surface of the endothelial cells, largely smooth and regular [2], is covered by a glycoprotein coat constituting the glycocalix responsible of the anti-thrombogenic properties of the endothelial surface.

The endothelial cells show many typical organelles, including mitochondria, microtubules and microfilaments ; the presence of abundant pinocytotic vesicles is in relation with the regulation of the permeability of the endothelial layer (Figure 3). Pynocitic vesicles are responsible for the movement of material from the lumen of the vessel into its wall.

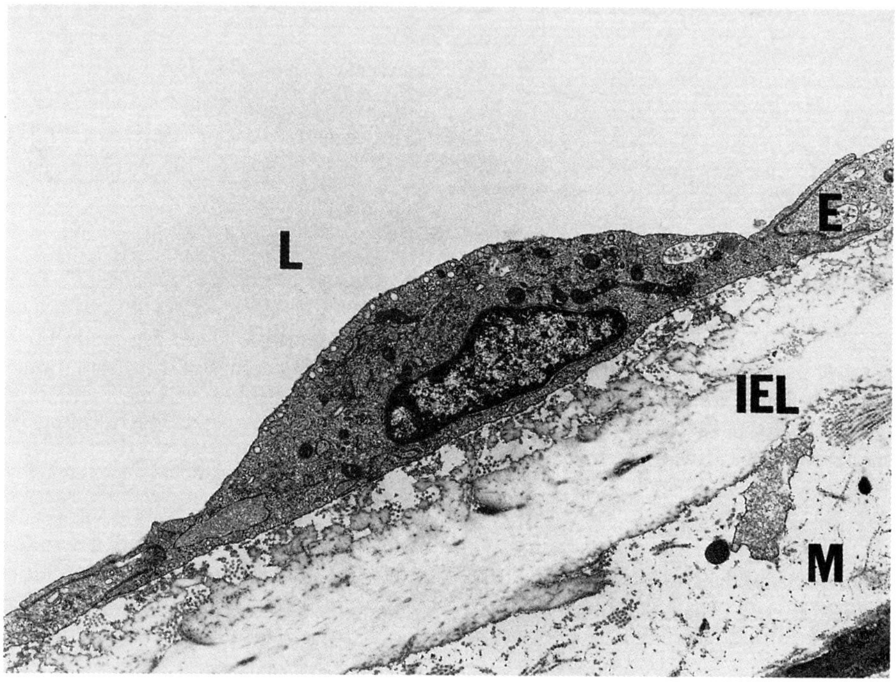

Figure 1. *Transmission electron microscopic view of a transverse section of the rabbit thoracic aorta showing the intima. L: lumen, E: endothelium, IEL: internal elastic lamina, M: media. Magn. x8000*

In general, the edges of the cells overlap and each one tends to override its immediately adjacent neighbors. The nature of the bonding between adjacent endothelial cells deserves consideration because of its relevance to both the mechanical strength and the permeability of the endothelial layer. The membrane of adjacent cells are mainly parallel and separated by an intercellular space of approximately 15-20 nm. The contents of the gap, or intercellular cleft, consist of glycosaminoglycans. There is in addition localized sites of firmer attachment, the junctional complexes : the tight junctions and the gap junctions or nexus.

Tight (occluding) junctions are frequent between adjacent cells; at these sites, the opposing cell membranes form a continuous band around the side of the cells, forming a *zonula occludens* providing a tight seal which prevents the passage of molecules between the lumen and cell base via the cleft. Gap (communicating) junctions are less frequent and are most evident in large arteries [3]. After cryofracture, the nexus represents a plaque containing 8 nm particles constituted of connexin and grouped around a central canal [4]. The main role of the gap junctions is to allow the inter-endothelial cells communication by movements of ions, metabolites and regulatory factors.

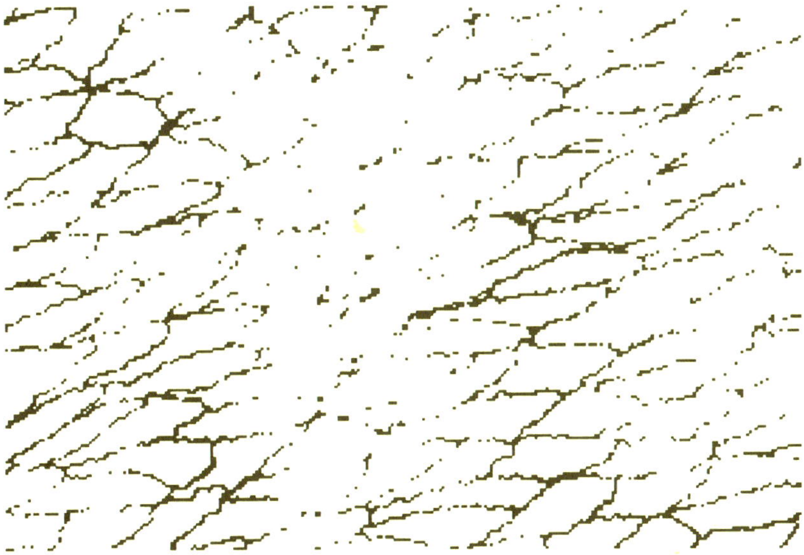

***Figure** 2. En face view of the endothelium. Endothelial junctions are stained with silver nitrate. Endothelial cells are oriented in the sense of flow (from left to right).*

Endothelial cells contain myofibrils and are able to constrict when submitted to catecholamines, histamine, serotonin, and angiotensin II. The constriction of the endothelial cells leads to an increase in the size of the gaps in the intercellular space and therefore in an increase in endothelial permeability [5].
It has been thought for a long time that endothelial cells, like neuronal cells, were unable to replicate; however autoradiographic methods evidenced a very low replication rate : 1 to 2 over 1000 cells per day in the rat thoracic aorta. In other words, the endothelium layer is completely renewed in 2-3 years [6]. In some arteries where the endothelial layer is submitted to high shear stress levels, as in the angle of bifurcations, the rate of renewal of endothelial cells is higher and can reach 10 cells over 100 per day. This endothelium regeneration occurs without denudation of the basal lamina; new cells grow around a senescent cell and move under it to recover the basal lamina, and reconstitute the endothelial layer integrity. This process corresponds to a non "denuding" desquamation of endothelial cells [7].

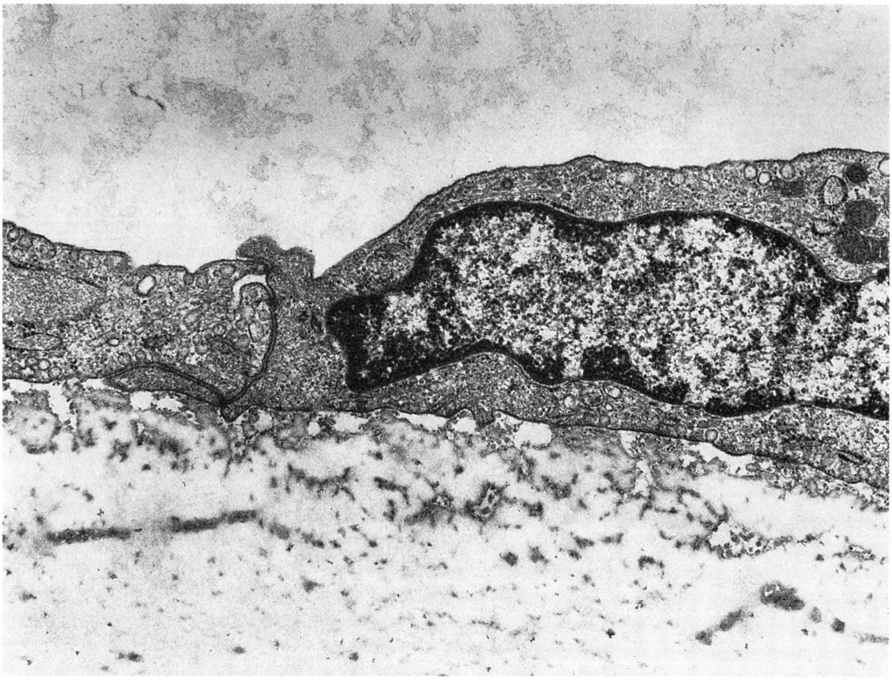

Figure 3*. Ultrastructure of the endothelial cell. Membrane vesicles on the basal and apical surfaces, as well as in the cytoplasm, are seen.(Magn. x15 000)*

For a long time, the endothelial layer was viewed as essentially a permeability barrier and an anti-thrombogenic surface. It is estimated that the endothelium allows only between 1 to 10 % of the intravascular proteins to penetrate into the vessel wall. In addition, the endothelial cell appears to be a site of synthesis of a number of important materials. Included are the prostaglandins and especially prostacyclin, factor VIII antigen, fibronectin, and histamine. Endothelium also participates in the metabolism of chylomicrons through the action of lipoprotein lipase, and endothelial cells can process a number of vasoactive factors such as bradykinin, serotonin, and norepinephrine. The major vasoactive role of the endothelial cell is the synthesis and release of nitric oxide, a soluble gas identified to be the major endothelium-derived-relaxing factor synthesized from arginine and endothelial nitric oxide synthase (see chapter *Endothelial Function and Dysfunction*). In addition the endothelial cells convert angiotensin I into angiotensin II by means of the angiotensin converting enzyme, an endo-ecto enzyme synthesized by the endothelial cell. The endothelial cell also has unique immunologic characteristics, possessing ABO antigens and the factor VIII antigen, the α2-macroglobulin antigen, and the tissue factor antigen. The wide diversity of antigens found on the endothelial surface may explain the susceptibility of the cell to immunologic injury. Therefore, it appears that the endothelial cell not only serves a multiplicity of primary vascular functions but it is also a highly diversified synthetic cell, the function of which may markedly influence the response of vessels to a number of agents or stimuli

Basal lamina

A basal lamina, which borders the endothelial cells on their abluminal surface forms a more or less continuous boundary between the endothelium and the immediately underlying intimal structure. It is formed of microfibrils of collagen (diameter ± 5 nm) and glycoproteins. Although multilayered and complex in some locations, each basal lamina layer consists of two zones when viewed in electron microscopy: the inner zone is clear, whereas the outer finely fibrillar zone is denser. The basal lamina seems to provide a pliable extensive bond which permits endothelial cells to comply with changes in configuration related to cardiac pulsation, and vessel torsion and flexion. In addition, discrete points of increased electron density and attachment have been noted between basal peripheral cytoplasmic dense bodies of endothelial cells and the internal elastic lamina. These attachments probably prevents slippage, buckling or excessive overriding of the lining cells when shearing stresses increase rapidly [8]. The basal lamina also plays the role of a support and a guide to allow regeneration of endothelial cells.

Although the *intima* is defined as the tissue which extends from the lumen to the media, the endothelium and basal lamina are, in many locations, directly applied to the internal elastic lamina of the media and therefore comprise the entire intima. Where cells and matrix fibers are between endothelium and elastic internal lamina, the intima is thicker than the endothelial layer and in some sites is as thick as the media. Intimal cells are predominantly smooth muscle cells; matrix contains elastic and collagen fibers, and proteoglycans.

Internal elastic lamina

The internal elastic lamina is a layer of elastic fibers (thickness of 70 to 100 nm). It is undulated in immersion-fixed histologic preparations but straightens to form a circular band on transverse sections when intraluminal pressure is maintained at normal levels during fixation. The internal elastic lamina is focally interrupted by gaps which correspond to fenestration.

Media

The media is constituted principally of smooth muscle cells, elastic fibers, and collagen fibers. Arteries are classified as elastic or muscular types according to the relative proportions of these cellular and fibrous components found in the media [9]. In elastic arteries, matrix fibers, in the form of well-defined elastic lamellae and collagen bundles, are abundant and prominent in the media. The conducting vessels of relatively large diameter in close proximity to the heart (such as aorta, brachiocephalic trunk, iliac an the main pulmonary arteries) are examples of elastic arteries.

Most of the muscular arteries arise as second or third order branches of the elastic arteries. The media of muscular arteries contains fewer connective tissue fibers than that of elastic arteries; smooth muscle cells being the predominant component. Except for the internal elastic lamina, which is prominent in muscular arteries, elastin is not organized in parallel lamellae but appears in the form of branching strands. The predominant muscular composition of these vessels corresponds to greater capacity to change diameter actively under the influence of neurohumoral stimulation.

Small arteries, i.e. those which comprise the distal subdivisions of the large muscular arteries, contains the lowest relative proportion of medial fibrous connective tissue. Such vessels change diameter markedly in response to stimuli and are with the arterioles the principal regulator of peripheral resistance. Although nervous influences have an important role in regulating vascular resistance in the peripheral circulation, moment to moment regulation of blood flow in most organs is primarily mediated by local control mechanisms which are independent of the nervous system. These local control mechanisms maintain the tissue blood flow constant by changing microvascular resistance in response to changes in perfusion pressure. Myogenic auto regulatory mechanisms are pressure-sensitive and produce arteriolar constriction in response to an elevated transmural pressure in the vessel : increases in blood pressure, at levels higher than 60-75 mmHg, do not result in increasing lumen size but in diameter diminution. This paradoxical phenomenon is due to the myogenic tone, specific to resistance arteries and appearing for diameters ranging from 300 to 50 μm, depending of specie and localization [10].

The relative proportions of elastin and collagen in the aortic media change with distance from the heart and also vary from vessel to vessel. The thoracic aorta contains a greater quantity of elastin than collagen, whereas the abdominal aorta has more collagen than elastin. Yet, the sum of the collagen and elastin along the aorta does not change with distance from the heart.

In elastic arteries, the mechanical function of the media is attributable to specific modes of interconnections of its structural elements allowing the constitution of "lammelar units" [9, 12] (Figure 4) : between two elastic lamellae, smooth muscle cells, collagen fibers, elastin microfibrils, and glycosaminoglycans of the fundamental substance (chondroitin-sulfates, heparane-sulfates, dermatane-sulfate, hyaluronic acid) are organized in functional units. The number of lamellar units in a given artery depends on the diameter of this artery. In humans, there are 40 to 60 units in the aorta, from new born to adults, 20 units in the rabbit, and 8 in the rat thoracic aorta. The number of lamellar units decreases from the heart to peripheral arteries; however, the distance between two elastic lamellae is roughly constant (12-17 μm) whatever the artery and the animal species.

The electron microscopy study of normally distended arteries reveals a further structural subdivision of arterial microarchitecture common to elastic and muscular arteries. Medias are composed of groups of fascicles of commonly oriented, elongated, smooth muscle cells lying mainly within tangential planes. Each fascicle is invested by a sheath of basal lamina and collagen fibrils, closely associated with a corresponding system of elastic fibers oriented in the same direction as the fascicle cells (Figure 5). Differences among arteries reside principally in the number, size and composition of these musculo-elastic structural groups. Although medias of muscular arteries are also composed of such musculoelastic fascicles, the associated matrix fiber systems are less prominent than in the aorta. On the basis of anatomic appearances, the size and alignment of individual fascicles appear to correspond to local differences in amplitude and resultant direction of tensile stress, respectively.

A

B

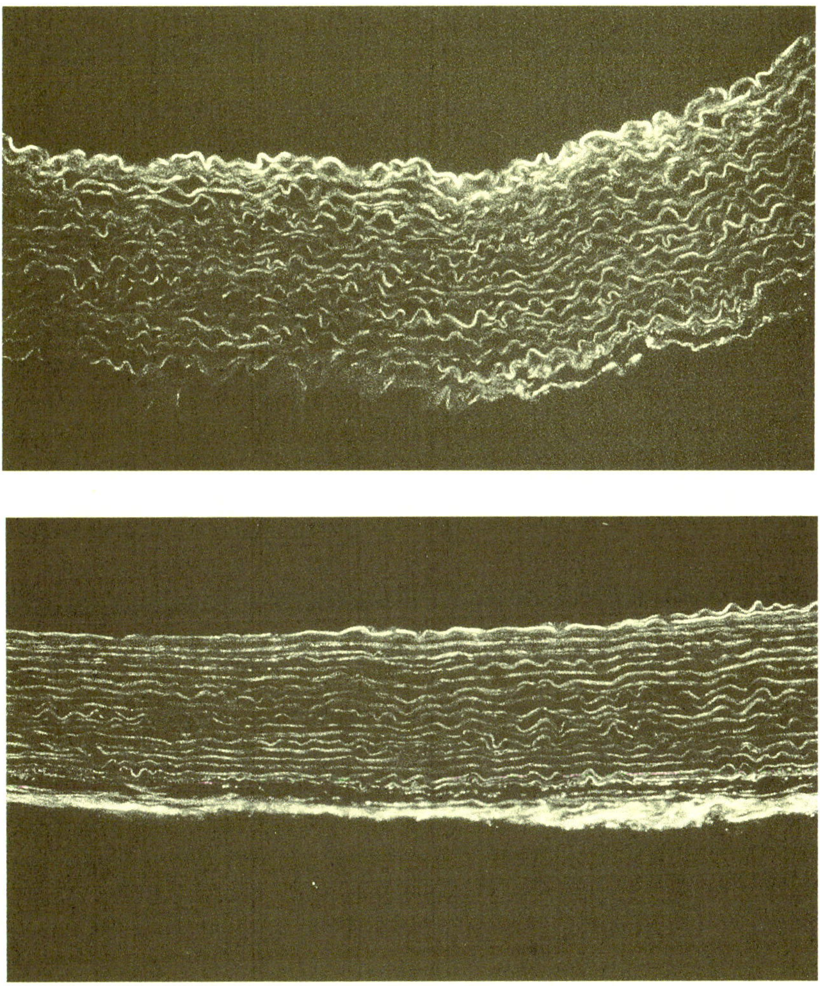

***Figure 4**. Lamellar unit· structure of elastic arteries under relaxed (A) and pressurized (B) conditions. Lumen is at the top and adventitia at the bottom.*

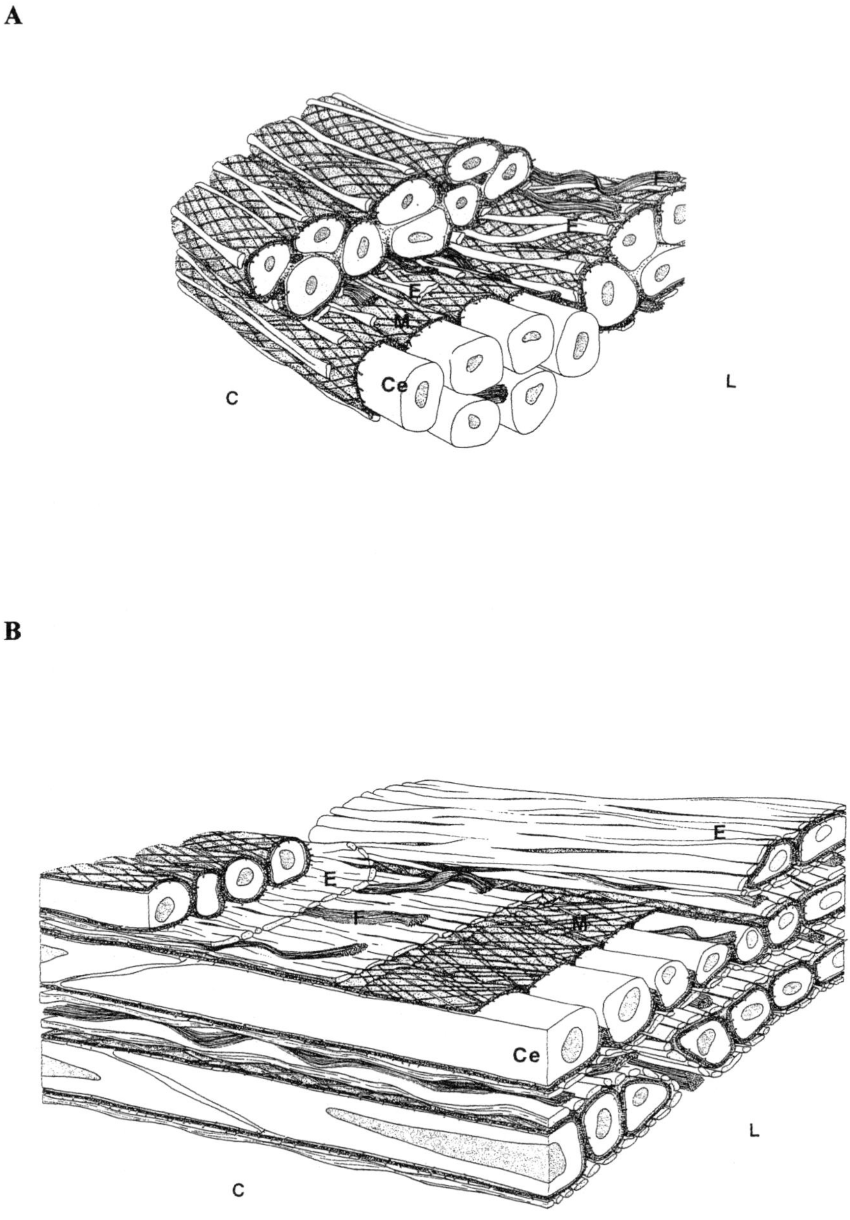

***Figure 5**. Representation of the organization in musculo-elastic fascicles of smooth muscle cells and fibers in mammalian elastic (A) and muscular (B) arteries. C: Transverse circumferential plane of section, L: lumen, E: elastic fibers, Ce: smooth muscle cell. (from Clark and Glagov [12])*

Adventitia

The external elastic lamina is the inner limit of the adventitia; in contrast, the outer limit of the adventitia is often difficult to define : it is usually contiguous with the perivascular connective tissue. The aorta has relatively slight adventitial condensation of fibrous connective tissue whereas the adventitia of the large muscular arteries consists of prominent elastic and collagen fibers in well-organized layers. In these locations, adventitia may be thicker than the media. However, adventitial cells are sparse and mainly fibroblasts; the adventitia also contains few thick elastic fibers. The adventitia contains vasa vasorum and nerves, the former providing nutrition to the adventitia and media; the latter contributing to the regulation of medial smooth muscle function. The vasomotor nerve fibers induce either vasocontriction of the vessel via the a adrenergic receptors, or vasodilatation by activation of b receptors. The nervous stimuli are transmitted to the outer layers of smooth muscle cells through neuro-muscular junctions; the excitation is then transmitted to the inner cells by electrical coupling between adjacent smooth muscle cells.

Vasa vasorum exist only in arteries with diameter larger than 200 μm; when the media layer contains more than 29 lamellar units, the outer part of the media is irrigated by the vasa vasorum. Thus, the oxygen and metabolites supply of the smooth muscle cells are provided both from the luminal blood flow (for the inner part of the media) and from the and vasa vasorum (for the outer layers of the media).

A lymphatic network contained in the adventitia collects the proteins, ions, soluble substances and water coming from the blood and transported through the vessel wall.

REFERENCES

1. Levesque MJ, Nerem RM. Elongation and orientation of cultured endothelial cells in response to shear. J Biomech Eng, 107:341-347, 1985.

2. Clark JM, Glagov S. Luminal surface of distended areries, eliminating configurational and technical artefacts. Br J Exp Pathol. 1976; 57: 129-135.

3. Simionescu M, Simionescu V, Palade GE. Segmental differentiations of cell junctions in the vascular endothelium Arteries and veins. J Cell Biol 1976; 68: 705-723

4. Rhodin JAG. Architecture of the vessel wall. In : Handbook of Physiology. The Cardiovascular System, vol II. Vascular Smooth Muscle, edited by Bohr D, Somlyo A, Sparks HV Jr. American Physiological Society, Bethesta, Maryland, 1980, p 1-31.

5. Tedgui A, Chiron B, Curmi PA, Juan L. The effect of nicardipine and verapamil on in vitro albumin transport in the rabbit thoracic aorta. Arteriosclerosis, 1987, 7:80-87.

6. Schwartz SM, Gajdusek CM, Reidy MA, Selden SC III, Haudenschild CC. Maintenance of integrity in aortic endothelium. Fed Proc 1980; 39: 2618-2625.

7. Schwartz SM. Dynamic maintenance of the endothelium. In : The Endothelial Cell. A Pluripotent Control of the Vessel Wall, edited by Thilo-Körner DGS, Freshny RI. Karger, Basel, 1983, p 113-125.

8. Ts'ao CH, Glagov S. Basal endothelial attachment: Tenacity at cytoplasmic dense zone in the rabbit aorta. Lab Invest. 1970; 23: 510-516.

9. Wolinsky H, Glacov S. A lamellar unit of aortic medial structure and function in mammals. Circ Res, 1967.; 20: 99-111.

10. Bevan JA, Garcia-Roldan JL, Joyce EH. Resistance artery tone is influenced independently by pressure and by flow. Blood Vessels 1990;27:202-207.

11. Feldman SA, Glagov S. Transmedial collagen and elastin gradients in human aortas: reversal with age. Atherosclerosis 1971;13:385-394.

12. Clark JM, Glacov S. Transmural organization of the arterial media. The lamellar unit revisited. Arteriosclerosis 1985; 5: 19-34.

2. MECHANICS OF THE LARGE ARTERY VASCULAR WALL

Bernard I. Levy
INSERM U141, Lariboisière Hospital, Paris, France

The aorta and large arteries are generally thought of as conduit vessels whose main function is to provide a conduit for blood flow to reach the peripheral tissues. However, because the pressure and flow curves are not a simple ratio, it has long been recognized that the cardiovascular system functions in more complex fashion than merely a simple resistance to blood flow. Blood pressure is highest at the beginning of the systemic circulation; the decrease of blood pressure is not linear with vessel diameter or distance in the vascular tree. Blood pressure decrease ranges from 30 to 40 % of the aortic pressure in vessels down from 250 to 50 µm in diameter [1-3] while most of the pressure drop occurs in the terminal arterioles with diameters smaller than 100 µm and which branch into numerous small capillaries. The site of the largest pressure drop may differ between tissues; however, in vessels smaller than 60 µm, no correlation has been found between the central arterial pressure and microvascular pressure which suggests that perfusion pressure is being controlled in these blood vessels and those with lower diameter [4].

Arterial blood flow is actually determined by blood pressure as well as by impedance of the vascular system. The latter is dependent on static and dynamic pressure-volume relations of blood vessels. Thus, any realistic representation of vascular properties must account for the visco-elastic properties of the vasculature [5].

PARAMETERS IN VESSEL WALL MECHANICS

Forces and stresses

Forces in the wall of a cylindrical tube can be oriented in three directions: circumferential, radial and longitudinal (Figure 1). Since the blood vessel may be considered to be incompressible [6, 7], dimension changes in one direction will result in dimension changes in at least one other direction. Moreover, arteries are anisotropic, i.e. stiffness is dependent on direction. It has been shown that vessel length increases when a vessel is inflated, but that this amount depends on the vessel type. Most of our knowledge about the anisotropy of arteries has been derived from studies on aortas [8] and carotid arteries [6]. In carotid arteries, longitudinal stiffness is about half of the tangential stiffness [6, 9].

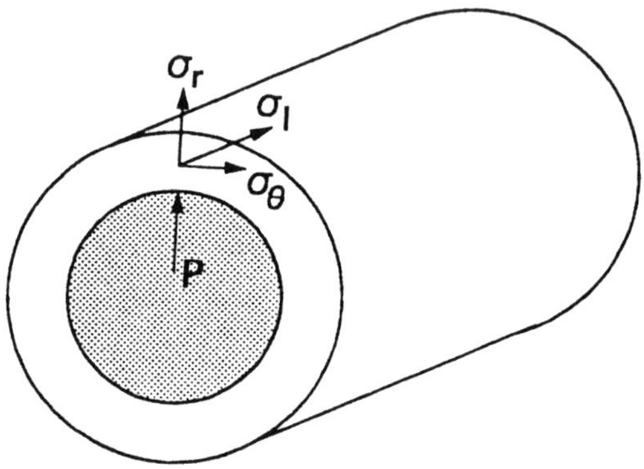

***Figure 1**. Pressurization of blood vessels results in stress development in three different directions: tangential (σ_θ), radial (σ_r), and longitudinal (σ_l).*

The "transverse anisotropy" was related to vascular tethering and to the distending pressure : arteries shorten as soon as they are released from their attachments in the body. Actually, most of the longitudinal distention of an arterial segment is related to the transmural pressure. When a segment of artery is submitted, in vitro, to increases in transmural pressure, its length increases by 10 to 20 % for pressure varying from 0 to 50 mmHg (Figure 2).
The longitudinal distention is less marked in older individuals, from 40% at age 12 to 12% or less after age 60 [10]. Recent results confirmed these results in rats' carotid arteries evidencing that aging alters arterial wall anisotropy [11].

The circumferential force developed in a tube is quantified by the tension T, which is calculated according to the Laplace's law [12]:

$$T = P\,r \qquad [N.m^{-1}]$$

where P [N.m $^{-2}$] is the transmural pressure and r [m,]the internal radius.

Stress (σ) refers to forces developed per unit area and is calculated as

$$\sigma = T/h \qquad [N.m^{-2}]$$

where h [m] is the wall thickness.

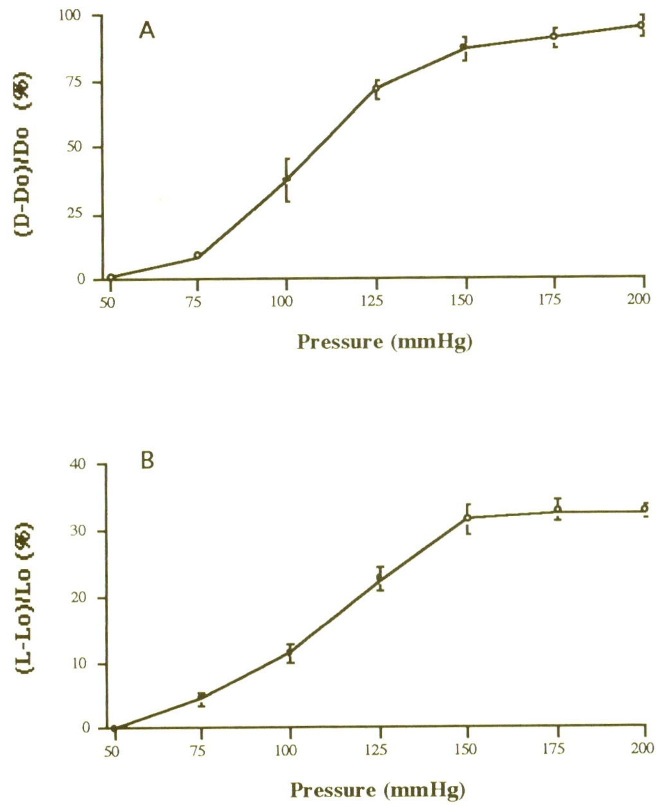

***Figure** 2. Changes (mean ± SEM) in diameter (A) and length (B) relative to their values at P=50 mmHg in normotensive Wistar Kyoto rats.*

The wall thickness and tension however depend on the transmural pressure i.e. on the circumferential strain of the artery. In linear mechanics, the relation between strain (ε), the relative increase in length (%), and stress (σ) is linear in a wide range and can be expressed by the elastic modulus E, also called Young's modulus.

$$E = \sigma / \varepsilon \qquad [N.m^{-2}]$$

In this case, E remains constant. Biological materials, however, do not possess a constant elastic modulus, and E is a function of strain. Yet, for each small increase in strain, an elastic modulus can be calculated, which holds only for one strain level. This elastic modulus is called incremental elastic modulus, Einc, and is thus the differential of the stress-strain relation. Einc is often expressed in the C.C.S unit dyne.cm^2, which equals 10^{-1} N. m^2.

Compliance

Though incremental elastic modulus is a useful parameter to describe mechanical properties of a tissue, in blood vessels it is also important to know the relation between pressure changes and volume changes. The compliance of a chamber is a quantitative description of its overall wall properties; compliance C is defined as the change in volume due to a change in pressure, that is:

$$C = \Delta V / \Delta P \qquad [m^5.N^{-1} \text{ or ml/mmHg}]$$

where ΔV is the change in volume [m^3, ml, or µl], and P, the change in transmural pressure [$N.m^{-2}$ or mmHg].

Compliance is determined not only by the mechanical properties of the wall but also by the geometry and size of the vessel. Moreover, because the pressure-volume relations of most biological systems are non-linear, compliance is a pressure-dependent quantity. Compliance should be clearly distinguished from a related quantity, capacity. Capacity refers to the amount of volume that a chamber can contain at any pressure; capacity depends on compliance and on the volume already contained in the chamber at zero transmural pressure. This introduces another quantity important and particularly difficult to measure, the unstressed volume, which is the volume contained in the vessel when the pressure is negligible.

In the study of isolated vessels and when using in vivo non invasive ultrasonic measurements of vessel diameters, it is convenient to calculate the compliance per unit vessel length, or cross sectional compliance Ccs, which corresponds with changes in lumen cross-sectional area :

$$Ccs = \Delta\, CSA / \Delta P \qquad [m^4.N^{-1} \text{ or } mm^2 \cdot mmHg^{-1}]$$

where CSA is the cross sectional area.

Finally, the compliance can be normalized to the vessel volume at a given transmural pressure to yield the distensibility D.

$$D = C / V_P \qquad [m^2.\ N^{-1}\ \ mmHg^{-1}]$$

DETERMINATION OF MECHANICAL PROPERTIES OF BLOOD VESSELS

The determination of the mechanical properties of blood vessels requires simultaneous measurement of blood vessel dimension (or volume) and internal pressure. Several experimental methods have been used to assess these parameters in vitro, in situ and in vivo.

In vitro measurements

Since the work of Mulvany and Halpern [14], the in vitro determination of the stress-strain relationship of vessel rings is the most commonly used experimental method to assess the mechanical properties. Arterial rings, as small as 150-200 μm diameter, are mounted by inserting two thin threads through the vascular lumen and attaching these threads to support plates at both sides of the vessel. One plate is connected to a force transducer and the second one to a displacement transducer; the vessel ring is immersed in an organ chamber containing a physiological solution. Various drugs can be added in the bathing solution to assess the vessel ring mechanical properties under different pharmacological conditions. The isometric myograph technique is commonly used and allows the measurement of the wall tension- ring circumference relationship under control, active (phenylephrine), and passive (sodium nitroprusside, or potassium cyanide poisoning).

Cylindrical segments of blood vessels can be cannulated, pressurized and perfused at controlled flow rates. This technique has been developed for use in large [15] and small vessels [16] and is widely used since the Halpern's system is commercially available. This method provides a more physiological loading of the vessel and the possibility for more precise pharmacological interventions. Halpern and Kelly [17] have summarized the differences between the isometric myograph and the cannulated vessel techniques. In cannulated vessels (and not in isometric myograph) 1) diameter is allowed to change, 2) shape is cylindrical, 3) a transmural pressure exists across the wall and thus the filtration rate through the vessel wall can be measured in nearly physiological conditions, 4) physiological axial length can be conserved, 5) pharmacological agents can be perfused and/or superfused.

The relation between the mechanical variables stress and strain and their distribution in the vessel wall depends on the experimental method used. In a comparative study, Cox [18] demonstrated that stress at equal levels of strain was higher in rings than in cylindrical segments from identical vessels. A comparison of dimensions from ring and cylindrical segments also showed significant differences between actual and calculated vessel radius. Furthermore, likely because of the presence of specific and non specific peptidases in the endothelial layer, the dose-effect relationship of constriction induced by peptide agents is quite different when the peptide is applied intraluminally or extraluminally [19]. Finally, the ring technique is relatively easy to perform and give valuable information concerning the physiology, the reactivity and the pharmacological properties of large arteries and resistance vessels. However, for precise in vitro measurement of the stress-strain relationship, it is recommended to use the method of pressurized and perfused cannulated vessels.

We developed an *in situ* experimental model allowing us to simply and accurately measure the static compliance of the carotid artery [20-22]. A segment of common carotid artery (22-25 mm length) free of collateral branches is isolated *in vivo* and *in situ*. A computer-controlled servo-system allows us to record the steady-state volume pressure relationship and therefore to evaluate the static mechanical properties of the isolated segment of common carotid artery. These measurements are performed under controlled conditions in the presence of a physiological vasomotor tone and after total abolition of the local smooth muscle tone by incubation with potassium cyanide. It is thus possible to compare the mechanical properties of the studied artery under active and passive smooth muscle conditions.

The advantages of such a model are the following: 1) to preserve the integrity of the endothelium in the isolated segment of artery. At the end of the experiment after excising and washing, the absence of fixation of Evan's blue dye by the wall indicated that the endothelial surface remained unaltered. 2) to keep the dissected, unexposed and non excised segment of carotid artery to its physiological length and in its normal fluid environment. 3) to avoid the collapse of the vessel. The assessment of only static and not dynamic mechanical properties of the studied segment of artery remains the main limitation of such experimental model.

Using ultrasonic echographic measurements of vessel diameter, we developed an *in vitro* model to study the reactivity and static and dynamic mechanical behavior of the rat carotid artery under controlled flow and pressure conditions [15]. The carotid artery is cannulated *in vivo* , before it is excised, and kept during the whole experiment at its original longitudinal stress and dimension. Ultrasonic echo-tracking techniques allow for the measurement of the internal diameter with an accuracy of less than 10 μm. The *in vitro* isolated artery can then be studied under controlled pressure and flow conditions. The isolated segment of artery can be submitted to steady or oscillatory pressure and flow.

Finally, the diameter-pressure static relationship of smaller muscular arteries can be assessed in situ by using optical measurement of diameter. Video-microscopy is used to measure diameters of mesenteric, arteries (500 to 100 μm) submitted to imposed levels of pressure or flow [23-25]. The limitation of such an experimental model does not depend on the dimension measurement but rather on the accuracy of pressure measurements. More sophisticated techniques of pressure measurement such as the servo-null technique [26] must be used to obtain reliable simultaneous measurements of phasic pressure in small arteries.

Figure 3 shows typical recordings of pressure-diameter recordings under basal condition, under "active" conditions (maximal vasoconstriction), and after poisoning of the smooth muscle cell with potassium cyanide (passive conditions) in a mesenteric artery. Using the previously described formulas and definitions, the compliance, distensibility, and elastic modulus can be derived from the pressure-diameter relationships. This experimental model also permits investigation of the flow-induced dilation of the isolated segment of vessel and the effect of the endothelium on the arterial mechanical properties.

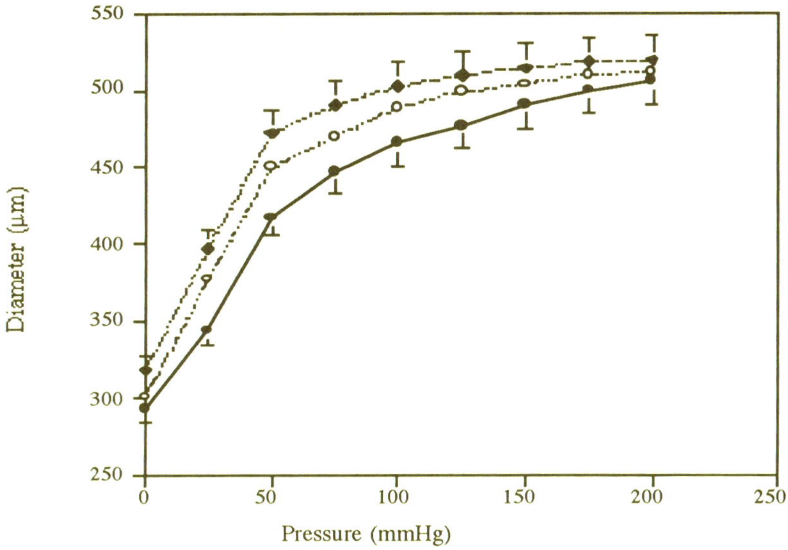

Figure 3. Pressure diameter relationship in a rat mesenteric artery under control conditions (open circles), after activation by phenylephrine (closed circles), abolition of the smooth muscle tone (diamonds).

In vivo measurements

New ultrasonic technologies have been developed allowing us to precisely measure, both in vitro and in vivo, instantaneous arterial diameter [27]. The probe consists of a 10 MHz strongly focused piezoelectric transducer operated in the pulse-echo mode. A stereotaxic arm permits motion of the transducer in x, y and z coordinates with micrometric steps to place the probe perpendicularly to the arterial axis in its largest cross-sectional dimension. The transducer is positioned so that its focal zone is located in the center of the artery ; thus, the back scattered echoes from both the anterior and posterior walls can be easily visualized. A typical radio frequency (RF) signal is then displayed on a computer monitor interfaced to the transducer system evidencing signals from the anterior and the posterior wall. These signals are only visible as the ultrasound beam crosses the axis of the vessel and are characterized by a first high-amplitude signal followed by a relatively silent acoustic period and then a second high-amplitude signal. The sample rate of the system is 5000 Hz and its resolution is close to 2 µm [28]. The highly non-linear characteristics of blood vessels require the study of dynamic elastic properties to be performed while applying small pressure perturbations superimposed on a constant mean level of pressure or strain. Under these conditions, linearization of the mechanical properties is allowed and modeling facilitated. By using ultrasonic devices, several authors reported differences in static and dynamic arterial mechanical properties evidencing smaller dynamic than static distensibility and, therefore, suggesting a significant viscosity of the arterial wall under in vivo conditions [29, 30]. It was constantly shown that the wall stiffness increased with frequency. At frequencies of 1 Hz, the increase is already high, but between 2 and 10 Hz, marked increases could be measured. Bergel [31] found a ratio of dynamic to static incremental elastic modulus at 1 Hz to range from 1.1 to 1.7 depending on the vessel type; the viscous component of the visco-elastic properties increases toward the peripheral arteries.

Pulse wave velocity

The older way of estimating the distensibility of the arterial system consists of measuring the arterial pulse wave velocity. The Moens-Korteweg equation relates pulse wave velocity (PWV) and elastic modulus E

$$PWV^2 = E\,h / 2\,\rho R \quad [m.s^{-1}]$$

where h is the wall thickness [m], R the vessel internal radius [m], r the specific mass of blood [kg.m^{-3}].

Assuming a non compressible vascular wall and small strains, PWV is equal to:

$$PWV^2 = V\,\Delta P / \rho\,\Delta V$$

The measurement of the pulse wave velocity in clinical practice is a simple and reliable method to assess global mechanical properties of the aorta and large elastic arteries [32].

Vascular impedance

Mechanical properties of the vascular bed can be determined *in vivo* as well. From aortic input impedance measurements, systemic arterial compliance can be determined [33]. However, the value of the calculated compliance strongly depends on the mathematical model for the arterial tree.

Vascular impedance expresses the relation of the force acting in the bloodstream to the resulting motion of blood. This relationship is a function of the physical properties of blood and the blood vessels. Vascular impedances are always ratios of pressure to flow; three different types have been defined :

Longitudinal impedance : $Z_L = (-dP/dx) / Q$

Input impedance : $Z_x = P / Q$

Transverse impedance : $Z_w = P / (-dQ/dx)$

P is the blood pressure and Q is the volumetric rate of flow in the longitudinal (x) direction.

Impedance spectrum is complex numbers, defined by pressure and flow or their derivatives. The impedance equations apply only to sinusoidal signals which can be derived from natural pulse waves *in vivo* by methods of frequency analysis. An impedance must be expressed as a spectrum of its complex values over a range of frequencies.

Longitudinal impedance (Z_L) is the ratio of the pressure *gradient* to flow and is thus the pulsatile analogue of vascular resistance. The use of the pressure *gradient* makes this impedance a property of a vascular segment of unit length, depending of the local properties of the vessel. Therefore, Z_L is usually expressed in dyne sec/cm ^{5} per cm vessel length.

Input impedance (Z_x) is the ratio of pressure to flow at a particular vascular cross section. It is thus a property of a particular site in the circulation, which is regarded as the *input* to all the vascular tree distal to that point; it is determined by the properties of the distal bed as well as local conditions. The input impedance at the end of a vessel is often referred to as its terminal impedance. If the distal bed is no more than a continuation of the local vessel, with exactly the same physical properties, the input impedance would be an expression of those uniform properties, called the characteristic impedance (Zc). Zx and Zc would thus be identical in a uniform tube terminated by a matching impedance. This condition does not occur in the circulation because the consecutive segments of the vascular tree differ in their dimension and their elasticity, and hence in their characteristic impedance. This mismatching generates reflected waves which influence the shape of the blood pressure and flow waves. Thus, the observed pressure and flow at a given point of arterial system are a mixture of incident and reflected waves. The relation of Zc to the properties of large vessels can be approximated as :

$$Zc = \rho \, PWV/A$$

where A is the vessel cross sectional area.

Aortic input impedance in the ascending aorta has been measured by many investigators, in man as well as in the dog and in the rat. A typical spectrum in the dog is shown in Figure 4.

The input impedance falls from 0 to 2 Hz and fluctuates with frequency thereafter. The modulus at zero frequency (mean pressure/mean flow) is 10 to 20 times greater than the rest of the frequency spectrum. The phase of the input impedance is negative at low frequencies, signifying that flow leads pressure. The phase angle approaches zero as frequency rises and becomes positive or turns downward again between 4 and 8 Hz. Characteristic impedance calculated by averaging the observed input impedances at frequencies from 2 to 12 Hz are ranging between 5 and 10% of the modulus at zero frequency (equal to the total peripheral systemic resistance).

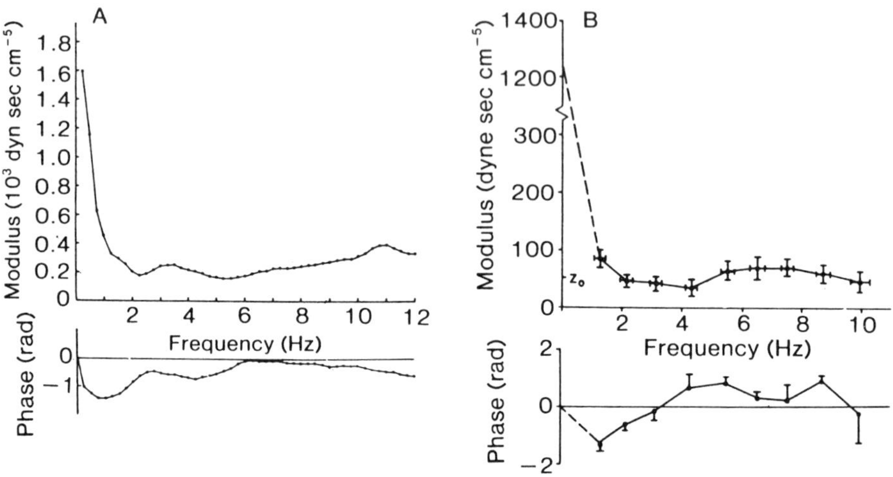

Figure 4. *Aortic input impedance in the ascending aorta in the dog.*

Significant changes of arterial input impedance can occur in the course of normal physiological adjustments of the circulation or in response to vasoactive drugs. Basically, three mechanisms may change the arterial impedance :

- Constriction or dilatation of arterioles modifies the peripheral reflection factors and thus alters the size of the frequency-dependent oscillations in the input impedance spectrum. Constriction enlarges the reflection factors whereas arteriolar dilation reduces reflection. The input resistance, or impedance at zero frequency, is more greatly altered by arteriolar vasomotion.

- An increase in elastic modulus increases phase velocity and thus shifts the Zx minima to higher frequencies. The smooth muscle activated in this case is that of the artery concerned (e.g. the aorta), rather than the peripheral vessels. The displacement of important reflection site to position nearer the heart has the same effect.

- Any alteration of cross sectional area of the vessel changes its characteristic impedance. According to the equation $Zc = \rho\ PWV/A$, the characteristic impedance, Zc, increases as the radius becomes smaller and *vice versa* . This response interacts with the previous one (variations in elastic modulus) because Zc is directly proportional to phase velocity. If wave velocity and radius both change exactly in the right proportions, the characteristic impedance remains unaltered. This occurs in many experimental conditions with opposite changes induced by vasoactive agents on the vessel diameter and its elastic modulus.

REFERENCES

1. Gore RW. Pressures in cat mesenteric arterioles and capillaries during changes in systemic blood pressure. Circ Res 1974; 34: 581-591

2. Zweifach BW. Quantitative studies of microcirculatory structure and function. I. Analysis of pressure distribution in the terminal vascular bed in cat mesentery. Circ Res. 1974; 34: 843-857.

3. Mulvany MJ, Aalkjaer C. Structure and function of small arteries. Physiol Rev. 1990; 70: 921-961.

4. Gore RW, Bohlen HG. Pressure regulation in the microciruclation. Fed Proc. 1975; 34: 2031-2037.

5. Frank O. Die theorie der Pulswellen. Z. Biol. 1926; 85: 91-130

6. Carew TE, Vaishnav RN, Patel DJ. Compressibility of the arterial wall. Circ Res. 1968; 23: 61-68.

7. Dobrin PB, Rovick AA. Influence of vascular smooth muscle on contractile mechanics and elasticity of arteries. Am J Physiol. 1969; 217: 1644-1652.

8. Patel DJ, Janicki JS, Carew TE. Static anisotropic elastic properties of the aorta in the living dogs. Circ res. 1969; 25: 765-779.

9. Lichtenstein O, Safar ME, Poitevin P, Levy BI. Biaxial mechanical properties of carotid arteries from normotensive and hypertensive rats. Hypertension 1995; 26: 15-19

10. W.R. Milnor. Hemodynamics. William and Wilkins Ed. Baltimore 1989. pp 71-73.

11. Gaballa MA, Jacob CT, Raya TE, Liu J, Simon B, Goldman S. Artery remodeling during aging passive and active stiffness. Hypertension 1998 (in press).

12. Caro CG, Pedley TJ, Schroter RC, Seed WA. The mechanics of the circulation, Oxford University Press, 1978.

14. Mulvany MJ, Halpern W. Mechanical properties of vascular smooth muscle cells in situ. Nature 1976; 260: 617-619

15. Caputo L, Tedgui A, Levy BI. Control of the carotid vasomotor tone by local renin angiotensin system in normotensive and hypertensive rats. Role of endothelium and flow. Circulation Res 1995; 77: 303-309

16. Duling BR, Rivers RJ. Isolation, cannulation and perfusion of microvessels. In Baker CH, Nastuk WL : Microcirculatory Technology. Academic Press, Orlando, 1986, pp. 265-280.

17. Halpern W, Kelly M. In vitro methodology for resistance arteries. Blood Vessels 991; 28: 245-251

18. Cox RH. Comparison of arterial wall mechanics using ring and cylindrical segments. Am J Physiol. 1983; 244 (Heart Circ Physiol. 13): H298-H303.

19. Fallon BJ, Stephens N, Tulip JR, Haegerty AM. Comparison of small artery contraction and morphology in pressurized and wire-mounted preparations. Am J Physiol 1995; 268: H670-H678.

20. Levy BI, Michel JB, Salzmann JL, Azizi M, Poitevin P, Safar ME, Camilleri JP. Effects of chronic inhibition of converting enzyme on the mechanical and structural properties of arteries in rat renovascular hypertension. Circulation Res, 1988, 63, 227-239

21. Levy BI, Benessiano J, Poitevin P, Safar ME. Endothelium dependent mechanical properties of the carotid artery in WKY and SHR : Role of angiotensin converting enzyme inhibition. Circulation Res 1990, 66: 321-328.

22. Levy BI, Duriez M, Phillipe M, Poitevin P, Michel JB. Effect of chronic dihydropyridine on the large arterial wall of spontaneously hypertensive rats. Circulation 1994; 90:3024-3033.

23. Qiu HY, Henrion D, Levy BI.. Endogenous angiotensin II enhances phenylephrine-induced tone in hypertensive rats. Hypertension, 1994, 24: 317-321.

24. Qiu HY, Henrion D, Levy BI. Alterations in the flow-dependent vasomotor tone in spontaneously hypertensive rats. Hypertension, 1994, 24: 474-479.

25. Qiu HY, Valtier B, Struijker-Boudier HAJ, Levy BI. Mechanical and contractile properties of in situ localized mesenteric arteries in normotensive and spontaneously hypertensive rats. J Pharmacol Toxicol Method, 1995, 33: 159-170.

26. Chilian WM, Eastham CL, Marcus ML. Microvascular distribution of coronary vascular resistance in beating left ventricle. Am J Physiol 1986; 25: H779-H788.

27. Hoeks APG, Brands PJ, Smeets FAM, Reneman RS. Assessment of the distensibility of superficial arteries. Ultrasound Med Biol 1990; 16:121-128.

28. Tardy Y, Hayoz D, Mignot JP, Richard P, Brunner HR, Meister JJ. Dynamic non-invasive measurements of arterial diameter and wall thickness. J Hypertens Suppl 1992; 10: S105-S109

29. Boutouyrie P, Bezie Y, Lacolley P, Challande P, Chamiot-Clerc P, Benetos A, de la Faverie JF, Safar M, Laurent S. In vivo/in vitro comparison of rat abdominal aorta wall viscosity. Influence of endothelial function. Arterioscler Thromb Vasc Biol 1997;17:1346-1355

30. Lichtenstein O, Safar ME, Mathieu E, Poitevin P, Levy BI. Static and dynamic mechanical properties of the carotid artery from normotensive and hypertensive rats. Hypertension 1998 32: 346-351

31. Bergel DH. The dynamic elastic properties of the arterial wall. J. Physiol (London), 1961; 156: 458-469

32. Asmar R, Benetos A, Topouchian J, Laurent P, Pannier B, Brisac AM, Target R, Levy BI. Assessment of arterial distensibility by automatic pulse wave velocity measurement. Validation and clinical studies. Hypertension 1995; 26: 485-490

3. NEUROHUMORAL CONTROL OF THE VASCULAR SYSTEM

Stéphane Laurent
Department of Pharmacology, Broussais Hospital, and INSERM U337, Paris, France

Through the modification of the neuronal discharge or changes in circulating catecholamines, the autonomic nervous system induces central or local vasomotor alterations and participates in the control of the internal environment and homeostasis. The vascular smooth muscle is the effector organ for these alterations. Indeed, with the exception of capillaries and some venules, the vessels have the ability to alter their calibre, to influence the regional (or total) peripheral resistance and capacitance, and to influence the cardiac output and its distribution. Several studies have demonstrated the extreme diversity of responses of different blood vessels to alteration in autonomic control [1].

The question of the neurohumoral control of blood vessels is further complicated by the fact that pressure and flow, which are partly influenced by the activity of the efferent nerve supply to the vasculature, modify arterial wall motion in addition to the direct effect of local innervation. The direct changes in artery tone brought about by neural activity are modified and diffused throughout the entire regional arterial system by the concomitant changes in the flow and pressure of the blood. Along this line, pressure and flow have been proposed as the true vascular neuroeffectors [2]. The purposes of this review are threefold : to recall the various neurohumoral components involved in the control of the vascular system, to emphasize the differences in the control of vascular tone between large conduit arteries and small resistive arteries, and to give evidence, from clinical investigation, for a neurohumoral control of large and small artery vasomotor tone.

NEUROHUMORAL COMPONENTS INVOLVED IN THE CONTROL OF VASOMOTOR TONE

Peripheral and central neurones

The degree of contraction of the smooth muscle cells of the resistance and capacitance vessels are governed by the activity of nerves connecting the central nervous system with the cardiovascular periphery. The sympathetic outflow to peripheral vessels originates in neurones located in the lateral parts of the reticular formation in the bulbar area of the brainstem. Postganglionic sympathetic nerves run to the blood vessels, with large variations in innervation. The resistance vessels, splanchnic veins, and cutaneous veins are densely innervated, leading to a strong vasoconstriction in response to sympathetic stimulation. By contrast, there is little response to sympathetic stimulation of the coronary arteries, the vessels of the brain, the veins draining skeletal muscle, the lung vessel, and large conduit arteries like the thoracic aorta.

The postganglionic sympathetic nerve endings form a network of unmyelinated slender processes (about 0.1 micron in diameter) which widen at regular intervals into varicosities (about 1 micron in diameter). The varicosities, which are denuded of Schwann cells, contain small granular vesicles (40 to 60 nm in diameter) in which the transmitter is stored. The varicosities are in close apposition to the smooth muscle cells they innervate, with a minimal distance of 80 to 100 nm (junctional cleft) [2].

Certain preganglionic sympathetic fibres continue to the central part (medulla) of the adrenals. They are essentially large sympathetic ganglia whose cells have lost their axons and have become specialized for the secretion of products directly into the blood-stream. The major biochemical difference between the adrenal medullar cells and the sympathetic neurones is that in the former most of the norepinephrine synthesized is transformed to epinephrine by the addition of a methyl group, a reaction catalysed by the enzyme phenylethanolamine-N-methyl-transferase. The mechanisms of storage and release of the adrenergic hormones obey the same rules as those occurring for the adrenergic nerve terminals. Particularly, the epinephrine released into the blood stream constricts most of resistance and capacitance vessels by stimulating α-adrenergic receptors, which are also activated by norepinephrine, dilates coronary arteries through β-adrenergic receptors, and dilates the resistance vessels of skeletal muscle and liver through activation of $\beta 2$-adrenoceptors, which are not sensitive to norepinephrine [3].

In the animal species which possess sympathetic cholinergic vasodilator fibres, activation of the cortico-hypothalamic pathway (defence reaction, fight-or-flight reaction) causes dilatation of the resistance vessels in skeletal muscle. In the skeletal muscles of the human, dilatation of the resistance vessels occurs during emotional stress, which is due not only to circulating norepinephrine but also to activation of vasodilators fibres travelling with the sympathetic nerves. The available pharmacological evidence suggests that these fibres are cholinergic. In tissues such as the salivary and sweat glands, the release of acetylcholine causes vasodilatation and activates the glandular cells. In the erectile tissues of the penis and the clitoris the released acetylcholine causes arteriolar vasodilation but strong constriction of the veins draining the sinusoids, which fill with blood and the corpora cavernosa harden [3].

Brain centers regulate the cardiovascular system, through a complex networks of interneurones, which, by liberating chemical transmitters to activate or inhibit each other, cause differential changes in the autonomic outflow.

The interneurones in the brainstem relay the afferent inputs from the peripheral sensors to appropriate areas of the brain, receive information from these areas and finally inhibit or excite discrete autonomic motor neurone pools to permit the various target organ to respond appropriately to the changing stresses imposed on the body. Chemical transmitters include norepinephrine, epinephrine, acetylcholine, dopamine, 5-hydroxytryptamine and the inhibitory amino-acid γ-aminobutyric acid (GABA). The nucleus tractus solitarius in the dorsal medulla is widely accepted as the major integrating center receiving information from the primary afferent baroreceptor neurones [4]. Studies from many workers have now established the importance of the rostral ventral medulla as the key pressor region from which bulbo-spinal presympathetic neurones descend to the intermediolateral cell column [5-6]. These studies have also established the caudal ventrolateral medulla as a depressor region, which exerts a direct influence on the rostral pressor region [5-6]. Ascending projections from these medullary centres, especially from the rostral pressor region, provide a major input to the locus coeruleus in the midbrain and to the hypothalamus. Descending projections from the medulla, especially from the rostral pressor region, provide possibly the single most important input for the regulation of sympathetic cardiovascular neurones, and of blood pressure [4].

A number of inputs modulate the activity of the vasomotor centers. Particularly, these inputs originate from the carotid and aortic mechanoreceptors, the cardiopulmonary mechanoreceptors, the arterial chemoreceptors, and the skeletal muscle receptors. These pathways, extensively analyzed in textbooks, will not be described in this chapter.

Neuromuscular transmission

Neural stimulation is associated with the production both of excitatory junction potentials (EJPs) and of force. The neurogenicity can be demonstrated by blocking neurotransmission with TTX and guanethidine [7-8]. The amplitude of the response is related to the frequency of the stimulation. Under *in vitro* conditions in small arteries, frequencies of 20-30 Hz are required for maximum response whereas 2-4 Hz are sufficient in large arteries, such as the aorta [8]. However, whether nerve-evoked EJPs summate to form action potentials is unclear [8].

Classical neurotransmission :

Norepinephrine : Norepinephrine is a transmitter of major importance in arteries. Although norepinephrine is known to be present in the perivascular nerve varicosities of small arteries, direct evidence that it is released during nerve stimulation of small arteries is lacking [8]. Indirect evidences are numerous, originating from the use of α-adrenergic antagonists [7-8]. The α-adrenoreceptors have been subdivided into α1- and α2-receptors, according to their response to various agonists and antagonists : Methoxamine, and phenylephrine, which are selective α1-stimulants; clonidine, guanfacine, and guanabenz which are α2-stimulants; epinephrine and norepinephrine, which stimulate both types; prazosin, doxazosin, and trimazosin, which are selective α1-blocking drugs; yohimbine, rauwolscine, and idazoxan, which are α2-blockers. The α1-receptors predominates postsynaptically in most vascular smooth muscle; α2-receptors are also important. In some vascular beds, like cerebral arteries of the dog, α2-receptors are more numerous, and there is considerable species variation. Post-synaptic α1-receptors in vascular smooth muscle are involved in the mediation of nerve impulse; α2-receptors, although mediating constriction in response to hormonal stimulation, are extrasynaptic. Presynaptic α2-receptors inhibit neuronal release of norepinephrine and participate to a negative feedback mechanism. The α1-receptors in vascular smooth muscle appear to be found in the adventitial layer, whereas the α2-receptors are found closer to the lumen, where they are acted upon catecholamines.

In small as in larger arteries, there is evidence for presynaptic regulation of the transmitters released from the perivascular varicosities. Most of available evidence originates from the consideration of norepinephrine uptake, which is inhibited by the amine pump inhibitors cocaine and desipramine. Although in large arteries blockade of α2-receptors enhances release of norepinephrine, evidence concerning presynaptic control of transmitter release in small arteries is limited [8].

The inability of a-antagonists to inhibit completely nerve-mediated responses has been attributed to transmitters other than norepinephrine, but also to the possibility that the post-synaptic receptors concerned are not conventional α-receptors [7]. Indeed, although the sensitivity of small arteries to most drugs is comparable to that seen in larger vessels, the sensitivity to norepinephrine is remarkably low : 0.3 to 1 μmol/l in small arteries *versus* 10 nmol/l in rat and rabbit aorta. The hypothesis has been raised (i) that the concentration of transmitter norepinephrine in the region of norepinephrine adrenoceptor could be high enough during nerve stimulation, because of the short distance between varicosity and smooth muscle, and (ii) that the postsynaptic norepinephrine adrenoceptors of small arteries have a particularly low affinity [7].

Acetylcholine : The pathways of neurogenic cholinergic influence on vascular smooth muscle are not completely understood [8-9]. There is some debate as to whether acetylcholine, released from the perivascular plexus, can reach receptors in the vessel wall and cause vasodilatation through the endothelium-derived relaxing factor (EDRF) nitric oxide (NO) [8]. In addition, mechanisms of cholinergic vasodilation that do not involve EDRF have been reported in arterial beds of certain species. The interaction between endothelium-derived factors and neuro-humoral factors are described in the chapter "*Endothelium function and dysfunction*".

Nonadrenergic, noncholinergic transmission :

Nonadrenergic, noncholinergic (NANC) mechanisms for the chemical transmission from autonomic nerves to their peripheral effector tissues have also been evidenced and are summarized in Table 1 [8-11].

Table 1 : Transmitter characteristics for nonadrenergic noncholinergic transmission. Adapted from Lundberg et al. [9].

	NPY	VIP	ATP	nitric oxide (NO)	CGRP	substance P
Synthesis	cell body precursor	cell body precursor	oxidative phosphorylation	NOS	sensory cell body	sensory cell body
Storage	LDV	LDV	SDV (LDV)	-	LDV	LDV
Second messenger	Ca^{2+}	$_{c}AMP$, K^{+}	Ca^{2+}	$_{c}GMP$, (PGs, $K_{Ca}2+$)	IP_3, Ca^{2+}, PKC	$_{c}AMP$, K_{ATP}
VSM response	vasoconstriction	vasodilation	vasodilation	vasodilation	vasodilation	vasodilation

NPY: neuropeptide Y; VIP: vasoactive intestinal peptide; ATP: adenosine 5' triphosphate ; CGRP: calcitonin-gene related peptide; NOS: nitric oxide synthase ; VSM: vascular smooth muscle; LDV: large dense cored vesicles ; SDV : small dense-cored vesicles ; IP_3 : inositol triphosphate ; PKC: protein kinase C ; $_{c}AMP$: cyclic AMP ; K_{ATP} : ATP-sensitive K^{+} channel;

For instance, electrical nerve stimulation evokes contraction of isolated blood vessels in vitro, even in the presence of adrenoceptor blocking agents [1-2,7]. Similarly, the classical muscarinic receptor antagonist atropine blocks the effects of exogenous acetylcholine but only marginally influences the parasympathetic nerve-mediated vasodilation in organs such as the submandibular salivary gland [9]. The perivascular nerves contain a variety of other substances than norepinephrine and acetylcholine, many of which are released in response to nerve stimulation and may behave as transmitters.

They include neuropeptide Y (NPY) [9-10], vasoactive intestinal peptide (VIP), adenosine 5'-triphosphate, nitric oxide (NO), calcitonin-gene related peptide (CGRP), and substance P. NPY and VIP occur in subpopulations of autonomic neurones, thus allowing further specialization of autonomic function. Thus, NPY is present in adrenergic cardiovascular nerves but not in noradrenergic fibres to exocrine glands [9]. In analogy, VIP is present in sphenopalatine ganglion cells innervating blood vessels but not in ciliary ganglion providing cholinergic innervation of iris smooth muscle [9]. Adenosine 5'-triphosphate is released together with norepinephrine in rat tail artery and in larger vessels [9]. The gas NO is not prestored as are other transmitters, but is synthesised upon demand from the terminal guanidino nitrogen atom of L-arginine by nitric oxide synthase (NOS). NO release does not involve vesicles and exocytosis but relies on diffusion, which is a slower process. NOS has been shown to be colocalized with VIP in presumably cholinergic parasympathetic post-ganglionic fibres, suggesting that these neurones can produce and release : the classical transmitter acetylcholine, peptides such as VIP, and the gas NO [9]. Attenuation of NO production by NOS inhibitors, such as N^G-mono-methyl-L-arginine (L-NMMA), markedly reduces parasympathetic NANC relaxation of vascular smooth muscle. Sympathetic neuroeffector mechanisms also seem to influence NO production because the vascular effects of NOS inhibition are reduced after sympathetic denervation of ganglionic blockade [11].

In summary, a large number of structural and chemical components are involved in the neurohumoral control of blood vessel, which is exceedingly complex, with considerable species and tissue variations. Specific combinations of chemical signals constitute "cocktails" of released transmitters, which vary in composition depending on strength and duration of activation.

LARGE CONDUIT ARTERIES VERSUS SMALL RESISTIVE ARTERIES

Due to the ability of small arteries and arterioles to adapt their myogenic activity to a bewildering variety of regional, local, and systemic factors, it is widely conceded that the major circulatory correlates of vascular smooth muscle contraction are to be found in the microvasculature, regulating local blood flow. Thus, the influence of the autonomic nervous system was usually analyzed in terms of pressure-flow relationships. In contrast with such diversity and specialization, large conduit arteries were frequently considered as simple conduits, the role of which was to deliver blood from the heart to the tissues in proportion to their metabolic needs and predominantly under the control of the regional resistance vessels. Indeed, there seems to be little evidence *in vivo* that vasomotor changes in large arteries occur independently of those in the resistive vessels [12]. The functional characteristics of conduit and resistive arteries, and their response to the activation of the autonomic nervous system are summarized in Table 2. However, it is important to consider also the response of large conduit arteries to changes in neurohumoral drive, because of their major role in arterial compliance and heart-vessel coupling, beside their conduit function.

Table 2. *Functional characteristics of conduit and resistive arteries, and their response to the activation of the autonomic nervous system.*

	Conduit Arteries		**Resistive arteries**
Function	conduit function	cushioning function	distribution of blood flow
Response to sympathetic activation	small reduction in calibre	• increased stiffness if vasoconstriction Di ↓ • decreased stiffness if Di #	
Consequences on circulation	contribution to the reduction in blood supply to peripheral organs	changes in heart-vessel coupling	• reduction in organ perfusion • increase in wave reflection → increase in central pulse pressure

Di : internal diameter

The large arteries have indeed two distinct interrelated functions: (i) to deliver an adequate supply of blood to the body tissues (the conduit function can by characterized by steady flow and steady pressure and their relationship, governed by the Poiseuille's law); and (ii) to smooth-out the pulsations occurring with intermittent ventricular ejection (cushioning function) characterized by oscillatory pressure and flow and their frequency-dependent relationships, which depends principally on the diameter of the artery and the physical properties of the arterial walls (elasticity). Changes in the neuro-humoral control of large arteries may affect not only their conduit function, but also their visco-elastic properties, through complex interaction between active changes in the vascular tone, transmural pressure and resulting cross-sectional area of arteries.

Changes in calibre only will be discussed first. Calibre changes in the large conduit arteries can occur in two ways: (i) passively, due to local or systemic changes in transmural pressure; and (ii) actively, due to changes in vascular smooth muscle contraction. Thus, constriction or dilatation of arterioles distal to the large arteries may modify the geometric and physical characteristics of the conduit vessel through two possible mechanisms. First, arteriolar vasomotion may induce systemic or local blood pressure changes, which could passively alter the diameter of the artery and its viscoelastic properties [12]. Second, arteriolar constriction or dilatation may modify the arterial wall smooth muscle tone actively throughout the endothelium-dependent mechanism of high-flow dilation [13-14]. However, there is evidence that the tone of large arteries may be regulated actively, and studies in isolated vessels, animals and humans suggest that large conduit arteries respond directly to both autonomic reflex and neurohumoral stimulation [15-17]. Finally, during changes in sympathetic activity, the alterations in the calibre and viscoelastic properties of conduit arteries depend on the balance between passive and active forces acting on the vessel wall.

Although the sympathetic system exerts a major influence in the control of the circulation, there are several differences in the response to adrenoreceptor stimulation and antagonism between large and small arteries, which have been emphasized in the previous paragraph. Particularly, dramatic changes occur between the initial part of the aorta and small distal arteries. The aorta, although poorly innervated as compared to peripheral arteries, contains highly sensitive vascular smooth muscle cells and an excess of a1-adrenoreceptors, and is sensitive to adrenergic stimulation [18-19]. An abrupt decrease in responsiveness to catecholamines occurs close to the roots of many aortic branches, and continues to fall-off as the arteries become smaller and further divide [20]. Furthermore, the sympathetic control of vascular tone of large arteries is more sensitive to α-adrenergic blockade than is that of smaller vessels [21].

The question of the physiological role and hemodynamic correlates of the diameter changes observed in central and large arteries *in vivo* needs to be addressed. Although these diameter changes have little effect on resistance to steady flow [12,22], they may have important consequences on the arterial visco-elastic properties. Several studies reported conflicting results regarding the effect of vasomotor tone on arterial stiffness. Whereas some authors found that stimulation of smooth muscle increased vessel stiffness [23], others reported a decrease [24] or no modification of elastic properties [16]. Whether vasoconstriction increases or decreases the stiffness of the arterial walls depends on whether the control and constricted vessels are compared at the same transmural pressure, circumference, strain and stress [22,25-26], because of the non-linear behavior of the stress-strain relationship and the methodological difficulties in assessing the unstressed diameter of the vessel, which is different in the constricted and relaxed artery [12,25-26]. Clinical studies using mental stress and cold pressor test to increase sympathetic tone, and taking into account transmural pressure at the site of a medium-sized muscular artery, the radial artery, have shown an increase in arterial stiffness when lumen diameter remained unchanged and a decrease in arterial stiffness when lumen diameter decreased [27-28]. These results suggest that the effects of vasomotor tone of large arteries visco-elastic properties are influenced by the changes in arterial geometry, which result from the balance between active and passive mechanisms.

In summary, even if the physiological variations in the diameter of large arteries are small in response to sympathetic activation, they can contribute to the reduction of blood supply to peripheral organs, induced by arteriolar vasoconstriction, and influence arterial distensibility, which is an important determinant of vascular impedance and cardiac afterload. In adapting their geometry and visco-elastic properties to haemodynamic changes of the resistive vessels (or to systemic influences), the large vessels participate in ventricular/vascular coupling and efficient control of the circulation.

We will examine separately the neurohumoral control of small and large arteries.

NEUROHUMORAL CONTROL OF SMALL ARTERIES AND ARTERIOLES

Experimental data

The numerous experimental evidence for a neurohumoral control of resistive arteries, extensively described in textbooks, will not be detailed in this paragraph. Some data have already been given above. A key argument is the reduction in arteriolar resistance by α-adrenergic inhibition, which demonstrates the tonic influence of the sympathetic system on the resistive component of the vasculature. This may be achieved at many sites, using either centrally acting drugs like clonidine, methyldopa, guanabenz and rilmenidine; ganglionic blockers or agents acting at sympathetic nerve endings like guanethidine; and α-blockers [29].

Clinical studies

The role of the autonomic nervous system on the control of systemic and/or local vascular tone has been extensively investigated. The respective contribution of β- and α-receptors and their subclasses were described in particular on the basis of the administration of selective agonists and antagonists [29-31]. The forearm vascular circulation was the most widely used model for the studies of the reflex and sympatho-adrenal regulation of the vascular tone [17,32-33]. This was because it was possible to infuse relatively high doses of drugs into the brachial artery without inducing any systemic haemodynamic effects or direct effects on the central nervous system, and also because of the accessibility of the forearm circulation to non-invasive techniques. Studies of peripheral blood vessels in humans have focused primarily on regulation of blood flow and vascular resistance, which are thought to reflect small vessel calibre and tonic activity [32-33].

The majority of the studies of forearm circulation used plethysmography to measure blood flow, and the observed changes in flow could be interpreted solely as due to changes in peripheral resistances, i.e. to dilatation or constriction of the small arteries and arterioles [17,22-23]. Forearm blood flow is an index of the input flow to the forearm, i.e. of brachial artery blood flow, which is determined as the product of blood flow velocity and cross-sectional area.

Recently, the pulsed Doppler system allowed the blood velocity to be measured separately and continuously at the site of a large artery, along with the determination of internal diameter, giving indirect information on the instantaneous fluctuations of the vasomotor tone of forearm arterioles [34,35]. For instance, during mental stress, brachial blood flow velocity increases two to three times [36]. As the increase in cardiac output measured simultaneously is small in comparison to the large increase in blood flow velocity, the most likely explanation for the velocity increase is that forearm vascular resistance decreases. This is indirectly evidenced by the significant relationship between brachial blood flow velocity measured with the pulsed Doppler system and the forearm vascular resistances, calculated from simultaneous plethysmographic measurements [37]. In some subjects the magnitude of spontaneous variations of brachial blood flow velocity represent 3- to 6-fold minimum values. Hand-exclusion attenuates these spontaneous fluctuations confirming that the cutaneous territory of the hand is principally responsible [36].

A recent important methodological advance in the clinical investigation of the neurohumoral control of blood vessels is the ability to record muscle and skin sympathetic nerve activity with microneurography [38,39], together with instantaneous blood flow. Thus, mental stress, which increases sympathetic tone, increases muscle sympathetic nerve activity (MSNA) and decreases forearm blood flow [38]. Using electrical stimulation of skin-sympathetic fibres of the left median nerve via microneurography needles, Stauss et al. [39] showed that the transmission from peripheral sympathetic nerves to cutaneous vascular smooth muscles in humans was strongest in the frequency band from 0.075 to 0.10 Hz [39].

Most of the physiology of the neurohumoral control of blood vessels has been established by clinical studies using plethysmography. Other clinical studies using pulsed Doppler systems are described in the paragraph devoted to large arteries. The vasodilator or vasoconstrictor response of an arteriolar territory depends on the type of stimulus. In addition, the vasomotor response to one type of stimulus depends on the arteriolar territory. For instance, emotional stress dilates arterioles of skeletal muscles, through activation of cholinergic vasodilator fibres and release of epinephrine, constricts arterioles of the splanchnic and kidney territories through an increase in sympathetic tone, and does not change skin blood flow, except for those subjects who develop emotional sweating in the hands where a consequent increase in flow occurs [3]. An increase in sympathetic tone, which can be provoked by various manoeuvres like cold pressor test or application of negative pressure to the lower body (LBNP, see below), increases arteriolar resistances and decreases regional blood flow [3,40].

The kidney, the coronary, and the cerebral circulation have specific response to stimuli. Normally, there is little activity in the sympathetic nerves to the kidney vessels. The afferent, but probably not the efferent, arterioles to the glomeruli have a sympathetic nerve supply. Changes in renal blood flow occur reflexly mainly by alterations in the activity of the arterial and cardiopulmonary mechanoreceptors, the arterial chemoreceptors, and afferents from the contracting skeletal muscles. Coronary blood flow may decrease in response to the reflex increase in sympathetic tone, but this decrease is usually offset by the metabolic vasodilatation resulting from the increased cardiac contractility. The cerebral vessels are innervated by autonomic nerves, but their role is unknown [40].

Antiadrenergic drugs, like the α-blocker prazosin, can unmask a tonic adrenergic tone on arterioles. Indeed, various agents like prazosin, doxazosin and urapidil lowers blood pressure by reducing systemic vascular resistances as a result of post-synaptic a1-receptor blockade and subsequent vasodilation [29-31]. After acute administration, β–blockers significantly reduce cardiac output and heart rate but do not decrease arteriolar resistances, which tend to increase as a consequence of baroreflex activation. Man in't Veld and Schalekamp [41] showed that the decrease in arteriolar resistances was most pronounced with β–blockers having a partial agonist activity or an intrinsic sympathomimetic activity (ISA) or, at a lesser degree, a b1-selectivity. The blockade of pre-synaptic β-adrenoreceptor, which could inhibit the release of norepinephrine from sympathetic nerve terminals and thereby reduce sympathetic activity to the heart, kidney and arterioles, has been proposed to explain the reduction in vascular resistance during long-term β–blocker therapy [31].

In summary, the response of small arteries and arterioles to neuro-humoral activation under physiological conditions largely depends on the type of the stimulus and the arterial bed. It also results from the potentiating or counteracting influences of circulatory changes on reflex mechanisms, and of mechanical factors (shear stress and transmural pressure) on the arterial wall.

NEUROHUMORAL CONTROL OF LARGE ARTERIES

Experimental data

Several experimental studies performed in anaesthetized animals *in vivo,* and arterial preparations *in vitro,* have indicated that changes in arterial smooth muscle activity (produced by catecholamines or the autonomic nervous system) affect the dimension as well as the visco-elastic properties of the large conduit arteries [10,15-16,18,19,22-26]. The extent of arterial diameter changes depend on the basal diameter, on the structure of their walls, on the resting tone [22,25-26], on the species [12,15], on age [12,26], and the method used [12].

The diameter changes in the large central arteries are small in comparison with changes in medium-sized and small arteries or arterioles. In dogs, the maximum norepinephrine-induced constriction of the aorta was 6 % of the outer diameter [15], while the haemorrhage or lumbar sympathetic stimulation produced a constriction of about 10 % [40]. The response of the artery was more pronounced in rabbits, where the range of diameter changes was between -6 % after norepinephrine and +17 % after phenoxybenzamine [43]. Active changes in the diameter of muscular, medium-sized arteries are greater in magnitude than those in central arteries [12]. The greatest range of active diameter variations are seen when the vasomotor agents are perfused from the lumen [25]. Thus, the range of diameter changes observed during norepinephrine perfusion vary from 40 % in the sheep carotid artery to 55 % in the dog iliac artery [44].

Clinical studies

Although pressure and flow techniques (blood velocity, plethysmography) were developed extensively for the study of resistive arteries, the studies of haemodynamics in the large conduit arteries in humans were until recently limited by the difficulty in measuring noninvasively diameter of the arteries [53].

New methods using pulsed Doppler velocimetry, introduced into clinical research, permitted the measurement of inner arterial diameter *in vivo* in man [34], particularly that of the brachial artery and superficial straight arteries. In addition to the sequential determination of the arterial diameter, the pulsed Doppler system allows the blood velocity to be measured separately and continuously, giving indirect information on the instantaneous fluctuations of the vasomotor tone of forearm arterioles [34-37]. Brachial artery diameter changes observed during various physiological and pharmacological conditions do not exceed 15 %, representing less than a 32 % increase in cross-sectional area [34-37,45-46]. Using a cylindrical representation of the artery, it can be assumed that variations of blood velocity account for the main part of the changes in volumic blood flow. Therefore, instantaneous fluctuations of blood flow velocity indicate that the vasomotor tone of arterioles is under permanent adjustment by local and systemic factors. As already mentioned, a flow-dependent mechanism can link changes in large artery calibre to changes in arteriolar vasomotor tone. For a clinical evaluation of this mechanism, flow velocity and arterial diameter may be studied before and after distal wrist occlusion (Figure 1).

When the wrist cuff is deflated, reactive hyperaemia increases blood velocity up to 8-fold with a parallel increase of arterial diameter of 12 ± 2 %. As myogenic relaxation and the vasodilating effect of recirculating metabolic compounds can be easily excluded, a velocity-dependent mechanism is the most likely explanation for the large artery relaxation [45].

Recently, it has been possible to determine the pulsatile changes in arterial diameter with high-resolution ultrasonic echo-tracking devices, either at the site of the radial artery (NIUS 02; SMH, Bienne, Switzerland; marketed by Capital Medical Services, Paris, France) [48] or at larger arteries like the common carotid, common femoral and brachial arteries, and the abdominal aorta (Wall Track System, marketed by Pie Medical, The Netherlands) [49]. By coupling the pulsatile changes in diameter to the pulsatile changes in blood pressure (finger BP determined by the Finapres system, or local pulse pressure determined by aplanation tonometry [50], it is possible to noninvasively determine the diameter-pressure curve of the artery over the systolic-diastolic range, and to derive the distensibility-pressure curve, in order to calculate the distensibility for a given blood pressure [51].

Most clinical studies on the neurohumoral control of large arteries have been performed at the site of the brachial and radial arteries, because of their accessibility and muscular nature.

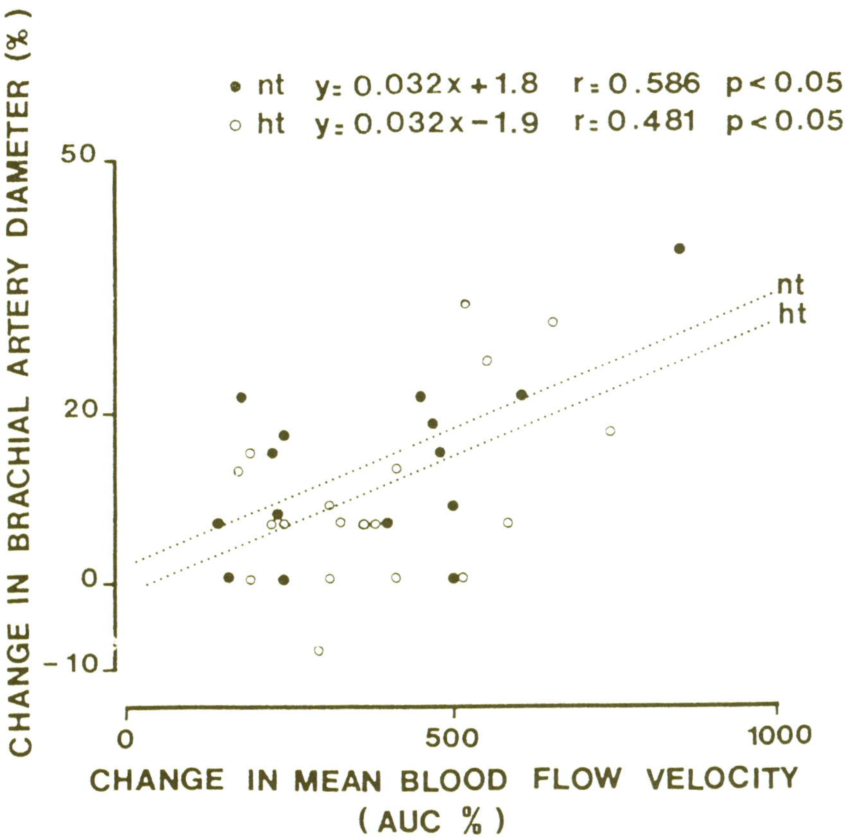

Figure 1. Flow-dependent dilatation of the brachial artery. Correlation between the increase in brachial artery diameter (expressed as percent change) and the increase in brachial blood flow velocity (expressed as percent change of the area under the curve -AUC- of velocity as a function of time) during the first 30 seconds of hand reactive hyperaemia. NT : normotensive subjects; HT : hypertensive patients. From Laurent et al. [45], with permission of the American Physiological Society

Response to catecholamines : Using pulsed Doppler velocimetry, the response of brachial artery diameter, flow velocity, blood flow and vascular resistances to subpressive doses of norepinephrine was studied in 28 subjects, including 9 normotensive people and 19 essential hypertensive patients [35]. Brachial artery haemodynamic parameters were studied before and after the administration of placebo (glucose) or increasing doses of norepinephrine (10, 20 and 40 ng/kg/min i.v.) given in a single-blind fashion. While the placebo infusion did not modify the brachial artery flow or diameter, the norepinephrine was accompanied by a significant decrease in brachial artery diameter, blood velocity, volumic flow and conductance (Figure 2).

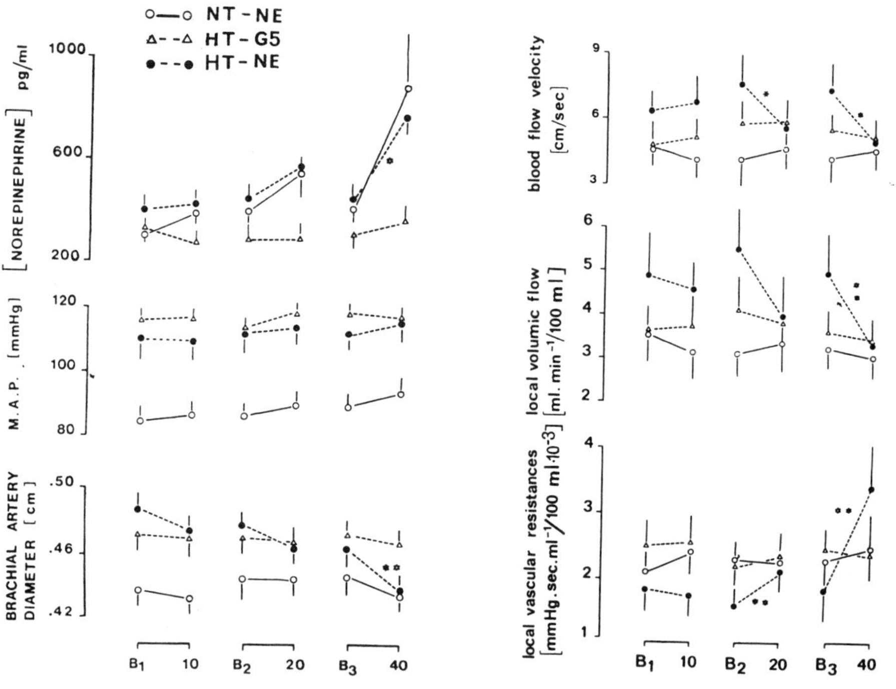

Figure 2. *Vasoconstriction of the forearm arterioles and the brachial artery in response to the perfusion of increasing doses of norepinephrine (NE: 10,20 and 40 ng/kg/min), with return to basal level between each perfusion sequence (B1, B2 and B3). Despite the lack of change in mean arterial pressure (M.A.P.), blood flow velocity, local volumic flow and brachial artery diameter decreased, and local vascular resistances increased in the group of hypertensive patients (HT) receiving NE (HT-NE). Plasma norepinephrine level increased significantly only at the highest perfusion rate. No significant change in plasma norepinephrine level, MAP, brachial artery diameter, blood flow velocity, local volumic blood flow and vascular resistance was observed in NT receiving NE (NT-NE) and in HT receiving placebo (glucose 5% : HT-G5). From Laurent et al. [35], with permission of the American Physiological Society*

The changes observed in the essential hypertensives were amplified in comparison with those observed in normotensive subjects. The changes in brachial artery diameter occurred in the absence of systemic arterial pressure variations. Change in brachial artery diameter was significantly correlated to change in plasma norepinephrine; the greater the rise in plasma norepinephrine, the greater the decrease in brachial artery diameter (Figure 3).

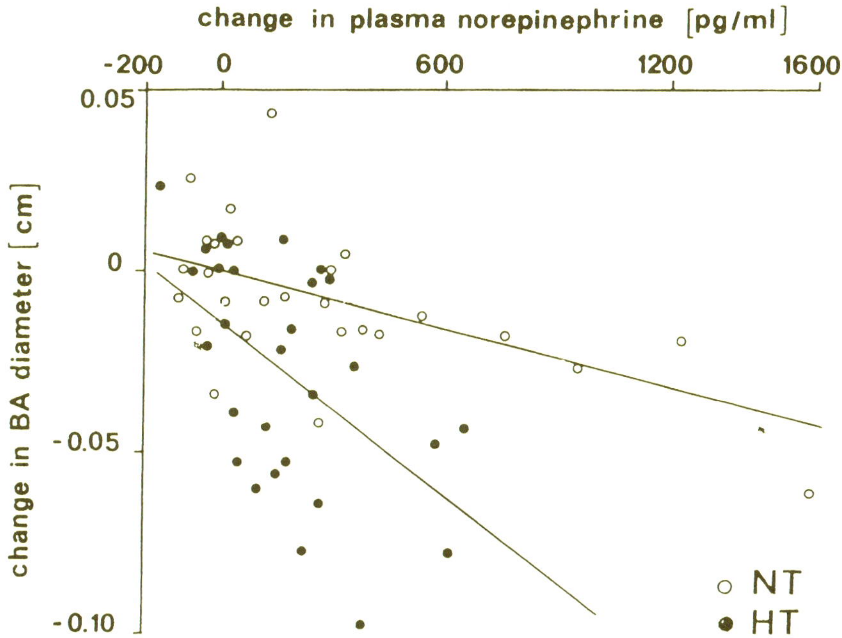

Figure 3. Relationship between change in plasma norepinephrine (NE) level and change in brachial artery diameter in 9 normotensive (NT) subjects and 10 hypertensive patients (HT) at 3 NE-infusion rates (10,20 and 40 ng/kg/min). Significant relationship was found in both populations ($P<0.01$ and $P<0.05$, respectively). The two regression lines were significantly different ($P<0.001$, ANCOVA), indicating that for any increase in plasma NE level, brachial artery (BA) diameter decreased to a greater extent in HT than in NT. From Laurent et al. [35], with permission of the American Physiological Society

Although the conclusion that the constriction of brachial artery is norepinephrine-dependent is logical, norepinephrine also induced a constriction of small arteriolar resistances vessels, and this can have direct repercussions on the upstream artery. A myogenic constriction cannot be excluded as a consequence of redistribution of local intravascular pressures. It seems improbable that myogenic response could cause the constriction of the brachial artery, as large arteries do not exhibit a substantial myogenic response. Changes in flow velocity can also be involved. Nevertheless, a flow-dependent endothelial mechanism was unlikely because, in response, to the largest doses of norepinephrine, blood velocity decreased to a lesser extent than the decrease in arterial diameter, and the changes in diameter and blood velocity were not correlated. Furthermore, in hypertensive patients, α-1-blockade by urapidil resulted in a marked increase in blood flow velocity but did not change brachial artery diameter [52], indicating that diameter changes are not necessarily the consequence of velocity alterations.

The action of epinephrine infusion on brachial artery haemodynamics was evaluated by Anderson and Mark [47]. They measured the brachial artery diameter and flow in three subjects before and during a 30-min infusion of epinephrine at 1.5 mg/ min. Epinephrine did not change brachial artery diameter, despite an increase in brachial artery blood flow velocity.

Response to mental stress : Mental stress can be utilized to study vascular responses to sympatho-adrenal stimulation, and to detect pressor hyper-responsiveness in hypertensive patients. The effects of mental stress has been evaluated on the brachial artery haemodynamics [36] and on the radial artery [27]. At both sites, a mental arithmetic stress test increased arterial pressure and blood flow velocity, but did not change arterial diameter (Figure 4).

The absence of increase in arterial diameter suggested that some mechanisms, e.g. myogenic or catecholamine-induced vasoconstriction, may have offset the increase in large conduit artery diameter, which would have been the consequence of (i) passive arterial pressure-dependent distension; (ii) velocity-dependent vasodilation; and (iii) cholinergic vasodilation.

Response to cold pressor test : Anderson and Mark measured changes in brachial artery diameter and flow in 9 subjects while a cold pressor test (CPT) was performed in the contralateral arm [38]. CPT reduced the brachial artery diameter by 11 % and brachial flow velocity by 54 %, together with a reduction in muscle sympathetic nerve activity (MSNA). This decrease in brachial artery diameter was observed in the presence of a significant increase in arterial pressure, and was therefore related to active smooth muscle contraction. CPT was also capable of causing brachial artery constriction after circulatory arrest induced by wrist occlusion. These findings were interpreted by the authors as suggesting a direct sympathetic stimulation with an exclusion of flow-mediated changes. The parallel increase in MSNA suggests a neurogenic sympathetic constriction. At the site of the radial artery, high-resolution tracking techniques have shown either no change [27] or a decrease [28] in arterial diameter, in response to CPT, depending on the balance between the passive increase in arterial diameter following the augmentation of intra-arterial pressure and the active decrease in diameter secondary to smooth muscle contraction (Figure 4).

Response to cardiopulmonary receptor deactivation and activation: Lower body negative pressure (LBNP) decreases the cardiopulmonary blood volume and pressure. According to the degree of negative pressure used, the LBNP can realize a condition of specific unloading of cardiac mechanoreceptors (from -5 to -20 mmHg) or can mimic a true orthostatic stress (-40 mmHg) [44]. In all circumstances the LBNP is a good circulatory model of the enhancement or stimulation of sympathetic activity. Anderson and Mark [47] have demonstrated that LBNP caused constriction of the brachial artery and a decrease in brachial artery flow.

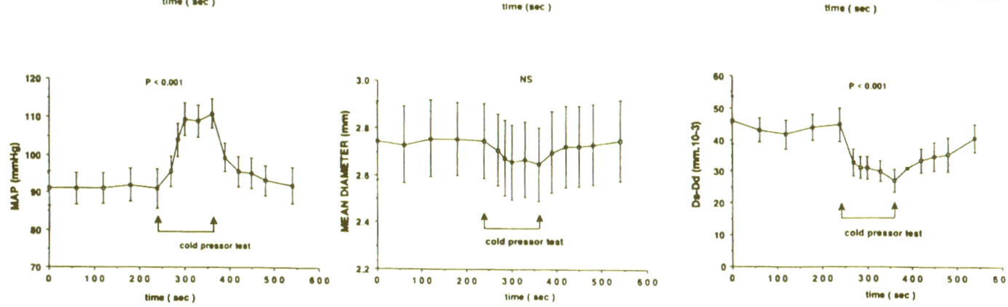

Figure 4 *. Time-course of the changes in MAP, radial artery diameter and pulsatile changes in radial artery diameter during cold pressor test and mental stress in 9 healthy subjects. Despite a significant increase in transmural blood pressure (increase in MAP), no significant change in diameter was observed, indicating an increase in smooth muscle tone in response to sympathetic activation, exemplified by the reduction in the pulsatile change in diameter. Adapted from Boutouyrie et al. (27), with permission of the American Physiological Society.*

Because the LBNP induced sympathetic constriction of small resistive vessels, the decrease in brachial artery diameter could be secondary to flow-mediated regulation. The absence of reduction in arterial diameter when LBNP was applied after distal circulatory arrest suggested that orthostatic manoeuvres influence large arteries diameter through a primary control of resistance vessels and velocity-mediated regulation. The problem is that after circulatory arrest the brachial artery diameter decreased, and superimposed changes induced by LBNP could be smaller than the resolution volume of the pulsed Doppler used. Increase in adrenergic activity with LBNP is not only associated with haemodynamic alterations in large peripheral arteries, but also induces changes in the aorta. Thus, Madkour et al. [54] have shown that LBNP induces a decrease in aortic compliance in parallel to an increase in plasma norepinephrine and a reduction of systolic blood pressure. The influence of aortic diameter changes on the compliance changes was not evaluated [44].

A passive elevation of the legs in a recumbent subject increases venous return, cardiopulmonary blood volume and cardiac filling pressure [55]. Roddie et al. [56] reported that such a leg elevation induced a reflex vasodilatation of forearm skeletal muscle arterioles through the stimulation of cardiopulmonary receptors and consequent inhibition of sympathetic vasomotor discharge. In a recent study using a dual-crystal pulsed Doppler system, we observed that the passive elevation of the legs in recumbent subjects, besides inducing a decrease in forearm vascular resistance and plasma renin activity, induced a sustained increase in brachial artery diameter and brachial artery flow velocity [38]. The initial maximum response after leg raising represented an increase of 13% of the baseline brachial artery diameter.

Changes in the arterial calibre occurred in the absence of arterial pressure modifications and the arterial relaxation was due to changes in vascular smooth muscle cell activity. Whether the arterial relaxation associated with cardiopulmonary baroreflex stimulation results directly from the inhibition of sympathetic activity remains uncertain. Indeed, leg raising induces a reflex vasodilatation of forearm skeletal arterioles and a decrease in vascular resistances [46,55]. Such a decrease in vascular resistance is associated with an increased distal outflow from the brachial artery and an increase in blood velocity [46]. Endothelial cells, when exposed to greater shear stress caused by an increase in blood velocity, induce relaxation of underlying smooth muscle and arterial relaxation. During the initial response to leg raising, brachial artery diameter and blood velocity increased. Moreover, during this peak response the variations in diameter and velocity were significantly correlated : the higher the increase in blood velocity, the higher the increase in arterial diameter. Therefore a velocity-dependent increase in brachial artery diameter, consecutive to reflex vasodilatation of forearm resistant vessels, could be the initiating mechanism. Nevertheless, the increase in arterial calibre observed after leg raising was long-lasting, while the increase in blood velocity was observed only during the initial period. The possibility of a direct reflex control of large conduit arteries should be considered [46,55].

Standard techniques have been applied to demonstrate the reflex nature of forearm resistance vessel changes in humans during cardiopulmonary receptor stimulation [47,55]. The use of these techniques to analyze the separate control of conduit arteries and resistive vessels is difficult. Indeed, there seems to be little evidence that vasomotor changes in large arteries and microvasculature occur independently from each other [12]. Under that condition, the manoeuvres that could block an eventual direct reflex control of large arteries would also block the reflex control of resistance vessels, and *vice versa*.

Response to autonomic nervous blockade : Brachial artery haemodynamics were studied in essential hypertensive patients before and after intravenous infusion of urapidil, a drug with hybrid action involving post-synaptic α-blockade [52]. Urapidil produced a significant drop in blood pressure, with a significant decrease in arteriolar resistance. Brachial artery blood flow increased, whereas brachial artery diameter did not change significantly. The interpretation remains uncertain because an absence of change in arterial diameter in the presence of decreased blood pressure could be due to an increase in unstressed diameter of the artery.

The action of beta-adrenoceptors on the large arteries depends on the class of the β-blocker and the artery studied. The forearm arterial effects of the non selective β-blocker propranolol and the angiotensin-converting enzyme inhibitor enalapril were compared by pulsed Doppler velocimetry in patients with essential hypertension. While the antihypertensive action of the two drugs was similar, only enalapril increased brachial artery diameter. Propranolol did not induce any arterial change [57]. Similar results have been reported after acute intravenous administration of propranolol [58]. This observation suggests that unopposed α-receptor 'vasoconstriction' of arterial wall is observed after non-selective beta-blockade.

In another study [59], patients with mild to moderate essential hypertension were separated into two age- and blood-pressure-matched groups: one receiving propranolol, the other pindolol. After a 3-month period the two drugs caused a similar decrease in blood pressure, but different effects on forearm arterial haemodynamics were observed. While propranolol did not induce any change, pindolol produced an increase in brachial artery diameter. As pindolol also induced a decrease in the local resistances, it was difficult to conclude whether the vasodilatation of the brachial artery was the consequence of the alteration of the microvasculature or of the partial agonist activity. However, on the basis of experimental studies, a relaxing effect of pindolol, independent of its β-blocking properties, can be suggested.

In summary, with techniques adapted to the measurement of large artery diameter in humans, it was possible to demonstrate significant changes in arterial lumen in response to catecholamines, to manoeuvres known to modify the sympathetic drive to large artery smooth muscle of skeletal territories, and to specific pharmacological blockade of adrenoceptors.

CONCLUSION

This review showed that a large number of structural and chemical components are involved in the neurohumoral control of blood vessel, which is exceedingly complex, with considerable species and tissue variations. It also emphasized the differences in the control of vascular tone between large conduit arteries and small resistive arteries. Thus, even if the physiological variations in the diameter of large arteries are small in response to sympathetic activation, they can contribute to the reduction of blood supply to peripheral organs, induced by arteriolar vasoconstriction, and influence arterial distensibility, which is an important determinant of vascular impedance and cardiac afterload. Finally, this review gave evidence, from clinical investigation, for a neurohumoral control of large and small artery vasomotor tone. The response of small arteries and arterioles to neuro-humoral activation under physiological conditions largely depends on the type of the stimulus and the arterial bed. It also results from the potentiating or counteracting influences of circulatory changes on reflex mechanisms, and of mechanical factors on the arterial wall. With techniques adapted to the measurement of large artery diameter in humans, it was possible to demonstrate significant changes in arterial lumen in response to catecholamines, to manoeuvres known to modify the sympathetic drive to large artery smooth muscle of skeletal territories, and to specific pharmacological blockade of adrenoceptors.

REFERENCES

1. Bevan JA, Bevan RD. « Changes in arteries as they get smaller. » *In Vasodilatation: Vascular smooth muscle, peptides, autonomic nerves and endothelium* PM Vanhoutte ed. New York: Raven Press Ltd, 1988.

2. Bevan JA. Pressure and flow: are these the true vascular neuroeffectors? Blood Vessels 1991;28(1-3):164-172

3. Burnstock G. Nervous control of smooth muscle by transmitters, cotransmitters and modulators. Experientia Basel. 1985;41:869-74.

4. Shepherd JT, Vanhoutte PM. *The human cardiovascular system. Facts and concepts.* New York, Raven Press,1979.

5. Chalmers J, Arnolda L, Llewellyn-Smith I, Minson J, Pilowsky P. "Central control of blood pressure". In *Textbook of Hypertension,* JD Swales, ed., Blackwell Scientific Publications, 1994.

6. Chalmers J, Pilowsky P. Brainstem and bulbospinal neurotransmitter system in the control of blood pressure. J Hypertens 1991;9:675-694.

7. Angus JA, Brougton A, Mulvany MJ. Role of alpha-adrenoceptors in constrictor responses of rat, guinea-pig and rabbit small arteries to neural activation. J Physiol Lond 1988;403:495-510.

8. Mulvany MJ, Aalkjaer C. Structure and function of small arteries. Physiol Rev 1990;70:921-61.

9. Lundberg JM. Pharmacology of co-transmission in the autonomic nervous system : integrative aspects on amines, neuropeptides, adenosine triphosphate, aminoacids, and nitric oxide. Pharmacol Rev 1996;48:113-178.

10. Gavazzi I, Boyle KS, Cowen T. Extracellular matrix molecules influence innervation density in rat cerebral blood vessels. Brain Res 1996;23:167-174.

11. Lacolley P, Lewis S, Brody MJ. Role of sympathetic nerve activity in the generation of vascular nitric oxide in urethane anesthetised rats. Hypertension 1991;17:881-887.

12. Gow BS. Circulatory correlates: vascular impedance, resistance, and capacitance. *In* DF Bohr et al (Eds.), The cardiovascular system, Vol II. Vascular smooth muscle (pp. 353-408). Baltimore: American Physiological Society (The Williams & Wilkins Company), 1980.

13. Furchgott RF, Zawadzki JV, Cherry PD. Role of endothelium in the vasodilator response to acetylcholine. *In* PM Vanhoutte and I Leusen (Eds.), Vasodilation (pp. 49-66). New York, Raven Press Publishers, 1981.

14. Pohl U, Holtz J, Buss R, Bassenge E. Crucial role of endothelium in the vasodilator response to increased flow in vivo. Hypertension 1986; 8: 37-44.

15. Barnett GO, Mallos AJ, Shapiro A. Relationship of aortic pressure and diameter in the dog. J Appl Physiol 1961; 16: 545-8.

16. Pagani M, Mirsky 1, Baig H, Manders WY, Kerkhof P, Vatner SF. Effects of age on aortic pressure-diameter and elastic stiffness stress relationships in unanesthetized sheep. Circ Res 1979;44: 420-9.

17. Hughes AD, Thom SAM, Martin GN. Size and site-dependent heterogeneity of human vascular responses in vitro. J. Hypertens, 1988;6 (Suppl. 4): S173-5.

18. Aars H. Diameter and elasticity of the ascending aorta during infusion of noradrenaline. Acta Physiol Scand 1971;83: 133-138.

19. Cox RH. Effects of norepinephrine on mechanics of arteries in vivo. Am J Physiol 1976; 2315: 420-425.

20. Laher I, Bevan JA. Alpha-adrenoreceptor number limits response of some rabbit arteries to norepinephrine. J Pharmacol Exp Ther 1985;233: 290-297.

21. Owen MP, Quinti CA, Bevan JA. Phentolamine-resistant neurogenic constriction occurs in small arteries at higher frequencies. Am J Physiol 1985; 249 (Heart Circ. Physiol., 18): H404-14.

22. McDonald DA. *Bloodflow in arteries*. London: Edward Arnold, 1974.

23. Peterson LH, Jensen RE, Parnell R. Mechanical properties ofarteries in vivo. Circ Res, 1960; 8:622-39.

24. Bagshaw RJ, Peterson LH. Sympathetic control of the mechanical properties of the canine carotid sinus. Am J Physiol 1972; 222: 1462-8.

25. Dobrin PB, Rovick AA. Influence of vascular smooth muscle on contractile mechanics and elasticity of arteries. Am J Physiol 1969; 217: 1644-51.

26. Gow BS. The influence of vascular smooth muscle on the visco-elastic properties of blood vessels. In *Cardiovascular fluid dynamics*, DH Bergel, ed. New York, Academic Press, 1972.

27. Boutouyrie P, Lacolley P, Girerd X., Beck L., Safar M., Laurent S. Sympathetic activation decreases radial artery compliance in humans. Am J Physiol 1994;267:H1368-76.

28. Joannides R, Richard V, Moore N, Godin M, Thuillez C. Influence of sympathetic tone on mechanical properties of muscular arteries in humans. Am J Physiol 1995 Feb;268 (2 Pt 2):H794-H801.

29. Prichard BNC, Weber MA. "Antiadrenergic drugs. Principles and practice of a-antiadrenergic therapy". In *Cardiovascular Drug Therapy,* Franz Messerli ed. Philadelphia, Saunders Publishing 1996.

30. Timmermans PMBWM, Van Zwieten PA. "Alpha-adrenoreceptors antagonists". In *Pharmacology of antihypertensive drugs, Handbook of hypertension.* Peter A Van Zwieten, ed., Amsterdam, Elsevier Science Publishers, 1984.

31. Frishman WH, Hershman D. "Beta-adrenoreceptor blockers. Principles and practice of beta-adrenoreceptor blockade.". In *Cardiovascular Drug Therapy,* Franz Messerli ed. Philadelphia, Saunders Publishing 1996.

32. Kiowski W, Hulthen UL, Ritz R, Buhler FR. Alpha 2-adrenoreceptor mediated vasoconstriction in human arterial vessels. Clin Pharmacol Ther 1983; 34: 565-9.

33. Jie K, Van Brummelen P, Vermey P, Timmermans PBMWM, Van Zwieten PA. Effects of exogenous adrenaline and noradrenaline on vascular post-synaptic alpha1 and alpha2 adrenoreceptors in man. J Hypertens 1984, 2 (Suppl 3): 119-21.

34. Safar M, Peronneau J, Levenson J, Simon A. Pulsed Doppler: diameter velocity and flow of brachial artery in sustained essential hypertension. Circulation 1981; 63: 393-400.

35. Laurent S, Juillerat L, London GM, Nussberger J, Brunner H, Safar ME. Increased response of brachial artery diameter to norepinephrine in hypertensive patients. Am J.Physiol 1988; 255: H36-43.

36. Laurent S, I,acolley P, Brunel P, Safar M. Effects of short lasting mental stress on systemic and brachial haemodynamics in essential hypertension. Circulation 1988; 78 (Suppl. IV): IV-175.

37. Safar ME, Daou JE, Safavian A, London GM. Comparison of forearm plethysmographic methods with brachial artery Doppler flowmetry in man. Clin Physiol 1988; 8: 163-70.

38. Anderson EA, Sinkey CA, Mark AL. Mental stress increases sympathetic nerve activity during sustained baroreceptor stimulation in humans. Hypertension 1991 Apr;17(4 Suppl):III43-III49

39. Stauss HM, Anderson EA, Haynes WG, Kregel KC. Frequency response characteristics of sympathetically mediated vasomotor waves in humans. Am J Physiol. 1998 Apr; 274(4 Pt 2): H1277-H1283.

40. Shepherd JT, Abboud FM. *Peripheral circulation and organ blood flow. The Cardiovascular System. Handbook of Physiology.* Bethesda, Maryland, 1983.

41. Man'in't Veld AJ, Shalekamp MADH. Effects of 10 different beta-adrenoreceptors antagonists in hemodynamics, plasma renin activity and plasma norepinephrine in hypertension. The key role of vascular resistance changes in relation to partial agonist activity. J Cardiovasc Pharmacol 5 (suppl 1) : S30-S36, 1983.

42. Gerova M, Gero J, Dolezel S, Blazkova-Huzulakova I. Sympathetic control of canine abdominal aorta. Circ Res 1973;33:149-152.

43. Aars H. Effects of altered smooth muscle tone on aortic diameter and aortic baroreceptor activity in anesthetised rabbits. Circ Res 1971;28:254-62.

44. Cox RH. Mechanics of canine iliac artery smooth muscle in vitro. Am J Physiol 1976;230:462-70.

45. Laurent S, Lacolley P, Brunel P, Laloux B, Pannier B, Safar M. Flow-dependent vasodilation of the brachial artery in essential hypertension. Am J Physiol 1990;258 (Heart Circ. Physiol. 27) : H1004-H1011.

46. London GM, Pannier BP, Laurent S, Safar ME. Cardiopulmonary baroreflex control of brachial artery diameter in sustained essential hypertension. J Hypertens 1989; 7 879-83.

47. Anderson EA, Mark AL. Flow-mediated and reflex changes in large peripheral artery tone in humans. Circulation 1989; 79: 93-100.

48. Tardy Y, Meister J-J, Perret F, Brunner H, Ardity M. Non-invasive estimate of the mechanical properties of peripheral arteries from ultrasonic and photoplethysmographic measurements. Clin Phys Physiol Meas 1991:12;39-54.

49. Hoeks APG, Brands PJ, Smeets GAM, Reneman RS. Assessment of the distensibility of superficial arteries. Ultrasound Med Biol 1990; 16: 121-128.

50. Kelly R, Daley J, Avollo A, O'Rourke MF. Arterial dilation and reduced wave reflection. Benefit of dilevalol in hypertension. Hypertension 1989; 14: 14-21.

51. Laurent S, Caviezel B, Beck L, Girerd X, Billaud E, Boutouyrie P, Hoeks A, Safar M. Carotid artery distensibility and distending pressure in hypertensive humans. Hypertension 1994;23[part 2]:878-883.

52. Levenson J, Simon ACh, Bouthier JD, Benetos A, Safar ME. Post-synaptic alpha-blockade and brachial artery compliance in essential hypertension. J Hypertens 1984;2: 37-41.

53. London GM, Laurent S, Safar M. "The autonomic nervous system and large conduit arteries". In *Vasodilation*, M O'Rourke, M Safar, V Dzau , Eds, London, Edward Arnold,1993.

54. Madkour A, Levenson J, Bravo El,, Simon A & Fouad-Tarazi FM (1989) Preload, adrenergic activity, and aortic compliance in normal and hypertensive patients. Am. Heart. J. 1183 1243-7.

55. London GM, Levenson JA, Safar ME, Simon AC, Guerin AP, Payen D. Hemodynamic effects of head-down tilt in normal subjects and sustained hypertensive patients. Am J Physiol 1983; 245:H194-H202.

56. Roddie IC, Shepherd JT, Whelan RF. Reflex changes in vasoconstrictor tone in human skeletal muscle in response to stimulation of receptors in low-pressure area of the intrathoracic vascular bed. J Physiol (Lond) 1957; 139: 369-76.

57. Simon ACh, Levenson JA, Bouthier JA, Safar ME. Comparison of MK 421 and propranolol in mild to moderate essential hypertension and their effects on arterial and venous vessels of the forearm. Am J Cardiol 1984; 53: 781-5.

58. Levenson JA, Simon ACh, Fiessinger JN, Safar ME, London GM, Housset EM (1982) Systemic arterial compliance in patients with arteriosclerosis obliterans of the lower limbs. Observation on the effect of intravenous propranolol. Arteriosclerosis, 2, 266-71.

59. Maarek BL, BouthierJA, Simon ACh, Levenson J, Safar ME. Comparative effects of propranolol and pindolol on small and large arteries and veins of the forearm circulation in hypertensive man. J Cardiovasc Pharmacol 1986; 8: (Suppl 4) S61-6.

4. ENDOTHELIAL FUNCTION AND DYSFUNCTION

Paul Vanhoutte[1], Chantal Boulanger[2],
1 IRIS, Courbevoie, France
2 INSERM Unit 141, Lariboisière hospital, Paris, France

The endothelial cells line the luminal surface of all blood vessels and are involved in numerous regulatory functions, such as the control of contraction and proliferation of vascular smooth muscle, adhesion of leucocytes and platelets, permeability and inflammatory responses. The endothelium also possesses thrombolytic and fibrinolytic properties. In addition, its metabolic activity regulates the oxidation of plasma lipids, the formation of angiotensin II and the degradation of circulating catecholamines and kinins. In 1980, Furchgott and Zawadzki [1] demonstrated that endothelial cells generate vasoactive subtances. This seminal observation has become crucial to vascular biology, leading ultimately to the understanding of the physio logical role for nitric oxide.

Although nitric oxide appears to be the major vasodilator released by endothelial cells in a vast majority of blood vessels, other substances, some of them still unknown, may play a role as well. Furthermore, soon after Dr Furchgott's discovery, it became clear that endothelial cells not only release relaxing factors but also produce contracting substances including endothelin. The release of endothelium-derived vasoactive subtances is controlled by a host of neuromediators and by shear forces exerted by the blood flowing in the blood vessel. Under physiological conditions, a precise and balanced release of relaxing and contracting factors contributes to appropriate organ perfusion. However, this balance is altered in diseases such atherosclerosis, diabetes, chronic heart failure, coronary artery disease or hypertension.

ENDOTHELIUM-DERIVED RELAXING FACTORS *(Figure 1)*

Acetylcholine and other mediators release endothelium-derived relaxing factor(s) (EDRF) in many arteries and veins from different species, including humans [1, 2]. Endothelium-dependent relaxations appeared early in evolution, which implies that these endothelial mediators fulfill an essential role [3]. Identification of the factors mediating such relaxations has been slowed down by their short half-lives. Bioassay experiments showing the diffusibility of EDRF from the endothelium to the smooth muscle also pointed out that the relaxing activity of the perfusate from a blood vessel was quickly lost when the transit time between the donor segment and the detector (an artery without endothelium) reached a few seconds. These bioassay experiments permitted the identification of the crucial role of superoxide anions in the inactivation of EDRF [4,5].

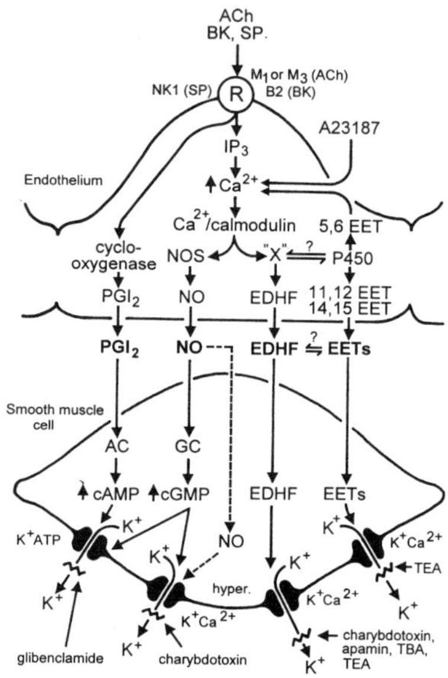

Figure 1. *Endothelial cells release multiple relaxing factors. M3, B2, NK1: M3 muscarinic, B2 bradykinin and NK1 neurokinin receptor subtypes respectively; R = receptor; A23187 = calcium ionophore; NOS = nitric oxide synthase; COX = cyclooxygenase; cyt P450 = cytochrome P450 monooxygenase; NO = nitric oxide; PGI2 = prostacyclin; EDHF = endothelium-derived hyperpolarizing factor; 5,6 EET = 5,6-epoxy-eicosatrienoic acid; 11,12 EET = 11,12-epoxy-eicosatrienoic acid; 14,15 EET = 14,15-epoxy-eicosatrienoic acid; GC = guanylate cyclase, cGMP = cyclic guanosine monophosphate ; cAMP = cyclic adenosine monophosphate; ATP = adenosine trisphosphate; IP3 = inositol trisphosphate; TEA = tetraethyl ammonium; TBA = tetrabutyl ammonium.*

Nitric oxide

Furchgott and Ignarro independently proposed that nitric oxide, or a related nitroso-compound, is the primary mediator of endothelium-dependent relaxation [6,7]. Moncada and coll then identified EDRF as nitric oxide (NO), synthesized following the conversion of endothelial L-arginine into L-citrulline [8,9]. Endothelium-dependent relaxations can be inhibited by analogues of L-arginine such as NG-monomethyl-L-arginine or nitroarginine-L-methylester which compete with the natural precursor L-arginine at the catalytic site of the enzyme. When infused intravenously, these inhibitors induce long-lasting increases in blood pressure [10, 11]. This suggests that the continuous basal release of nitric oxide by the endothelium contributes to the regulation of peripheral resistance [12]. Indeed, mice lacking the gene for the endothelial isoform of nitric oxide synthase (type III) have a slightly higher arterial blood pressure than wild type animals [13].

Endothelial NO synthase

The endothelial enzyme that produces NO (type III NO synthase) is a NADPH-dependent oxygenase involving the oxidation of one of the guanidine-nitrogen atom of L-arginine. The enzyme is localized in plasma membrane caveolae of the endothelial cells [14-18]. The production of nitric oxide by the endothelium requires the membrane-bound localization of the enzyme as well as the presence of cofactors (such as tetrahydrobiopterin, flavin adenine dinucleotide and flavin mononucleotide) and substrate (L-arginine) [19]. The enzyme binds calmodulin in a calcium-dependent manner and can therefore be activated by stimuli that increase the concentration of intracellular free calcium (such as acetylcholine or bradykinin) [20]. However, the release of NO can also occur in the absence of an increase in intracellular free calcium, as observed during shear stress [21].

Under normal conditions, the availablity of the semi-essential amino acid L-arginine in the plasma (about 100 to 200 μmmol/L) does not appear to be a limiting factor for the production of NO by the type III NO synthase (Km arg. : 5 to 10 μmmol/L) [22]. Endothelial cells not only take up circulating L-arginine by the mean of the y+ transporter but also can recycle L-citrulline into L-arginine [23]. Endogenous inhibitors of NO synthase such as asymmetrical dimethyl-L-arginine (ADMA) or NG-monomethyl-L-arginine (L-NMA) may interfere with L-arginine uptake and activation of the NO synthase(s) but the degree of NO synthase activation will depend upon the ratio L-arginine:(ADMA+L-NMA). The circulating levels of these endogenous inhibitors is in renal failure and in animal model of atherosclerosis [24, 25].

Although type III NO synthase is also called "constitutive" NO synthase, the level of its messenger RNA can be modulated. Indeed, type III NO synthase mRNA levels appear to be augmented by shear stress and by exposure to vascular endothelial cell growth factor [26, 27]. Estrogens may also regulate the expression of the endothelial isoform. In addition, type III NO synthase mRNA is lowered by oxidized-LDL while tumor necrosis factor shortens its half-life [28, 29]. This could contribute to pathological changes in nitric oxide production by the blood vessel wall. In addition, the subsequent production of endothelial NO may be affected by changes in membrane localization of the enzyme (which is regulated by post-translational palmytoylation and myristoylation) or its interaction with caveolin-1 [30], as well as by alterations in the availability of L-arginine or the unavailability of cofactors. Finally, the amount of active NO released by endothelial cells depends upon the level of superoxide anions which will degrade the mediator [4,5]. Thus, alteration of these different steps involved in the activation of endothelial nitric oxide release may contribute to the development of vascular disease.

Other isoforms of NO synthase

Another calcium-dependent NO synthase (type I) is expressed in nitrergic neurons innervating the adventitia of some cerebral arteries. Nitric oxide released following activation of type I NO synthase is the mediator of the non-adrenergic non-cholinergic innervation (NANC nerves) [31].

Exposure to cytokines or lipopolysaccharides induces the expression of a calcium-independent NO synthase (type II) in the blood vessel wall. Once induced, this isoform releases spontaneously large and sustained amounts of the mediator which may in part explain the hyporeactivity to contractile agents observed in septic shock (for review see 32].

Effect of endothelial NO (Figure 2)

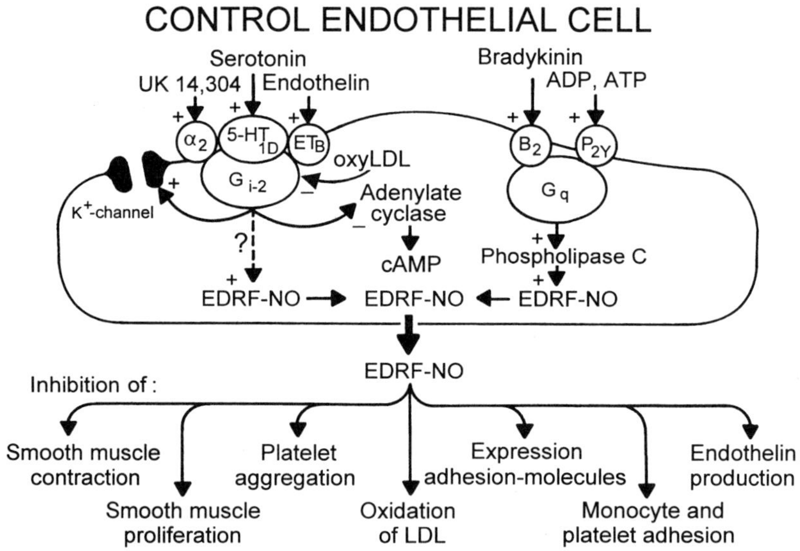

Figure 2.*Postulated signal transduction processes in a normal endothelial cell. Activation of the cell causes the release of EDRF-NO which has important protective effects in the vascular wall. α = alpha-adrenergic; 5-HT = serotonin-receptor; ET = endothelin receptors; B = bradykinin receptor; P = purinoceptor; G = coupling proteins; cAMP = cyclic adenosine monophosphate; NO = nitric oxide; LDL = low density lipoproteins; + = activation; - = inhibition.*

Relaxation of vascular smooth muscle

Endothelial NO causes the relaxation of the underlying vascular smooth muscle and can be regarded as the physiological antagonist of endogenous vasoconstrictors like catecholamines, angiotensin II or endothelin-1. Nitric oxide stimulates soluble guanylate cyclase and augments the levels of cyclic GMP in vascular smooth muscle cells [33]. This increase in cyclic GMP could contribute to the relaxation of the smooth muscle by decreasing Ca++ influx, augmenting calcium uptake by Ca++ ATPases or by a direct interaction at the level of contractile proteins. In some blood vessels, cyclic GMP specifically inhibits phosphodiesterase activity and may prevent cyclic AMP from degradation, thus augmenting cyclic AMP-mediated relaxations. In addition, NO may cause relaxation by interacting with K+ channels either directly or through an increase in cyclic GMP [34].

In addition to directly relaxing vascular smooth muscle, NO also regulates the production of the potent vasoconstrictor endothelin-1 in endothelial cells [35]. This effect is mediated by an increase in cyclic GMP and is also observed for atrial natriuretic factor.

NO can also be made available to vascular smooth muscle following a complex interaction with hemoglobin from red blood cells, glutathion and small erythrocytic thiols. Release of oxygen to the tissue is accompanied by the transfer of NO, bound to a cysteine group in hemoglobin, to small erythrocytic thiols where it can be released to cause relaxation of blood vessels [36]. These data suggest that NO may be more than a purely local regulator.

Antiproliferative effects?

Endothelial nitric oxide can inhibit the proliferation of vascular smooth muscle cells [37]. However, the antiproliferative effect of nitric oxide and NO donors is observed in cultured vascular smooth muscle cells at high concentrations which appear to be more compatible with the release of NO from an inducible NO synthase [38, 39]. Moreover, the antiproliferative effect of NO may depend upon the growth factors present in the medium since culture experiments suggest that NO in the presence of fibroblast growth factor may actually stimulate the proliferation of vascular smooth muscle [40].

NO may have indirect antiproliferative effects, since the mediator downregulates the production by endothelial cells of growth factors such as endothelin-1 and PDGFB chain [35, 41]. This effect appears to be mediated by an increase in cyclic GMP.

Protective role and antiatherogenic effect

Soon after the demonstration that NO could mediate endothelium-dependent relaxations, it was observed that in synergy with prostacyclin it prevents adhesion of platelets at the endothelial surface and their aggregation [42]. By inhibiting the degranulation of the platelets, NO prevents the release of vasoconstrictors and growth factors such as thromboxane A2, serotonin and PDGF.

By impairing the activation of nuclear factor NF-kB, NO negatively regulates the expression of adhesion proteins (such as VCAM-1, ICAM-1) and MCP-1, a chemokine mediator of inflammation involved in the recruitment of circulating monocytes [43-45]. Therefore, nitric oxide has a protective role against the first events involved in cellular adhesion which precedes the formation of foam cells and the development of subsequent atherosclerotic lesions. In addition, NO prevents the oxidation of low-density lipoproteins, another crucial event in the development of atherosclerosis [46-47].

Hyperpolarizing factor(s) EDHF

Endothelium-dependent hyperpolarizations have been observed in a large variety of blood vessels, including human coronary arteries [48]. Although endothelium-derived NO, itself or prostacyclin can cause hyperpolarization of vascular smooth muscle cells, endothelium-dependent hyperpolarizations are mediated mainly by a distinct diffusible substance termed endothelium-derived hyperpolarizing factor (EDHF). The identity of EDHF is unknown, although several candidates have been proposed such as ammonium anions, cytochrome P450 metabolites, hydrogen peroxide and cannabinoids (Fig 1). The intracellular pathway leading to EDHF synthesis must involve an increase in intracellular calcium since the calcium ionophore A23187 causes endothelium-dependent hyperpolarization and the latter is absent in calcium-free medium. Since, depending on the species, endothelium-dependent hyperpolarizations are mediated by the activation of either KCa- or KATP-channels in vascular smooth muscle, several endothelium-dependent hyperpolarizing factors may exist [see 48].

Prostacyclin

Prostacyclin is formed from arachidonic acid following the activation in turn of phospholipase A2, cyclooxygenase and prostacyclin synthase. It is released during activation of endothelial cells by shear stress or by agonists which also cause the release of NO. In certain cases, endothelium-derived prostacyclin causes endothelium-dependent relaxation of vascular smooth muscle cells by stimulating adenylate cyclase (increasing the level of cyclic AMP) or by opening KATP-channels. In most cases prostacyclin has weak vasodilator properties [see 48,49].

Other relaxing factors

Carbon Monoxide

Carbon monoxide (CO) is synthesized during activation of hemoxygenase-1 and -2 in the blood vessel wall [50-52]. It may contribute to endothelium-dependent relaxations of blood vessels which are insensitive to NO synthase inhibitors. However, since CO augments the levels of cyclic GMP, the relaxations it causes are sensitive to inhibitors of soluble guanylate cyclase. CO also downregulates the expression of endothelin-1 and growth factors in endothelial cells [50].

C-natriuretic peptide

Endothelial cells produce C-natriuretic peptide (CNP) [53, 54]. Exogenous CNP causes relaxation of isolated arteries and veins by activation of particulate guanylate cyclase and in some blood vessels, by activation of soluble guanylate cyclase and calcium-dependent K^+ channels [53]. Evidence that endothelial CNP mediates endothelium-dependent relaxations *in vivo* is lacking. Experiments on co-cultures of endothelial and smooth muscle cells suggest that the former may release sufficient amounts of CNP to inhibit the proliferation of the latter [54].

Parathyroid hormone related peptide

Parathyroid hormone related peptide can be synthesized both by endothelial and by vascular smooth muscle cells [55]. This peptide relaxes isolated blood vessels and causes hypotension when infused *in vivo*. As with CNP, an actual role of this peptide in endothelium-dependent regulation of vascular smooth muscle has to be demonstrated.

Mechanisms of release of relaxing factors

A number of physiological stimuli can cause endothelium-dependent relaxations.

Shear stress

The shear stress exerted by the blood flowing on the arterial wall is one of the main factors in the release of relaxing factors. This conclusion was reached from studies showing that an increase in flow rate through an isolated artery substantially increased the release of EDHF [56, 57]. A similar result was obtained if a stable flow rate was replaced by a pulsatile one. This explains why flow-induced dilatation is endothelium-dependent in the intact organism [57]. Thus, if resistance vessels in a peripheral organ (heart or skeletal muscle) suddenly dilate, the resulting surge of blood causes dilatation of the large arteries irrigating that organ. These flow-induced, endothelium-dependent changes in arterial diameter tend to normalize wall shear stress to baseline values. The augmented release of endothelium-derived relaxing factors induced by shear stress results -at least in part- from the activation of the kinin-kallikrein system and the local formation of bradykinin in the blood vessel wall [58, 59].

Endothelium-dependent relaxation mediated by shear stress involves integrins (transmembranar glycoproteins involved in cell-matrix interaction) and the endothelial cytoskeleton [60, 61]. Indeed, both inhibition of integrin binding to the extracellular matrix and disruption of the endothelial cytoskeleton specifically inhibit flow-induced dilatation without affecting endothelium-dependent relaxations in response to pharmacological agents. Increases in shear stress can stimulate the immediate release of nitric oxide, prostacyclin and possibly EDHF by the endothelial cells [62]. In addition, long term exposure to shear stress upregulates the expression of type III NO synthase in cultured endothelial cells [26]. Therefore, the continuous stimulation of endothelial NO release by flowing blood likely contributes to the protective role of the endothelium against the adhesion of platelets, neutrophils and monocytes in the intact organism. In addition, when the blood flow is augmented for a sustained period of time, NO participates in the long-term increase in diameter of the arteries [63].

Activation of endothelial receptors (Figure 2)

The endogenous substances stimulating the release of relaxing factors are either circulating hormones, autacoids formed by the blood vessel wall or substances released during the coagulation of blood.

Hormones

Epinephrine (adrenaline), norepinephrine (noradrenaline) and synthetic alpha$_2$-adrenergic agonists cause endothelium-dependent relaxations which are prevented by alpha$_2$-adrenergic antagonists, thus demonstrating the presence of alpha$_2$-adrenoceptors on endothelial cells [64]. These endothelium-dependent relaxations may participate in the vasodilator effect of catecholamines in blood vessels such as the coronary arteries. In rat cerebral arterioles, alpha$_2$-adrenergic agonists cause relaxation by augmenting the direct relaxing effect of basally released NO on vascular smooth muscle cells and rather than augmenting the release of NO from the endothelium [65].

Vasopressin and oxytocin cause endothelium-dependent responses by acting on endothelial V$_1$-vasopressinergic receptors [66]. This effect is particularly striking in cerebral arteries where vasopressin causes potent endothelium-dependent relaxations while it does not have this effect in the peripheral circulation. The endothelium-dependent dilatation induced by vasopressin may help to contribute to the preferential redistribution of blood flow to the brain when large quantities of the hormone are secreted (e.g. during hemorrhage).

Autacoids

The autacoids releasing EDRF include histamine, bradykinin, and various neuropeptides.

Histamine causes potent endothelium-dependent relaxations in many blood vessels by activating H_1-histaminergic receptors [67]. The endothelial action of histamine in arterioles likely accounts for the vasodilatation and hence for the rubor, characteristic of the local allergic reaction.

Various neuropeptides, in particular substance P, cause endothelium-dependent relaxations [68]. This effect may be involved in the local vasodilatation during antidromal (axon reflex) stimulation of sensory nerves.

Bradykinin is a potent stimulator of the release of EDRF and causes precapillary vasodilatation, possibly by its action on endothelial B2-kinin receptors. Bradykinin releases both NO and at higher concentrations EDHF, as observed in human coronary arteries [69]. Since the precursors of bradykinin (kininogens) are found throughout the body and particularly in platelets, and given that the vessel wall contains the enzymes required to transform these kininogens into bradykinin, locally generated bradykinin may have an important role in the regulation of vasomotor tone [70, 58, 59]. In particular, the local production of bradykinin may mediate endothelium-dependent dilatations to increase in flow. Reduction in the breakdown of bradykinin by angiotensin converting enzyme inhibitors may contribute to the beneficial effect of these compounds, in addition to preventing the formation of angiotensin II [71].

Platelet products and thrombin *(Figure 3)*

The endothelial action of thrombin and of the platelets products 5-hydroxytryptamine (serotonin) and adenosine diphosphate (ADP) contributes to the protective role of the endothelium against undesired coagulation. Thus, in normal blood vessels platelet aggregation, with the inevitable release of ADP and serotonin, as well as the local formation of thrombin leads to massive endothelium-dependent vasodilatation [72]. This helps to eliminate the microaggregate. In addition, NO in synergy with prostacyclin also inhibits further platelet adhesion and aggregation thereby eliminating the danger of vascular occlusion. Conversely, if the endothelial lining is interrupted (e.g. by trauma) aggregation proceeds with the continous release of serotonin and thromboxane A_2 which have unrestricted access to the vascular smooth muscle, which contracts. The blood vessel closes down to constitute the vascular phase of hemostasis .

ENDOTHELIUM-DEPENDENT CONTRACTIONS

Endothelial cells also can initiate contraction of the underlying smooth muscle by releasing constricting substances. The endothelium-derived contracting factors (EDCF) include the peptide endothelin, vasoconstrictor prostanoids such as thromboxane A2 and prostaglandin H2 as well as superoxide anions and components of the renin-angiotensin system.

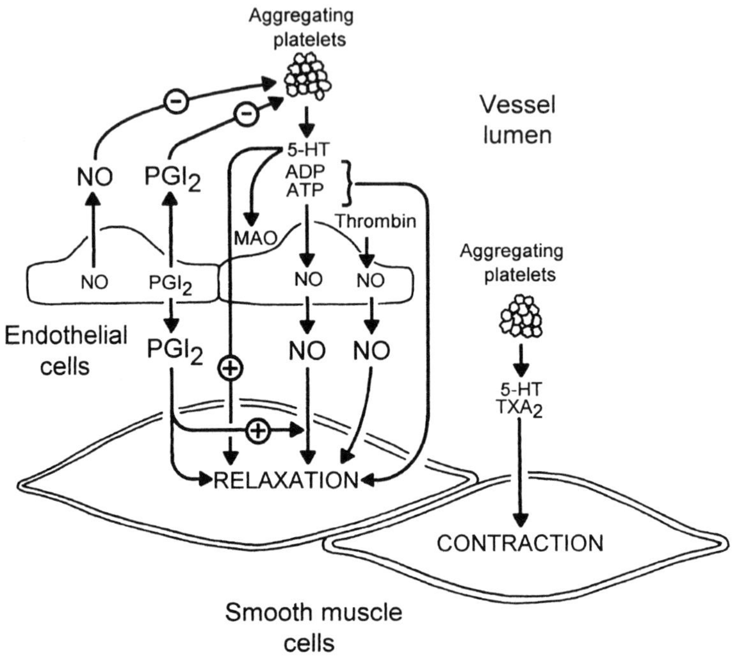

Figure 3. *Interaction between platelet products, thrombin and endothelium. If the endothelium is intact, several of the substances released from the platelets [in particular, the adenine nucleotides (ADP and ATP) and serotonin (5-HT)] cause the release of NO and prostacyclin (PGI2). The same is true for any thrombin formed. The released NO will relax the underlying vascular smooth muscle, opening up the blood vessel, and thus flushing the microaggregate away ; it will also be released towards the lumen of the blood vessel to brake platelet adhesion to the endothelium and, synergistically with prostacyclin, inhibit platelet aggregation. In addition, monoamine oxidase and other enzymes will break down the vasoconstrictor serotonin, limiting the amount of the monoamine that can diffuse toward the smooth muscle. Finally, the endothelium acts as a physical barrier that prevents the access to the smooth muscle of the vasoconstrictor platelet products serotonin and thromboxane A2 (TXA2). These different functions of the endothelium play a key role in preventing unwanted coagulation and vasospastic episodes in blood vessels with a normal intima. If the endothelial cells are removed (e.g. by trauma) or dysfunctional, the protective role of the endothelium is lost locally, platelets can adhere and aggregate, and vasoconstriction follows ; this contributes to the vascular phase of hemostasis. + = activation ; - = inhibition.*

Contractions blocked by inhibitors of cyclooxygenase

A group of endothelium-derived contracting factors is generated by the metabolism of arachidonic acid involving cyclooxygenase. In peripheral veins, but also in the cerebral circulation and in some arteries from hypertensive animals, endothelium-dependent contractions are mediated by thromboxane A2 or prostaglandin H2 which activate the same thromboxane-endoperoxide receptor [73-75]. Among the stimuli that cause endothelium-dependent contractions sensitive to inhibitors of cyclooxygenase, a physiological important response probably is that to stretch. The endothelium-dependent contraction of a cerebral artery in response to stretch closely resembles the autoregulatory response. A possible explanation for autoregulation of the cerebral circulation initiated by a sudden stretch of the vessel wall in response to an increase in blood pressure, is the release of endothelium-derived contracting factors which would activate the underlying smooth muscle to restore a normal flow rate.

In addition, cyclooxygenase is a source of superoxide anions which can cause contraction directly or indirectly by inactivating EDRF-NO (Fig 4). Endothelial cells also produce reactive oxygen species either by means of the xanthine oxidase, NADPH-oxidases system or by nitric oxide synthases when available concentrations of tetrahydrobiopterin or L-arginine are too low [76, 77].

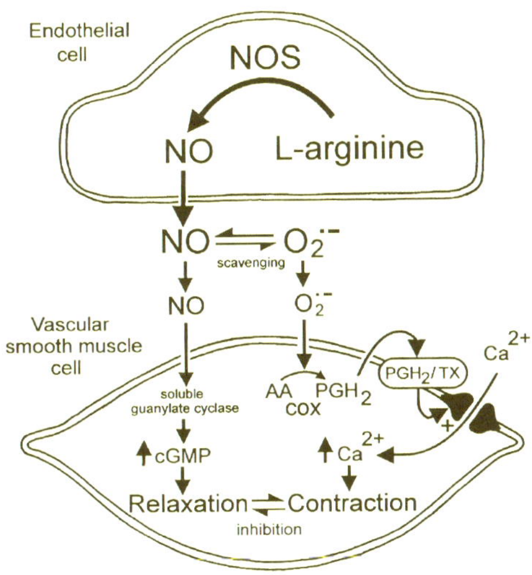

***Figure 4**. Interactions between endothelium-derived nitric oxide and superoxide anions. AA = arachidonic acid, COX = cyclooxygenase, cGMP = cyclic GMP; PGH2 = endoperoxides, TX = thromboxane, NOS = NO synthase.*

Hypoxia-induced contractions

Coronary, cerebral and pulmonary arteries rapidly contract when exposed to sudden hypoxia. This endothelium-dependent contraction is caused by the transfer of a diffusible substance that remains unknown. It is not dependent on cyclooxygenase, and is exacerbated by a reduced release of nitric oxide [78, 79].

Endothelin-1

Endothelial cells produce exclusively endothelin-1. Translation of messenger RNA generates preproendothelin, which is converted to big-endothelin; its conversion to the mature peptide endothelin-1 by endothelin converting enzymes is necessary for the development of its vascular activity [80-82]. The expression of messenger RNA and the release of the peptide from cultured endothelial cells is stimulated by thrombin, transforming growth factor b1, interleukin-1, epinephrine, angiotensin II, arginine vasopressin, calcium ionophore and phorbol ester. It is inhibited by nitrix oxide [35,41]. Endothelin-1 causes vasodilation at lower concentration by activating endothelial ETB receptors coupled to the release of NO, prostacyclin and EDHF. At higher concentrations, it causes marked and sustained contractions by activation of ETA, and in some blood vessels of ETB receptors on vascular smooth muscle (Fig 5; 83-85). The circulating levels of endothelin-1 are low suggesting either a discrete endogenous production under physiological conditions, the presence of potent inhibitory mechanisms (such as the negative control induced by NO) or a preferential abluminal release of the peptide towards vascular smooth muscle cells [35, 86, 87].

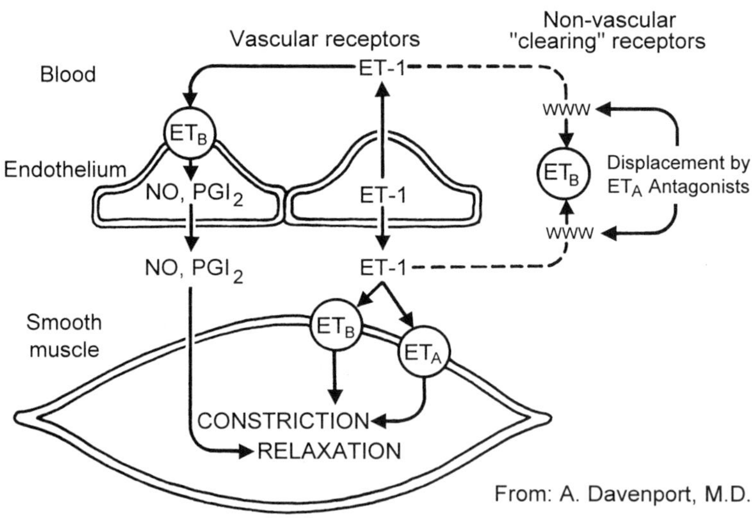

Figure 5. *Action of endothelin-1 on the vascular wall. ET = endothelin, endothelin-receptor, PGI2 = prostacyclin (courtesy of Dr. A. Davenport).*

ENDOTHELIUM-DERIVED FACTORS AND PATHOLOGY

Regenerated endothelium

Endothelial cells proliferate at a very low rate under normal conditions. This rate of proliferation is increased with ageing, after angioplasty or under pathological conditions such as hypertension. However, the regenerated endothelial are dysfunctional. Thus, in a model of regenerated endothelium following balloon abrasion of the porcine coronary artery, a selective reduction in release of relaxing factors is observed in response to agonists activating receptors linked to pertussis toxin sensitive Gi-proteins (such as serotonin 5HT1D, endothelin ET_B and alpha$_2$-adrenergic receptors). However, the response of regenerated endothelial cells to agonists activating receptors not coupled to pertussis toxin sensitive Gi-proteins is maintained (e.g. bradykinin or ADP) [88, 89]. The presence of such dysfunctional endothelial cells promote the occurrence of coronary vasospasm in response to ergonovin or platelet-derived product such as serotonin, by leaving unopposed the direct contractile effect of the serotonergic agonist on the vascular smooth muscle cells [90]. This vasospastic event is associated with signs of severe cardiac ischemia. The fact that the response to all receptors coupled to pertussis toxin sensitive Gi-proteins is reduced suggests that these Gi-proteins may be dysfunctional in regenerated endothelial cells. Regenerated endothelial cells express a normal amount of alpha-subunits of Gi proteins but their function may be altered [91]. Gi-proteins are expressed preferentially in endothelial cells from large human epicardial coronary arteries as compared with those from small coronary arterioles which are less prone to the occurrence of atherosclerosis [92, 93].

Reperfusion injury

When the blood flow is restored following temporary occlusion of a coronary artery, the reperfusion injury involves not only the myocardium but also the blood vessel wall. One hour after reperfusion, almost all endothelium-dependent responses have disappeared or are severely inhibited, because the reperfusion involves profound metabolic changes due to the reintroduction of oxygen and massive formation of oxygen-derived free radicals. Three months after reperfusion injury, the reactivity of the reperfused artery is still abnormal, particularly as regards the response to aggregating platelets and thrombin [94]. As a result, an hyperconstriction is observed and platelets and other circulating cells adhere to the reperfused segment. This abnormal adhesion can lead to the release of vasoconstrictors, growth factors from adherent cells or favor the transmigration of monocytes to the subendothelium, initiating a local inflammatory response. If the reperfusion is done more carefully, the subsequent chronic endothelial dysfunction is not so dramatic [95].

Atherosclerosis

A blunting of endothelium-dependent relaxations has been observed in various models of atherosclerosis and in human atherosclerotic coronary arteries [96, 97]. The first endothelium-dependent response to be lost is that to serotonin [98]. The most important mechanism in the decrease of endothelium-dependent relaxations is a reduced release of NO. As the disease progresses and the artery thickens and stiffens, it become increasingly difficult for NO to reach the vascular smooth muscle cells before it is inactivated. In animal models of atherosclerosis, inducible NO synthase is expressed in the blood vessel wall although the amount of active NO available is reduced due to an increased formation of superoxide anions [99, 100]. The beneficial effect of a chronic treatment with L-arginine in some animal models of atherosclerosis may be explained by an increased availability of the substrate to both isoforms resulting in a greater generation of NO, although a decreased formation of superoxide anions certainly plays a role as well [101, 102]. Inhibition of superoxide anions in the atherosclerotic wall may also explain the beneficial effect of chronic treatments with antioxidants [103]. The expression of endothelin-1 is augmented in the human atherosclerotic plaque but is unlikely to directly contribute to the impaired endothelium-dependent vasodilatation since endothelin receptor antagonists do not improve endothelium-dependent responses [104, 105].

Endothelial dysfunction may be one of the initial steps involved in the development of atherosclerosis. As endothelial cells regenerate with aging or following exposure to risk factors such as hypertension, smoking or hyperlipidemia, their protective role against the adhesion of platelets, monocytes or neutrophils lessens. This sets up the stage for the development of an inflammatory response which, together with the presence of oxidized LDL in the vessel wall, will further worsen the initial endothelial dysfunction (Fig 2).

Hypertension

In most animal models of hypertension, endothelium-dependent relaxations to acetylcholine (and other agonists) are impaired [73]. The characteristics of the endothelial dysfunction differs depending on the model studied. It is associated either with a decreased production of NO (and EDHF) and/or with the concommittant release of endothelium-derived contracting factors which interact with NO or impair its effect. This duality may explain the apparent discrepancy between the results from studies evaluating endothelial function and the contribution of NO in hypertensive subjects. Indeed, endothelium-dependent dilatation are either decreased (in most studies) or unchanged in patients with arterial hypertension [106-110].

In salt-sensitive hypertensive rats, endothelium-dependent relaxations are decreased because of an impaired release of NO, because of a decreased activity of the calcium-dependent NO synthase in blood vessels and an increased level of the endogenous inhibitor ADMA [111, 113]. Exposure to L-arginine restores endothelium-dependent relaxations [114]. In large arteries of spontaneously hypertensive rats (SHR), the endothelium-dependent relaxations to acetylcholine,

ADP or serotonin are impaired because of the release of endoperoxides, in particular prostaglandin H2 which activate thromboxane-endoperoxides receptors on the vascular smooth muscle cells [75]. Endothelial function is normalized when cyclooxygenase is inhibited or when thromboxane-endoperoxides receptors are blocked. The endothelial dysfunction associated with the activation of thromboxane-endoperoxide receptors may result not only from an exaggerated release of endoperoxides [75] but also from an impaired effect of NO on the vascular smooth muscle cells or from an alteration in the release of endothelial NO [109]. The calcium-dependent nitric oxide activity is increased in homogenates from SHR blood vessels suggesting either an increased expression of endothelial NO synthase or the presence of another isoform such as type I (neuronal) NO synthase [111, 115]. The presence of a cyclooxygenase-dependent EDCF in hypertension is supported by the fact that a cyclooxygenase inhibitor improves endothelium-dependent response to acetylcholine in patients with essential hypertension [109]. In the microcirculation of the SHR, endothelium-dependent relaxations to acetylcholine appear decreased because of the concommitant release of superoxide anions [116].

The endothelial dysfunction observed in hypertension probably is a consequence of high blood pressure since antihypertensive treatments normalize these responses. However, it may amplify the increase in vascular resistance at the origin of the hypertension since the inhibition of NO release (as in type III NO synthase knockouts or after enzyme inhibition by L-arginine analogs) causes an increase in blood pressure.

Heart Failure

Heart failure is accompanied by a generalized increase in peripheral vascular resistance resulting from a number of compensatory mechanisms (e.g. neural, hormonal, local factors). Dysfunction of the endothelium also contributes to the peripheral vasoconstriction. Indeed, decreased endothelium-dependent relaxations have been reported in several experimental models of heart failure. Likewise, in patients with the disease, the vasodilator response to acetylcholine is reduced [117]. The endothelial dysfunction may be specific for certain responses (e.g. acetylcholine) but has not been observed with all endothelium-dependent vasodilators (e.g. substance P, the calcium ionophore A23187). In addition, plasma levels of endothelin-1 increase in proportion to the severity of the cardiac failure and chronic treatment with a combined ET_A - ET_B endothelin-antagonist improves long term survival, at least in animals [118,119].

REFERENCES

1. Furchgott RF, Zawadzki JV: The obligatory role of endothelial cells in the relaxation of arterial smooth muscle by acetylcholine. Nature 1980; 299: 373-376.

2.Lüscher TF, Vanhoutte PM: The Endothelium: Modulator of Cardiovascular Function. Boca Ranton, CRC Press, 1990; pp 1-230.

3. Miller VM, Vanhoutte PM. Endothelium-dependent responses in isolated blood vessels of lower vertebrates. Blood Vessels 1986; 23: 225-235.

4. Rubanyi GM, Vanhoutte PM. Superoxide anions and hyperoxia inactivate endothelium-derived relaxing factor. Am. J. Physiol 1986; 250: H822-H827.

5. Gryglewski RJ, Palmer RMJ, Moncada S. Superoxide anions is involved in the breakdown of endothelium-derived relaxing factor. Nature 1986: 320: 454-456.

6. Furchgott RF: Studies on the relaxation of rabbit aorta by sodium nitrite: the basis for the proposal that acid-activatable inhibitory factor from bovine retractor penis is inorganic nitrite and the endothelium-derived relaxing factor is nitric oxide; in Vanhoutte PM (Ed): Vasodilatation: Vascular Smooth Muscle, Peptides, Autonomic Nerves and Endothelium, Raven Press, New York 1988, pp 401-414.

7. Ignarro LJ, Byrns RE, Wood KS: Biochemical and pharmacological properties of endothelium-derived relaxing factor and its similarity to nitric oxide radical; in Vanhoutte PM (Ed): Vasodilatation: Vascular Smooth Muscle, Peptides, Autonomic Nerves and Endothelium,Raven Press, New York, 1988; pp427-436.

8. Palmer RMJ, Ferridge AG, Moncada S: Nitric oxide accounts for the biological activity of endothelium-derived relaxing factor. Nature 1987; 327: 524-526.

9. Palmer RM, Ashton DS, Moncada S: Vascular endothelial cells synthesize nitric oxide from L-arginine. Nature 1988; 333: 664-666.

10. Moncada S, Palmer RMJ, Higgs EA. Nitric oxide: physiology, pathophysiology and pharmacology. Pharmacol Rev 1991; 43: 109-142.

11. Rees DD, Palmer RMJ, Schulz R, et al.: Characterization of three inhibitors of endothelial nitric oxide synthase in vitro and in vivo. Br J Pharmacol 1990; 101: 746-751 .

12. Rees DD, Palmer RMJ, Moncada S: Role of endothelium-derived nitric oxide in the regulation of blood pressure. Proc Natl Acad Sci 1989; 86: 3375-3382.

13. Huang PL, Huang Z, Mashimo H, Block KD, Moskowitz MA, Bevan JA, Fishman MC: Hypertension in mice lacking the gene for endothelial nitric oxide synthase. Nature 1995; 377: 239-242.

14. Pollock JS, Förstermann U, Mitchell JA, Warner TD, Schmitt HHH, Nahane M, Murad F: Purification and characterization of particulate endothelium-derived relaxing factor synthase from cultured and native bovine aortic endothelial cells. Proc. Natl. Acad. Sci. USA 1991; 88: 10480-10484.

15. Lamas S, Marsden PA, Li GK, Tempst P, Michel T: Endothelial nitric oxide synthase: molecular cloning and characterization of a distinct constitutive enzyme isoform. Proc. Natl. Acad. Sci. USA. 1992; 89: 20496-20501.

16. Nishida K, Harrison DG, Navas JP, Fisher AA, Dockery SP, Uematsu M, Nerem RM, Alexander RW, Murphy TJ: Molecular cloning and characterization of the constitutive bovine aortic endothelial cell synthase. J Clin Invest 1992; 90: 2092-2096.

17. Marsden PA, Heng HHQ, Scherer SW, Stewart RJ, Hall AV, Shi XM, Tsui LC, Schappert KT: Structure and chromosomal localization of the human constitutive endothelial nitric oxide synthase gene. J. Biol. Chem. 1993; 268: 17478-17488.

18. Ju H, Zou R, Venema VJ, Venema RC. Direct interaction of endothelial nitric-oxide synthase and caveolin-1 inhibits synthase activity. J Biol Chem 1997; 272(30):18522-18525.

19. Schmidt K, Werner ER, Mayer B, Wachter H, Kukovetz WR: Tetrahydrobiopterin-dependent formation of endothelium-derived relaxing factor (nitric oxide) in aortic endothelial cells . Biochem J. 1992; 281: 297-300.

20. Busse R, Mulsch A, Fleming I, Hecker M: Mechanisms of nitric oxide release from the vascular endothelium. Circulation 1993; 87: V18-V25.

21. Ayajiki K, Kindermann M, Hecker M, Fleming I, Busse R: Intracellular pH and Tyrosine phosphorylation but not calcium determine shear-stress-induced nitric oxide production in native endothelial cells. Circ Res 1996; 78: 750-758.

22. Stuehr DJ, Griffith OW. Mammalian nitric oxide synthases. Adv Enzymol 1992; 65: 287-346.

23. Hecker M, Sessa WC, Harris HJ, Anggard EE, Vane JR. The metabolism of L-arginine and its significance for the biosynthesis of endothelium-derived relaxing factor: cultured endothelial cells recycle L-citrulline in L-arginine. Proc Natl Acad Sci USA, 1990; 87: 8612-8616.

24. Vallance, P., Leone, A., Calver, A., et al.: Accumulation of an endogenous inhibitor of nitric oxide synthesis in chronic renal failure. Lancet 1992; 339: 572.

25 . Boger RH, Bode-Boger SM, Brandes RP, Phivthong-ngam L, Bohme M, Nafe R, Mugge A, Frolich JC. Dietary L-arginine reduces the progression of atherosclerosis in cholesterol-fed rabbits: comparison with lovastatin. Circulation 1997; 96(4):1282-1290

26. Uematsu M, Ohara Y, Navas JP, Nishida K, Murphy TJ, Alexander RW, Nerem RM, Harrison DG. Regulation of endothelial cell nitric oxide synthase mRNA expression by shear stress. Am J Physiol 1995; 269(6 Pt 1):C1371-C1378.

27. Papapetropoulos A, Garcia-Cardena G, Madri JA, Sessa WC. Nitric oxide production contributes to the angiogenic properties of vascular endothelial growth factor in human endothelial cells. J Clin Invest 1997; 100(12):3131-3139

28. Rosenkranz-Weiss P, Sessa WC, Milstien S, Kaufman S, Watson CA, Pober JS. Regulation od nitric oxide synthesis by proinflammatory cytokines in human umbilical vein endothelial cells. J Clin Invest 1994; 93: 2236-2243.

29. Liao JK, Shin WS, Lee WY, Clark SL: Oxidized low-density lipoproteins decrease the expression of endothelial nitric oxide synthase. J. Biol. Chem. 1995; 270: 319-324.

30. Garcia-Cardena G, Oh P, Liu JW, Schnitzer JE, Sessa WC. Targeting of nitric oxide synthase to endothelial cell caveolae via palmytoylation: implication for nitric oxide signalling. Proc Natl Acad Sci USA. 1996; 93: 6448-6453.

31. Bult H, Boeckxstaens GE, Pelckmans PA, Jordaens FH, Van Maercke YM, Herman AG. Nitric oxide as an inhibitory non-adrenergic non-cholinergic neurotransmitter. Nature 1990; 345 (6273):346-347

32. Stoclet JC, Andriantsitohaina R, Kleschyov A, Muller B. Nitric oxide and cGMP in regulation of arterial tone. Trends in Cardiov Med 1998; 8: 14-19.

33. Rapoport RM, Murad F: Agonist-induced endothelium-dependent relaxation in rat aorta may be mediated through cyclic GMP. Circ. Res. 1983; 52: 352-357.

34. Bolotina VM, Najibi S, Palacino JJ, Pagano PJ, Cohen RA. Nitric oxide directly activates calcium-dependent potassium channels in vascular smooth muscle. Nature 1994, 368: 850-853.

35. Boulanger C, Lüscher TF: Release of endothelin from the porcine aorta. Inhibition of endothelium-derived nitric oxide. J Clin Invest 1990; 85: 587-590.

36. Jia L, Bonaventura C, Bonaventura J, Stamler JS: S-nitrosohaemoglobin: a dynamic activity of blood involved in vascular control. Nature 1996; 380: 221-226.

37. Scott-Burden T, Vanhoutte PM: The endothelium as a regulator of vascular smooth muscle proliferation. Circulation 1993; 87: V-51-V-55.

38. Garg UC, Hassid A: Nitric oxide-generating vasodilators and 8-bromo-cyclic guanosine monophosphate inhibit mitogenesis and proliferation of cultured rat vascular smooth muscle cells. J Clin Invest 1989; 83: 1774-1780.

39. Scott-Burden, T., Schini, V.B., Elizondo, E, Junquero, D.C., Vanhoutte, P.M. Platelet-derived growth factor suppresses and fibroblast growth factor enhances cytokine-induced production of nitric oxide by cultured smooth muscle cells: effect on proliferation. Circ Res 1992; 71: 1088-1100.

40. Hassid, A., Arabshahi, H., Bourcier, T., Dhausi, G.S., Matthews, C. Nitric oxide selectively amplifies FGF-2-induced mitogenesis in primary aortic smooth muscle cells. Am. J. Physiol. 1994; 267: H1040-H1048.

41. Kourembanas S, McQuillan LP, Leung GK, Faller DV. Nitric oxide regulates the expression of vasoconstrictors and growth factors by vascular endothelium under both normoxia and hypoxia.J Clin Invest 1993; 92(1):99-104.

42. Radomski, M. W., Palmer, R. M., Moncada, S. The anti-aggregating properties of vascular endothelium: interactions between prostacyclin and nitric oxide. Br J Pharmacol 1987; 92: 639.

43. Tsao PS, Wang B, Buitrago R, Shyy JY, Cooke JP. Nitric oxide regulates monocyte chemotactic protein-1. Circulation 1997; 96(3):934-940.

44. Zeiher AM, Fisslthaler B, Schray-Utz B, Busse R. Nitric oxide modulates the expression of monocyte chemoattractant protein 1 in cultured human endothelial cells. Circ Res 1995; 76(6):980-986.

45. Spiecker M, Peng HB, Liao JK Inhibition of endothelial vascular cell adhesion molecule-1 expression by nitric oxide involves the induction and nuclear translocation of IkappaBalpha. J Biol Chem 1997; 272(49):30969-30974.

46. Yates MT, Lambert LE, Whitten JP, Mc Donald I, Mano M, Ku G, Mao JT. A protective role for nitric oxide in the oxidative modification of low-density lipoproteins by mouse macrophages. Febs Lett 1992; 309: 135-138.

47. Hogg N, Kalyanaraman B, Joseph J, Struck A, Parthasarathy S. Inhibition of low density lipoprotein oxidation by nitric oxide: potential role for atherogenesis. Febs Lett, 1993; 334: 170-174.

48. Mombouli JV, Vanhoutte PM. Endothelium-derived hyperpolarizing factor(s): updating the unknown. TiPS 1997; 18:251-256.

49. Moncada S, Vane VR. Pharmacology and endogenous roles of prostaglandin endoperoxides, thromboxane A2 and prostacyclin. Pharmacol Rev 1979; 30: 293-331.

50. Morita T, Kourembanas S. Endothelial cell expression of vasoconstrictors and growth factors is regulated by smooth muscle cell-derived carbon monoxide. J Clin Invest. 1995; 96: 2676-2682.

51. Christodoulides N, Durante W, Kroll MH, Schafer AI. Vascular smooth muscle cell heme oxygenase generate guanylyl cyclase-stimulatory carbon monoxide. Circ 1995; 91: 2306-2309.

52. Wang R, Wang Z, Wu L. Carbon monoxide-induced vasorelaxation and the underlying mechanisms. Br J Pharmacol 1997; 121: 927-934.

53. Wright RS, Wei CM, Kim CH, Kinoshita M, Matsuda Y, Aarhus LL, Burnett JC Jr, Miller VM. C-type natriuetic peptide-mediated coronary vasodilation: role of the coronary nitric oxide and particulate guanylate cylcase systems. J Am Coll Cardiol 1996; 28: 1031-1038.

54. Komatsu Y, Itoh H, Suga S, Ogawa Y, Hama N, Kishimoto I, Nakagawa O, Igaki T, Doi K, Yoshimasa T, Nakao K; Regulation of endothelial production of C-type natriuretic peptide in coculture with vascular smooth muscle cells. Circ Res 1996; 78: 606-614.

55. Philbrick WM, Wysolmerski JJ, Galbraith S, Holt E, Orloff JJ, Yang KH, Vasavada RC, Weir EC, Broadus AE, Stewart AF. Defining the roles of parathyroid hormone-related protein in normal physiology. Physiol Rev 1996; 76: 127-173.

56. Rubanyi GM, Romero C, Vanhoutte PM. Flow-induced release of endothelium-derived relaxing factor. Am J Physiol 1986; 250: H1145-H1149.

57. Pohl U, Holtz J, Busse R, Bassenge E. Crucial role of endothelium in the vasodilator response to increase in flow in vivo. Hypertension 1986, 8: 37-44.

58. Mombouli JV, Vanhoutte PM.Kinins and endothelium-dependent relaxations to converting enzyme inhibitors in perfused canine arteries. J Cardiovasc Pharmacol 1991;18(6):926-927.

59. Groves P, Kurz S, Just H, Drexler H. Role of endogenous bradykinin in human coronary vasomotor control. Circulation 1995;92(12):3424-3430.

60. Hutcheson IR, Griffith TM. Mechanotransduction through the endothelial cytoskeleton: mediation of flow- but not agonists-induced EDRF release. Br J Pharmacol. 1996; 118: 720-726.

61. Muller JM., Chillian WM, Davis MJ. Integrin signalling transduces shear stress-dependent vasodilation of coronary arterioles. Circ Res 1997; 80: 320-326.

62. Davies PF. Flow-mediated endothelial mechanotransduction. Physiol Rev. 1995; 75: 519-560.

63. Tronc F, Wassef M, Esposito B, Henrion D, Glagov S, Tedgui A. Role of NO in flow-induced remodeling of the rabbit common carotid artery. Arterioscler Thromb Vasc Biol 1996; 16 (10):1256-1262.

64. Miller VM, Vanhoutte PM. Endothelial alpha2-adrenoceptors in canine pulmonary and systemic blood vessels. Eur J Pharmacol. 1985; 118: H432-H437.

65. Bryan RM Jr, Steenberg ML, Eichler MY, Johnson TD, Swafford MW, Suresh MS. Permissive role of NO in alpha 2-adrenoceptor-mediated dilations in rat cerebral arteries. Am J Physiol 1995; 269(3 Pt 2):H1171-H1174.

66. Katusic ZS, Sheperd JT, Vanhoutte PM. Vasopressin causes endothelium-dependent relaxations in canine basilar arteries. Circ Res 1984; 55: 575-579.

67. Toda N. Mechanism of histamine actions in human coronary arteries.Circ Res 1987; 61: 280-286.

68. Berkenboom G, Depierreux M, Fontaine J. The influence of atherosclerosis on the mechanical responses of human isolated coronary arteries to substance P, isoprenaline and noradrenaline. Br J Pharmacol 1987;92(1):113-120.

69. Nakashima N, Mombouli JV, Taylor AA, Vanhoutte PM. Endothelium-dependent hyperpolarization caused by bradykinin in human coronary arteries. J. Clin. Invest 1993; 92: 2867-2871.

70. Mombouli JV, Vanhoutte PM. Kinins and endothelial control of vascular smooth muscle. Annu Rev Pharmacol Toxicol 1995;35:679-705.

71. Hornig B, Kohler C, Drexler H. Role of bradykinin in mediating vascular effects of angitoensin-converting enzyme inhibitors in humans. Circulation 1997; 95: 1115-1118.

72. Cohen RA, Sheperd JT, Vanhoutte PM. Inhibitory role of the endothelium in the response of isolated coronary arteries to platelets. Science 1983; 221: 273-274.

73. Vanhoutte PM, Boulanger CM. Endothelium-dependent response in hypertension. Hypert Res 1995; 18: 87-98.

74. Miller VM, Vanhoutte PM. Endothelium-dependent contractions to arachidonic acid are mediated by products of cyclooxygenase in canine veins. Am J Physiol 1985; 248: H432-H437.

75. Ge T, Hughes H, Junquero DC,Wu KK, Vanhoutte PM, Boulanger CM. Augmented expression of prostaglandin H synthase and contraction to prostaglandin H2 in the aorta of spontaneously hypertensive rats. Circulation Research 1995; 76: 1003-1010.

76. Pou S, Pou WS, Bredt DS, Snyder SH, Rosen GM. Generation of superoxide by purified brain nitric oxide synthase. J Biol Chem 1992; 267: 24173-24176.

77. Heinzel B, John M, Klatt P, Bohme E, Mayer B: Ca2+ calmodulin dependent formation of hydrogen peroxide by brain nitric oxide synthase. J. Biol. Chem 1992; 281: 627-630.

78. Graser T, Vanhoutte PM.Hypoxic contraction of canine coronary arteries: role of endothelium and cGMP. Am J Physiol 1991; 261(6 Pt 2):H1769-H1777.

79. Pearson PJ, Lin PJ, Schaff HV, Vanhoutte PM Augmented endothelium-dependent constriction to hypoxia early and late following reperfusion of the canine coronary artery. Clin Exp Pharmacol Physiol 1996;23(8):634-641.

80. Yanagisawa, M., Kurihara, H., Kimura, S., et al.: A novel potent vasoconstrictor peptide produced by vascular endothelial cells. Nature 1988; 332: 411.

81. Ohnaka K, Takayanagi R, Nishikawa M, et al.: Purification and characterization of a phosphoramidon-sensitive endothelin-converting enzyme in porcine aortic endothelium. J Biol Chem 1993; 268: 26759.

82. Xu D, Emoto N, Giaid A, et al.: ECE-1: a membrane-bound metalloprotease that catalyzes the proteolytic activation of big endothelin-1. Cell 1994; 78: 473.

83. Arai, H., Hori, S., Aramori, I., et al.: Cloning and expression of a cDNA encoding an endothelin receptor. Nature 1990; 348: 730.

84. Sakurai, T., Yanagisawa, M., Takuwa, Y., et al.: Cloning of a cDNA encoding a non-isopeptide-selective subtype of the endothelin receptor. Nature 1990; 348: 732.

85. Seo, B., Oemar, B. S., Siebenmann, R., et al.. Both ETA and ETB receptors mediate contraction to endothelin-1 in human blood vessels. Circulation 1994; 89: 1203.

86. Wagner, O. F., Christ, G., Wojta, J., et al.: Polar secretion of endothelin-1 by cultured endothelial cells. J Biol Chem 1992; 267: 16066.

87. Stewart, D. J., Langleben, D., Cernacek, P., et al.: Endothelin release is inhibited by coculture of endothelial cells with cells of vascular media. Am J Physiol 1990; 259: H1928.

88. Flavahan NA, Shimokawa H, Vanhoutte PM. Pertussis toxin inhibits endothelium-dependent relaxations to certain agonists in porcine coronary arteries. J. Physiol. 1989; 408: 549-560.

89. Shimokawa H, Flavahan NA, Vanhoutte PM. Natural course of the impairment of endothelium-dependent relaxations in regenerating porcine endothelial cells: role of a pertussis toxin sensitive G-protein. Circ Res 1989; 65; 740-753.

90. Shimokawa H, Flavahan NA, Shepherd JT, Vanhoutte PM. Endothelium-dependent inhibition of ergonovine-induced contraction is impaired in porcine coronary arteries with regenerated endothelium. Circulation 1989; 80: 643-650.

91. Borg-Capra C, Fournet-Bourguignon MP, Janiak P, Villeneuve N, Bidouard JP, Vilaine JP, Vanhoutte PM . Morphological heterogeneity with normal expression but altered function of G proteins in porcine cultured regenerated coronary endothelial cells. Br J Pharmacol 1997; 122(6):999-1008.

92. Tsutsui M, Shimokawa H, Tanaka S, Kuwoaka I, Hase K, Nogami N, Nakanishi K, Okamatsu S: Endothelial Gi protein in human coronary arteries. Eur. Heart J 1994; 15: 1261-1266.

93. Shimokawa H, Tsutsui M, Mizuki T, Hase K, Kuwaoka I, Nogami N, Okamatsu S, Nakanishi K: Endothelial Gi protein expression is markedly low in human coronary microvessels. J Cardiov. Pharmacol 1996; 27: 297-302.

94. Pearson PJ, Schaff HV, Vanhoutte PM. Acute impairment of endothelium-dependent relaxations to aggregating platelets following reperfusion injury in canine coronary arteries. Circ Res 1990; 67(2):385-393

95. Lee JJ, Olmos L, Vanhoutte PM. Recovery of endothelium-dependent relaxations four weeks after ischemia and progressive reperfusion in canine coronary arteries. Proc Assoc Am Physicians 1996;108(5):362-367.

96. Ludmer PL, Selwyn AP, Shook TL, Wayne RR, Mudge GH, Alexander RW, Ganz P. Paradoxical vasoconstriction induced by acetylcholine in atherosclerotic coronary arteries. N Engl J Med 1986;315(17):1046-1051.

97. Golino, P, Piscione F, Willerson JT, Cappelli BM, Focaccio A, Villari B, Indolfi C, Russilillo E, Condorelli M, Chiarello M. Divergent effects of serotonin on coronary artery dimensions and blood flow in patients with coronary atherosclerosis and control patients. N Engl J Med 1991; 324: 641-648.

98. Shimokawa H, Vanhoutte PM: Impaired endothelium-dependent relaxation to aggregating platelets and related vasoactive substances in porcine coronary arteries in hypercholesterolemia and atherosclerosis. Circ Res 1989; 64: 900-914.

99. Wilcox JN, Subramanian RR, Sundell CL, Tracey WR, Pollock JS, Harrison DG, Marsden PA Expression of multiple isoforms of nitric oxide synthase in normal and atherosclerotic vessels. Arterioscler Thromb Vasc Biol 1997; 17(11):2479-2488.

100. Minor RL Jr, Myers PR, Guerra R Jr, Bates JN, Harrison DG Diet-induced atherosclerosis increases the release of nitrogen oxides from rabbit aorta. J Clin Invest 1990; 86(6):2109-2116.

101. Wever RM, Luscher TF, Cosentino F, Rabelink TJ Atherosclerosis and the two faces of endothelial nitric oxide synthase. Circulation 1998; 97(1):108-112.

102. Stroes E, Kastelein J, Cosentino F, Erkelens W, Wever R, Koomans H, Luscher T, Rabelink T Tetrahydrobiopterin restores endothelial function in hypercholesterolemia. J Clin Invest 1997; 99(1):41-46.

103. Ting HH, Timimi FK, Haley EA, Roddy MA, Ganz P, Craeger MA. Vitamin C improves endothelium dependent vasodilation in forearm resistance vessels of human with hypercholesterolemia. Circulation, 1997; 95: 2617-2622.

104. Lerman A, Edwards BS, Hallett JW et al.: Circulating and tissue endothelin immunoreactivity in advanced atherosclerosis. New Engl J Med 1991; 325: 997-999.

105. Hasdai D, Best PJ, Cannan CR, Mathew V, Schwartz RS, Holmes DR Jr, Lerman A. Acute endothelin-receptor inhibition does not attenuate acetylcholine-induced coronary vasoconstriction in experimental hypercholesterolemia. Arterioscler Thromb Vasc Biol 1998; 18(1):108-113.

106. Linder L, Kiowski W, Buhler FR, Luscher TF. Indirect evidence for release of endothelium derived relaxing factor in human forearm circulation in vivo. Blunted response in essential hypertension. Circulation 1990 Jun;81(6):1762-1767.

107. Panza JA, Quyyumi AA, Brush JE Jr, Epstein SE. Abnormal endothelium-dependent vascular relaxation in patients with essential hypertension. N Engl J Med 1990 5;323(1):22-27.

108. Creager MA, Roddy MA. Effect of captopril and enalapril on endothelial function in hypertensive patients. Hypertension 1994 Oct;24(4):499-505.

109. Taddei S, Virdis A, Ghiadoni L, Magagna A, Salvetti A Cyclooxygenase inhibition restores nitric oxide activity in essential hypertension. Hypertension 1997; 29(1 Pt 2):274-279.

110. Cockcroft JR, Chowienczyk PJ, Benjamin N, Ritter JMPreserved endothelium-dependent vasodilatation in patients with essential hypertension. N Engl J Med 1994; 330(15):1036-1040.

111. Lüscher TF., Raij L. and Vanhoutte P.M. Endothelium-dependent vascular responses in normotensive and hypertensive Dahl-rats. Hypertension 1987; 9: 157-163.

112. Hayakawa H, Raij L. Nitric oxide synthase activity and renal injury in genetic hypertension. Hypertension 1998 Jan;31(1 Pt 2):266-270.

113. Matsuoka H, Itoh S, Kimoto M, Kohno K, Tamai O, Wada Y, Yasukawa H, Iwami G, Okuda S, Imaizumi T. Asymmetrical dimethylarginine, an endogenous nitric oxide synthase inhibitor, in experimental hypertension. Hypertension 1997; 29(1 Pt 2):242-247.

114. Patel A, Layne S, Watts D, Kirchner KA. L-arginine administration normalizes pressure natriuresis in hypertensive Dahl rats. Hypertension 1993; 22(6):863-869.

115. Boulanger CM, Heymes C, Benessiano J, Geske RS, Lévy BI, Vanhoutte PM. Type I Nitric oxide synthase is expressed in vascular smooth muscle cells: activation by angiotensin II in hypertension. The Physiologist, 1998; 41, 4: 271.

116. Tschudi MR, Mesaros S, Luscher TF, Malinski T. Direct in situ measurement of nitric oxide in mesenteric resistance arteries. Increased decomposition by superoxide in hypertension. Hypertension 1996 ;27(1):32-35.

117. Cannan CR, McGoon MD, Holmes DR Jr, Lerman A. Altered coronary endothelial function in a patient with asymptomatic left ventricular dysfunction. Int J Cardiol 1996; 53(2):147-151.

118. Lerman A, Kubo SH, Tschumperlin LK, Burnett JC Jr. Plasma endothelin concentrations in humans with end-stage heart failure and after heart transplantation. J Am Coll Cardiol 1992; 20(4):849-853.

119. Mulder P, Richard V, Derumeaux G, Hogie M, Henry JP, Lallemand F, Compagnon P, Mace B, Comoy E, Letac B, Thuillez C. Role of endogenous endothelin in chronic heart failure: effect of long-term treatment with an endothelin antagonist on survival, hemodynamics, and cardiac remodeling. Circulation 1997; 96(6):1976-1982.

5. MECHANICAL FACTORS AND VASCULAR BIOLOGY

Alain Tedgui, Stéphanie Lehoux, Bernard Levy
INSERM U141 and IFR Circulation, Hôpital Lariboisière, Paris, France

"Mechanoreception is the most widely distributed sensory modality in animals. It subserves the conscious senses of hearing, orientation to local gravity, and touch. It provides the voluntary musculature with information about distension and tension which is required for coordinated movement. It provides the viscera with the ability to sense blood pressure, lung, gut, bladder, and mammary inflation. It may also be used to transduce osmotic pressure: from sex to gas pain mechanoreceptors are there"

Sachs F. Biophysics of mechanoreception. *Membr Biochem.* 1986;6:173-195.

Blood vessels are permanently subjected to mechanical forces in the form of stretch, which, due to the pulsatile nature of blood flow, exposes vessels to cyclic mechanical strain, and shear stress. Blood pressure is the major determinant of vessel stretch. It creates radial and tangential forces which counteract the effects of intraluminal pressure, and which affect all cell types in the vessel. In comparison, fluid shear stress results from the friction of blood against the vessel wall, and it acts in parallel to the vessel surface. Accordingly, shear is sensed principally by endothelial cells, strategically located at the interface between the blood and the vessel wall. Alterations in stretch or shear stress invariably produce transformations in the vessel wall that will aim to accommodate the new conditions and to ultimately restore basal levels of tensile stress and shear stress [1-3]. Hence, while acute changes in stretch or shear stress correlate with transient adjustments in vessel diameter, mediated through release of vasoactive agonists or change in myogenic tone, chronically altered mechanical forces usually instigate important adaptive alterations of vessel wall shape and composition. The concept of vascular remodeling has therefore been used to describe the transformations that occur in vessels undergoing mechanical stresses.

MECHANICAL FORCES

Pressure, tension and tensile stress

The blood pressure exerts forces on the vessel wall in a direction perpendicular to the endoluminal surface. These forces are balanced by intraparietal tangential forces in the longitudinal and circumferential directions exerted by the different elements of the vessel wall, opposing the distending effects of blood pressure in the circumferential direction, the force per unit length of the vessel (the parietal tension) is related to the blood pressure and the vessel radius by Laplace's law:

$$T = P.r$$

The relation between circumferential tension and deformation of the vessel as intraluminal pressure increases depends both on the geometry and the elastic characteristics of its wall (see chapter *Mechanics of the Large Artery Vascular Wall*). The circumferential tension is actually borne by the whole thickness of the arterial wall. Each element of the wall only bears part of this tension. The tension per unit of thickness represents the stress exerted on the wall in the circumferential direction. It is expressed as:

$$\sigma = Pr/h$$

where h is the thickness of the wall.

Blood flow and shear stress

As blood flows, it exerts a frictional force on the endothelial surface. This force is expressed as a shear stress (τ) on the endothelium, defined as the product of the blood viscosity and the blood velocity gradient measured at the vessel wall. The shear stress transmitted to the endothelium by the blood flow tends to displace the endothelium and the intimal layer in the direction of flow (one might equally say that it is because the endothelium is fixed that friction occurs). In the case of laminar flow (where the profile of blood velocity is parabolic), the shear stress is expressed as:

$$\tau = 4\ \mu\ Q/\pi\ r^3$$

where μ is the viscosity, Q the flow rate and r the vessel radius. Note that the radius appears at the third power in the denominator; thus, at a constant volume flow, a slight reduction in vascular diameter produces a much greater increase in shear stress.

The arteries are normally exposed to pulsatile pressure and flow, and consequently to changes in pressure and blood velocity over the course of the cardiac cycle. The mechanical stresses related to those variations lead to both short-term and long-term reactions. The dynamic characteristics of the blood flow and the cyclic movements of the vessel wall that accompany pulsatile flow play a crucial role in atherogenesis [1]. The effects of pulsatile flow may be stronger in some regions than others due to the presence of oscillatory shear stress (as occurs, for example, on the external surface of the carotid sinus). The number of pulsations over time can influence both the degree and the localization of intimal lesions in susceptible vascular territories [4].

In summary, the vascular wall is normally exposed to two types of mechanical forces: (a) circumferential stress acting tangentially on the wall and directly related to pressure and dimensions (diameter and thickness) of the vessel, and (b) shear stress acting in a longitudinal direction at the blood-endothelium interface and directly related to the flow-velocity profile.

MECHANICAL STRESS AND VASCULAR REMODELING

On the basis of observations in chick embryos, Thoma in 1893 produced the hypothesis that the size of blood vessels is regulated by the magnitude of blood flow, while the thickness of the vessel wall depends on the magnitude of the tension forces generated by the blood pressure. This hypothesis has subsequently been experimentally confirmed. It has been demonstrated, for example, that the diameter of the abdominal aorta of a lamb undergoes a significant reduction between the 4th and 14th days *post partum* [5]. This reduction can be accounted for by a fall of approximately 70% in the blood velocity in the abdominal aorta at the time of delivery, due to the disappearance of the placenta circulation, and is associated with apoptosis of vasular cells [6]. At the same time, the diameter of the thoracic aorta increases in parallel with the increase in systemic blood flow. Similarly, the thicknesses of the pulmonary artery and aorta, which are almost identical at birth since pressures in utero are similar in both vascular territories, evolve differently after birth. The pulmonary artery atrophies during development, following the fall in pulmonary pressure post partum, while the thoracic aorta thickens in relation to the increase in systemic pressure [7].

Effects of pressure

Numerous studies have demonstrated a direct relationship between the circumferential stress to which the vessel wall is exposed and the structure of the wall itself. When the stress increases due an increase in arterial pressure, smooth muscle cell hypertrophy and increased of collagen and elastin contents follow. Inversely, when the circumferential stress falls, the wall undergoes atrophy [8]. Several physiologic and experimental arguments confirm the relationship between circumferential stress and the thickness and composition of the vessel wall:

1) From one animal species to another, as the diameter of a particular blood vessel increases, the number of lamellar units and the total thickness of the wall increase proportionately, so that the circumferential stress remains constant irrespective of the size of the animal, from the rat to the horse. This " ideal " value is of the order of 2.10^6 dyne/cm^2 in the descending thoracic aorta [9, 10]. It varies according to the arterial territory and essentially depends on the structure of the blood vessel concerned.

2) In all experimental models of arterial hypertension, a close correlation is observed between the level of arterial pressure and the frequency of polyploidy and hypertrophy of the smooth muscle cells of the arterial wall [11].

3) Smooth muscle cell hypertrophy in the wall of the major arterial trunks only develops when the distending pressure has reached a threshold level, and never precedes the onset of hypertension, even when the neurohumoral abnormalities responsable for hypertersion are already present.

The conceptually simplest experimental model used to underscore the effect of blood pressure on the arterial wall structure is arterial stenosis. In this model high blood pressure and wall stresses are observed proximal to the coarctation, whereas normal or low pressure and stresses occur distal to the stenosis. In rats and in monkeys with thoracic coarctations [12, 13], the arterial wall was reported to be thickened in the section submitted to high blood pressure and normal in the lower part of the arterial network of the same animals, in relation with the level of blood pressure and tensile wall stresses.

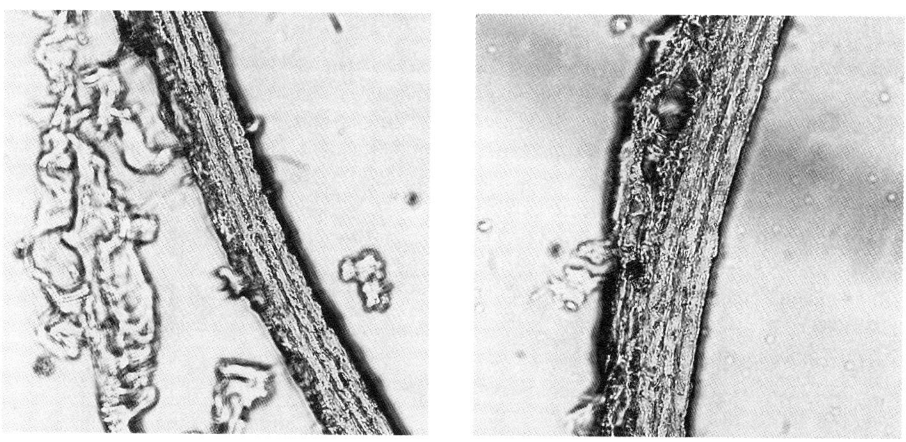

***Figure 1**. Increased thickness of the aorta of hypertensive rats (right) compared to normotensive rats (left)*

The effects of mechanical tensile stress on the arterial wall have been extensively described and have been applied to the understanding of hypertension. Tensile stress is a strong determinant of the vascular structure among other factors including sympathetic activity and autocrine and paracrine factors, especially the renin-angiotensin system. In the early phase of essential arterial hypertension, it is generally admitted that the vessel wall is submitted to increased pressure because of abnormal peripheral resistances related to genetic, humoral, nervous and/or structural factors. Hypertension has long been considered to be a disease of the cardiovascular system that affects almost exclusively the arterioles and the heart. However, damage to the large arteries are clearly involved in cardiovascular morbidity and mortality associated with hypertension [14]. Numerous animal and human studies have shown that sustained hypertension is associated with structural and functional alterations to both large arteries and arterioles. There is good evidence that hypertension is associated with increased arterial wall thickness [9, 15] and alterations to the structural composition of the arterial wall [16, 17] (Figure 1), leading to altered arterial function [18, 19].

It has been suggested that smooth muscle cell (SMC) hypertrophy, accompanied by polyploidism, contributes to the increase in SMC mass in several animal models of hypertension [20]. The chronic increase in blood pressure and therefore in circumferential tensile stress, as occurs in chronic hypertension, increases the arterial wall thickness [21-23]. According to Laplace's equation (σ = Pr/h), the hypertrophy of the arterial wall compensates for the increase in blood pressure and contributes to maintain a normal level of circumferential stress (Figure 2).

In elastic and large conduit arteries, the presumably adaptive response to hypertension serves to reduce and eventually to normalize the tensile stress [24]. In spontaneously hypertensive rats (SHRs), the medial thickening of the thoracic aorta is due to smooth muscle hyperplasia and hypertrophy [25]. An increase in the DNA content and polyploidy have also been reported in hypertensive aorta [20]. In contrast, it appears that in the carotid arterial wall, SMC hyperplasia is the predominant cause of medial thickening [25]. The marked increases in the SMC size and/or number is likely to be related to increased protein synthesis by SMCs in the vascular wall.

In resistance arteries, increases in pressure and therefore in wall tensile stress are associated with medial thickening mostly due to SMC hyperplasia and proportional changes in contractile and matrix proteins. Medial hypertrophy is associated with a decrease in the lumen diameter of resistance arteries, resulting in an increase in the media/lumen ratio, and thus increased hemodynamic resistance. Longitudinal age studies of small arteries both in SHRs and in renal hypertensive rats [26] show that the media-to-lumen ratio of the vessels increases concomitant with the increase in pressure, such that the media stress (force exerted by the vascular smooth muscle per unit area) does not increase.

To simplify our understanding of these phenomena, we can suppose that everything occurs as though the thickness of the artery, the denominator of the equation determing the tensile stress (σ = Pr/h), increases in such a way as to balance out the increase in the numerator (wall tension), whether the latter is due to an increase in pressure or an increase in the radius of the vessel. However, some studies have tended to demonstrate that for a given hypotensive effect, different classes of antihypertensive drugs may have different effects on the arterial wall. In particular, it has been demonstrated that angiotensin II-converting enzyme inhibitors are more effective in reducing arterial hypertrophy and fibrosis in hypertensive animals [27].

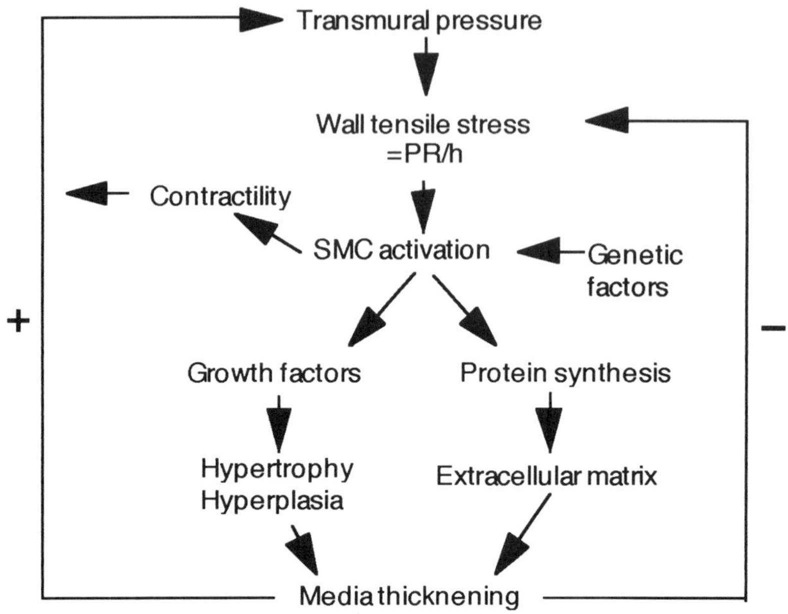

Figure 2. *Regulation of vascular remodeling by the transmural pressure: an increase in the wall tensile stress resulting from hypertension activates vascular smooth muscle cells (SMC) and induces both SMC hypertrophy and/or hyperplasia and extracellular matrix synthesis in relation with changes in SMC phenotype (from contractile to secretory). The medial thickening resulting from vascular wall remodeling continues until tensile stress is normalized. In essential hyperpertension, genetic factors induce over-activation of SMC and an abnormal increase in contractility. Concomitant increase in wall thickness and hypercontractility of resistance arteries participate in arterial hypertension.*

The renin-angiotensin system could play an important role in the vascular wall remodeling process. Similar wall stresses were found in muscular mesenteric arteries from normotensive rats and SHRs despite marked differences in arterial pressure and wall tension [28]. In contrast, in hypertensive transgenic rats over-expressing renin, a disproportional wall hypertrophy was observed in muscular mesenteric arteries. Circumferential wall stresses were therefore markedly lower in Ren-2 transgenic rats than in SHRs despite similar levels of hypertension [29]. In a study of blood vessels maintained in organ culture for several days, we have demonstrated that protein synthesis by smooth muscle cells is greatly increased when the vessels are subjected to elevated intraluminal pressures [30]. Furthermore, pressure and angiotensin II are capable of inducing fibronectin expression in the arterial wall in a synergistic manner [31]. The effector mechanisms for the increased expression of fibronectin in response to pressure involve local activation of the vascular renin-angiotensin system [31]. Mechanical and humoral factors can therefore interact to promote vascular remodeling.

Effects of radius changes

According to Laplace's law, tensile stress can also be modified by changes in lumen size and/or vascular wall thickness. Under physiological conditions in young adults, vascular wall tension decreases in proportion to arterial diameter such that it is approximately 10,000 times greater in the aorta than in a capillary [32]. Taking into account the relative thickness of the vascular wall narrows the difference in wall stress between the aorta and the capillaries from 10,000-fold to approximately 6-fold [33, 34].

The effect of aging on arterial geometry is characterized by the constant enlargement observed in humans [35] and in animals [36] with increasing age. It is noteworthy that arteries continue to grow during the whole life while the body size does not change for a long time. Arterial growth affects both the diameter and the length of vessels: the aorta for example becomes tortuous with normal aging because of increased length, and it dilates to such an extent that its intimal surface area doubles between the second and sixth decade in men. The aortic circumference increases with age independently of the aortic location from the ascending aortic arch to the bifurcation of the abdominal aorta (Figure 3). Age appears to have minimal effect on medial thickness, whereas the effect on intimal thickness is pronounced, such that the increase in intimal thickness with age is the predominant factor responsible for the well-known age-related increase in total wall thickness of the aorta. Interestingly, the changes in lumen radius and total wall thickness (media + intima) occurring with age are in agreement with the concept that tensile wall stress is maintained at an ideal value during the whole life under non-pathological conditions.

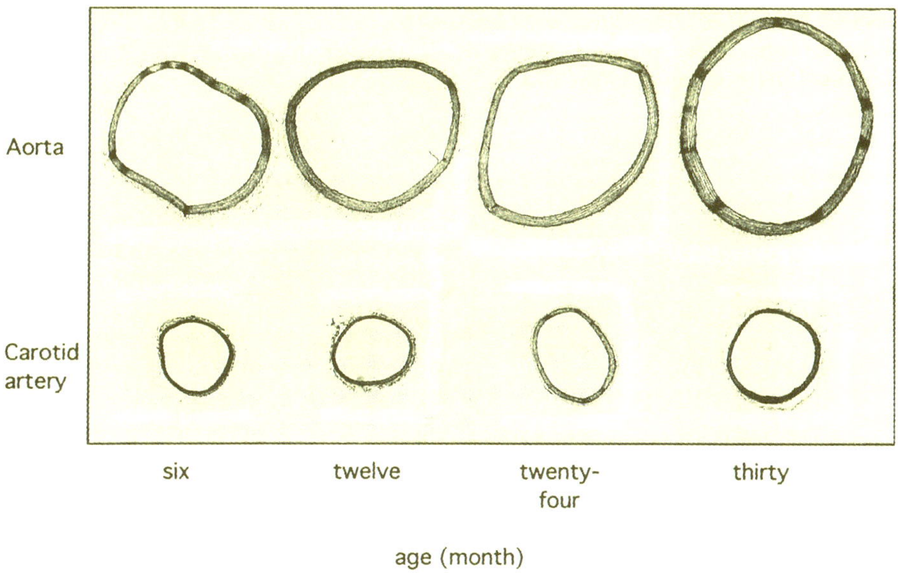

Figure 3. *Increased vessel caliber in rat arteries with age (from Michel et al [36]).*

Hence, alterations in the structural and physical properties of the arterial wall are present in the normal aging population. In an experimental study involving rats 6 to 30 months old, significant increases in the aortic and carotid wall elastin and collagen contents were reported concomitant with aortic enlargement and thickening. Furthermore, qualitative changes in the internal elastic lamellae were also observed in the form of ruptures or lack of continuity. These structural changes resulted in a dramatic decrease in arterial wall distensibility: the wall of the large arteries was markedly stiffer in older animals. Because vessel compliance is the product of arterial wall distensibility and arterial lumen volume, the decrease in distensibility with aging is partially compensated for by the concomitant increase in vascular volume.

In a non-invasive human study (more than 13,000 subjects), Crouse et al [37] reported that wall thickness of the common carotid artery was greater by 21% in humans aged 60 to 64 years compared with those aged 45 to 49 years. However, lumen diameters were also greater in older individuals, resulting in no lumen constriction associated with thickened carotid wall. In the same way, in experimental [38] and humans studies [39] concerning arterial remodeling occurring in atherosclerosis, it appears that the artery compensates for intimal formation by vessel enlargement. Finally, at least in experimental atherosclerosis, the degree of vessel enlargement might be more important than intimal thickening in determining the chronic lumen size. In most conditions examined, a tendency towards the normalization of the wall tensile stress can be hypothesized.

A clear and typical illustration of the role of the vessel radius changes in determining the wall thickening is the use of restrictive external supports to limit the wall distension of vein grafts submitted to arterial blood pressure (Figure 4). Several experiments show that the maintenance of the external radius of such veins by an external non-distensible stents diminishes the hyperplasic response of the vein wall [40, 41].

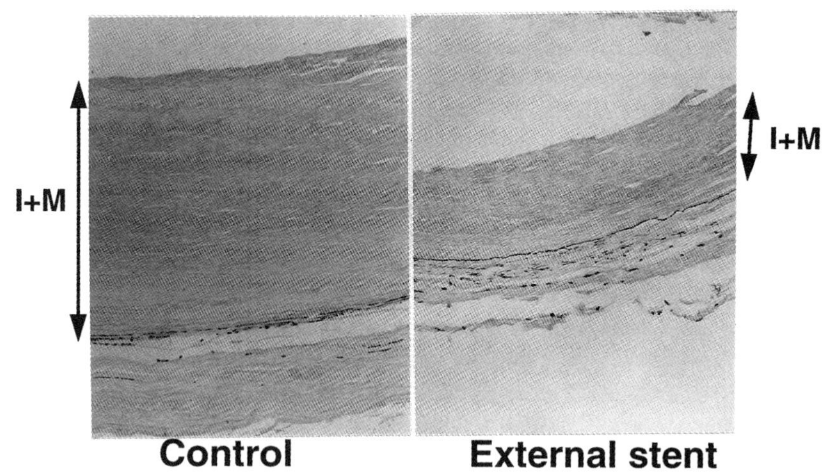

Figure 4. Rigid external stenting applied to rabbit jugular vein graft to prevent vein distension reduces intima-media thickness (from Batellier et al. [40])

Effects of pressure-induced crush

In addition to the circumferential tensile stress responsible for wall stretch, the intraluminal pressure engenders two other stresses within the vessel wall: an axial stress responsible for vessel elongation and a radial stress causing wall crush. This later effect might be important if wall stretching is limited or prevented. In *in vivo* conditions, pressure-induced vascular stretch is limited due to the mechanical properties of the vessel wall and surrounding tissues, and/or to the pressure-induced myogenic tone. As a result, in the presence of limited vascular stretch, graded increases in intraluminal pressure are associated with incremental levels of pressure-induced arterial wall crush. This is particularly evident in percutaneous balloon angioplasty in which the arterial wall is subjected to very high intraluminal pressures associated with limited vascular stretch leading to marked pressure-induced vascular crush. It has been reported that sustained pressure applied to rat cultured SMCs was able to promote DNA synthesis and cell proliferation [41]. Similarly, human vascular smooth muscle cultured in a three-dimensional collagen gel system increases DNA synthesis in response to transient mechanical compression, in part via autocrine or paracrine fibroblast growth factor-2 [42]. The effect of arterial compression on DNA synthesis might involve early proto-oncogene expression. Indeed, acute pressure-induced crush of isolated rabbit thoracic aorta stimulates sustained expression of *c-fos* and *c-jun* genes [43]. This is particularly interesting inasmuch as Fos and Jun proteins form a dimeric complex (AP-1) that binds to specific sequences in the 5' flanking region of several target genes [44, 45] and are involved in the regulation of DNA synthesis and cell proliferation. Moreover, an increase in *c-fos* expression proportional to the degree of balloon inflation pressure has been reported in a rat model of arterial injury [46]. The increase in *c-fos* expression appears to be associated with a proportional increase in SMC proliferation. Therefore, compression might be an additional mechanical factor generated by hemodynamic forces which must be taken into account when attempting to understand the mechanisms involved in vascular remodeling.

Effects of blood flow

Shear stress arising from the mechanical effects of blood flow on the vascular endothelium has an effect on arterial growth. Restricting blood flow by placing a stenosing ring around the common carotid artery in young rats leads to a significant delay in growth of the carotid during the subsequent weeks, with a reduction of 25% in its diameter as compared with the contralateral carotid [47]. The content of fibrous proteins in the arterial wall is also controlled by blood flow and shear stress. In rabbits, the reduction in caliber of the developing carotid associated with a reduction in its blood flow is accompanied by a reduction in the elastin content of the carotid arterial wall [48].

The most spectacular illustration of the phenomenon of flow-dependent growth is the arteriovenous fistula model. In carotid-jugular arteriovenous fistulas created in dogs [49] or rabbits [3], the flow rate in the developing carotids can be multiplied by a factor of up to 8. As long as there is no excessive increase in blood flow, the shear stress is normalized by a compensatory increase in carotid diameter (Figure 5). Under physiologic conditions, the mean shear stress to which the vascular endoluminal surface is exposed is remarkably constant, close to 10-15 dyn/cm^2, irrespective of which part of the vascular system is examined: arterial, venous or capillary.

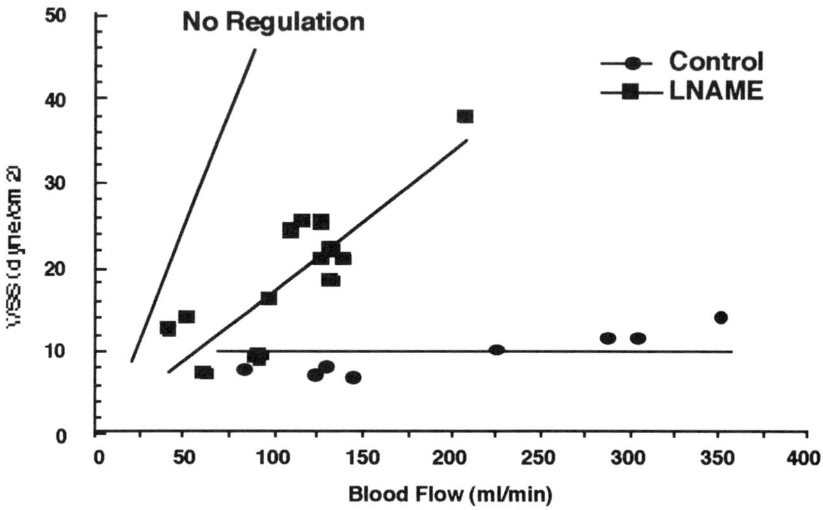

Figure 5. *Wall shear stress (WSS) is remins unchanged whatever the blood flow in flow-loaded carotid upstream of an arteriovenous fistula. The regulation is partially lost after chronic inhibition of NO synthesis by L-NAME treatment (from Tronc et al. [3]).*

A chronic increase in shear tends to enhance the L-arginine/NO pathway in the endothelial cell. The endothelial NO synthase mRNA expression is enhanced [50], and cGMP in flow-loaded vessels is markedly increased [51]. We have shown that chronic inhibition of NO production by L-NAME treatment inhibits, at least partially, the adaptive wall shear stress regulation in flow-loaded vessels [3] (Figure 6). However, simple relaxation of vascular smooth muscle is not enough to account for the very significant increase in vascular caliber, which may almost double in response to large increases in flow. Previous microscopic and ultrastructural studies of the arterial wall proximal to an arteriovenous fistula have shown extensive tears and fragmentation of the internal elastic lamina (IEL) [3, 52, 53]. Recently, Wong and Langille [54] have demonstrated enlarged fenestrae in the IEL of flow loaded rabbit carotid arteries using laser scanning confocal microscopy. Disruption of IEL in flow-loaded arteries suggest a potential role for matrix metalloproteinases (MMPs) in matrix digestion and reorganization leading to arterial wall remodeling [3].

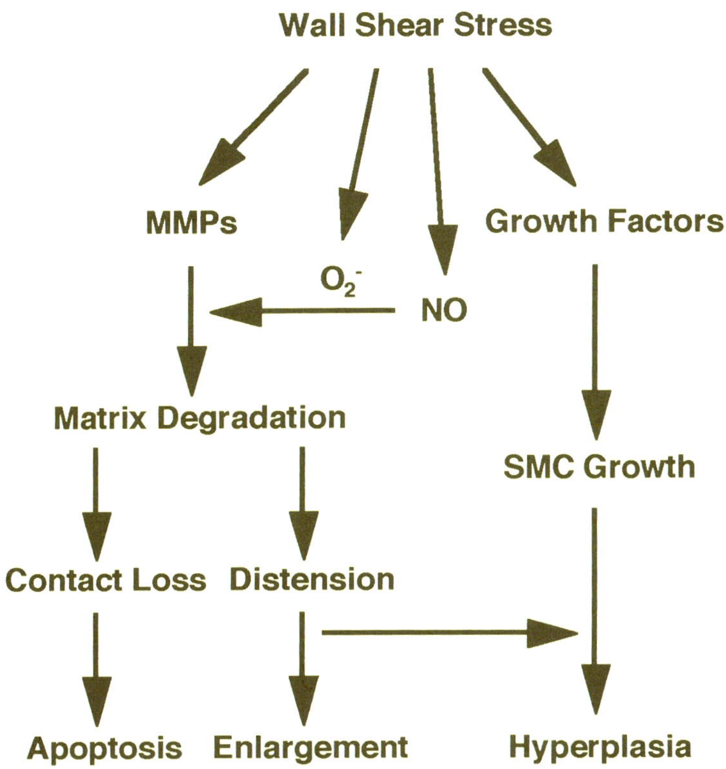

***Figure 6**. Diagram showing blood-flow mediated mechanisms leading to arterial remodeling*

Role of MMPs in blood-flow induced vascular remodeling

In agreement with this hypothesis, we recently found that MMP-2 and MMP-9 are induced in the vessel wall in response to increased in blood flow and play a major role in the adaptive remodeling of the flow-loaded artery, because this process can be prevented by inhibiting MMP activity by chronic treatment with doxycycline, a tetracycline with inhibitory effects on MMP expression or activation [55]. We therefore believe that increased shear stress stimulates the production of MMPs in the arterial wall resulting in matrix degradation, especially tears in the IEL, ultimately leading to arterial enlargement. Indeed, if MMP-induced IEL fenestrations are formed following increased blood flow, arterial distensibility increases resulting in an enhanced arterial diameter. As arterial caliber gradually increases, wall shear stress diminishes and the stimulus for MMP production/activation fades. This process will continue until arterial caliber is such that WSS is normalized. Interestingly, it has been shown that NO or peroxynitrite, which results from simultaneous production of NO and superoxide anions, activates MMPs [56-58], and fluid flow might stimulate MMP production by endothelial cells [59]. Therefore NO and MMPs stimulated by increased shear stress can cooperate to induce a vascular adaptive response to blood flow (Figure 6).

MECHANICAL FORCES AND VASCULAR CELL PHENOTYPE

Effects of stretch

Mechanical stretching of SMC produces a variety of responses that account for vessel morphology. On the one hand, stretch may be at the very core of the SMC differentiated state. Indeed, in a model of cultured rabbit aorta, we determined that a certain level of stretch is crucial for the maintenance of the differentiated phenotype of the SMC. Vessels placed in conditions of abnormally low intraluminal pressure (10 mmHg) showed decreased content, over three to six days, of smooth muscle marker proteins h-caldesmon and filamin content, despite the presence of fetal calf serum, a known mitogen, in the culture medium. In comparison, loss of these proteins was prevented in aortic segments kept at physiological intraluminal pressure (80 mmHg)[60]. The presence of endothelium was not essential for maintenance of SMC marker proteins. These results corroborate earlier experiments, where cyclic stretching of cultured SMC was shown to increase (we should say prevent the decrease) the expression of smooth muscle myosin heavy chains and myosin light chain kinase [61]. Furthermore, cyclic stretching (12-72h) of SMC augmented smooth muscle myosin heavy chain SM-1 and SM-2 protein content and decreased nonmuscle myosin NM-A, compared to static SMC cultures [62]. Since only SM-1 showed enhanced mRNA expression under cyclic strain [62], the relative increase in SM-2 was presumably due to reduced degradation of this marker protein. Therefore, constant mechanical stimulation appears to be required for maintenance of normal contractile phenotype of SMC in the arterial wall. Loss of stretch together with loss of extracellular matrix contacts are probably major causes of differentiation of SMC in culture. However, while a certain level of stretch is required to maintain SMC in a quiescent state, overstretching triggers adaptive processes resulting in increased protein synthesis and hypertrophy.

In fact, organ culture of aortas maintained during three to six days at high intraluminal pressure (150 mmHg) revealed that both total protein synthesis and fibronectin content are enhanced by stretch [30]. Interestingly, DNA synthesis remains unchanged in the same preparations (Figure 7), in keeping with observations of pressure-induced hypertrophy but not hyperploidy in large vessels *in vivo* [9], and contrary to DNA synthesis induction by stretch in SMC culture [63,64]. In the latter case, absence of interaction with the extracellular matrix environment is certainly responsible at least in part for the hyperplasic response to stretch.

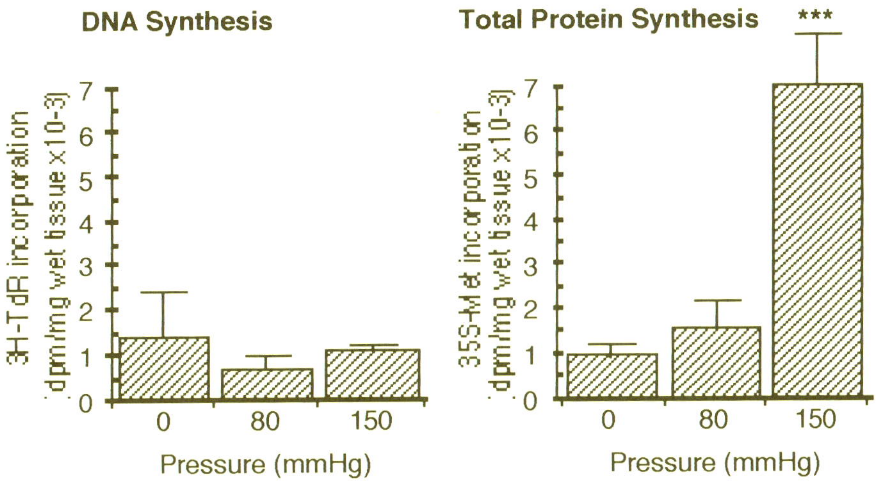

***Figure** 7. Effects of transmural pressure on DNA and total protein synthesis in vessel organ culture (from Bardy et al. [30])*

Effects of shear stress

The fact that shear acts mainly on endothelial cells, while stretch has repercussions on the entire vessel wall, does not rule out the likelihood that long-term changes in blood flow will bring about vascular remodeling. Accordingly, in endothelial cells subjected to oscillatory flow, fibronectin and laminin content was found to be greater than in static cultures [65]. However, while laminar shear stress may lead to a reorganization of cytoskeletal proteins and change of cell shape, it apparently does not change protein levels in cultured endothelial cells [66]. On the other hand, a variety of genes encoding for growth factors, (PDGF, TGF) [67,68], vasodilators (NO, prostacyclin) [3,50,69,70], vasoconstrictors (endothelin)[71] and adhesion molecules (intercellular adhesion molecule) [72] are regulated upon shear stimulation. While a number of these inductions are transient, some persist and may mediate long-term alterations in vessel structure and function that occur through regulation of protein and gene expression [3]. Finally, pressure-induced circumferential cyclic strain increases endothelial cell sensitivity to shear stress, resulting in a lowered threshold level of shear to provoke structural responses [73]. Ultimately, concomitant stimulation of vascular cells by both stretch and shear stress may hence produce maximal remodeling responses in the vessel.

TRANSMISSION OF MECHANICAL STRESSES AT THE CELL MEMBRANE

Integrins

Strategically situated at the boundary between the extracellular matrix and the cytoskeleton, integrins may act not only as mediators of cell adhesion but also as transduceers of biochemical signals across the cell membrane. Indeed, integrins are present at sites of close apposition of the cell surface and the extracellular matrix, and they form a bridge between matrix proteins and the cytoskeleton, mediating binding and attachment of the cell to components of the extracellular matrix, such as fibronectin, vitronectin and collagen, and creating focal adhesions [74,75]. A role for integrins as mediators of vascular strain is supported by observations that shear stress-induced tyrosine phosphorylation of endothelial cells in isolated arterioles exposed to intraluminal flow is abolished by inhibition of integrin binding to extracellular matrix proteins containing the RGD amino acid sequence [76], RGD being the key combination via which extracellular matrix proteins are bound by integrins. Response to strain is also abrogated by antibodies to β3 or αvβ5 integrins, which bind fibronectin [64], whereas flow-dependent vasodilatation is RGD and β3 integrin sensitive [76], and is blocked by removal of the glycocalyx with neuraminidase [77]. Besides, β3 integrin expression may actually be enhanced subsequent to cyclic stretching of endothelial cells [78]. In addition, mechanical strain of SMC grown on fibronectin or vitronectin induces cell proliferation, whereas elastin- or laminin-grown cells in the same conditions do not proliferate [64]. On the other hand, neonatal SMC grown on laminin express SM-1 and SM-2 smooth muscle myosin heavy chains in response to mechanical strain, contrary to cells grown on the poly RGD substrate pronectin [79], and cyclic stretch-induced SM-1 expression is greater in cells grown on laminin than collagen or fibronectin [62]. It therefore appears that both the ability of cells to sense mechanical strain and the ensuing biochemical response depend on the nature of the interaction of specific integrins with the extracellular matrix. Ultimately, the hyperplasic versus hypertrophic response to stretch of SMC exposed to extracellular matrix will depend on their phenotype, as suggested by organ culture experiments in which stretch produces matrix synthesis without inducing cell proliferation [30].

A recent report brings evidence that some extracellular matrix proteins actually serve as ligands to receptors. This is at least the case of collagen, which binds and activates the receptor-like tyrosine kinase discoidin domain receptor 1 (DDR1), a first example of a ligand shared by integrins and receptor tyrosine kinases. In this case, the receptors are phosphorylated with a protracted time course [80]. Collagen might thus signal via its two distinct receptor classes to regulate cellular responses to changes in the surrounding environment [80]. Hence, integrin-mediated cell signaling could implicate either a cytoskeletal alteration removing a constraint on the signal transduction machinery, or a direct action on the signaling systems, or a combination of both [81].

Ion channels

Submitting endothelial cells to either shear stress or stretch spurs a transient increase in intracellular calcium and divalent cations [82,83]. As it turns out, endothelial cells possess stretch-activated ion channels which are relatively nonselective across cations [84]. These were identified using patch-clamp techniques. In isolated endothelial cells, membrane stretching by applying suction through the patch electrode increased the opening frequency of channels permeable to calcium [85]. Mechanosensitive ion channels were further characterized by their sensitivity to gadolinium but not to classical calcium channel blockers such as nifedipine [83]. The presence of mechanotransducing ion channels in endothelial cells may help explain how the endothelium mediates vascular responses to hemodynamic stresses. For example, the mechanism of stretch-activated phospholipase C (PLC) activity in SMC was found to involve influx of calcium via gadolinium-sensitive ion channels, but not via nifedipine-sensitive ion channels [86]. MAP kinase activation by AngII also shows a calcium dependency in SMC [87]. On the other hand, potassium channels distinct from the non-selective gadolinium sensitive cation channels may also participate in transduction of mechanical stress [88]. This was brought to evidence by the observation that shear-induced increases in gene expression and cGMP concentrations are inhibited by tetraethylammonium ion (TEA), a non-specific potassium channel blocker [89]. Furthermore, membrane stretch and fatty acids directly activate large conductance calcium-activated potassium channels in vascular smooth muscle cells [90]. Finally, RSK, a downstream target of ERK 1/2, phosphorylates or may actually itself be the NHE-1 isoform of the Na^{+}-H^{+} exchanger [91]. This is particularly interesting in light of the observation that cellular sodium entry via a tetrodotoxin-inhibited, mechanosensitive channel modulates ERK 1/2 activation by shear [92]. Either way, there appears to be a definite role for ion channels in the response of vascular cells to mechanical forces.

INTRACELLULAR TRANSMISSION OF MECHANICAL STRESSES

Focal adhesion kinases

At the cellular level, subjecting endothelial cells to oscillatory flow spurs a clustering of $\alpha 5\beta 1$ integrins and a concomitant gathering of cytoskeletal proteins talin and vinculin [65]. In fact, during cell stimulation by mechanical factors such as stretch or shear stress, several signaling events are associated with the formation of focal adhesions, which consist of clustered integrins and accumulated cytoskeletal proteins. The recruitment of integrins into focal adhesions is mediated by the cytoplasmic domains of the bridging proteins, as deletion of the $\beta 1$ subunit cytoplasmic domain inhibits integrin association [93]. In turn, the cytoplasmic domains bind cellular cytoskeletal proteins that are present in the focal adhesions, α-actinin and talin [94,95]. α-Actinin is directly connected to actin microfilaments [96], whereas talin is linked via vinculin [97] which in turn binds α-actinin or tensin [98], both of which associate with actin [99].

Proteins present at focal adhesions, in particular the 125 kDa cytoplasmic tyrosine kinase FAK, become tyrosine phosphorylated when cells are stimulated by integrin antibodies, cell adhesion or RGD-containing compounds [100-102]. Shear and adhesion activate FAK, but stimuli are not additive [103-105]. FAK associates with paxillin [106] and talin [107], and both FAK and paxillin can bind to the cytoplasmic tail of integrins independently [108]. Focal adhesions containing talin, vinculin and paxillin have been reported to form in endothelial cells despite the absence of FAK association and in conditions of reduced tyrosine phosphorylation, suggesting that FAK activation is downstream of focal adhesion and stress fiber formation, and that its role is one of a signaling protein in focal adhesions rather than one of focal adhesion assembly [109]. In confirmation, there is evidence that aggregation of FAK with $\alpha 5\beta 1$ integrin, RGD or fibronectin occurs even in the presence of tyrosine kinase inhibition or actin filament assembly disruption by cytochalasin D. However, cytoskeletal protein recruitment and activation of downstream kinases is prevented [110,111]. Rho, a small G-protein, is also implicated in the regulation of formation of stress fibers and focal adhesions, through phosphorylation of FAK, p130Cas (an adaptor protein bound by FAK), and paxillin, and, independently, actin polymerization [112]. FAK may be downstream of Rho, and Rho may be involved in activation of FAK by 7 transmembrane domain receptors of AngII, bombesin, and lysophosphatidic acid [81].

Shear in endothelial cells increases the tyrosine phosphorylation and activity of FAK, and its association with Grb2 [103]. In fact, it is the attachment of c-Src, a membrane associated non-receptor tyrosine kinase, at a key region of autophosphorylation on FAK which creates a binding site for the Src-homology-2 (SH2) domain of Grb2 [113]. C-Src is active in its dephosphorylated state, which seems to be modulated by stretch [114], and is inactivated by C-terminal Src kinase (Csk). Following its activation, c-Src is translocated to focal contacts (Figure 8).

The MAP kinase cascade: upstream events

The MAP kinase cascade is a major pathway through which signals coming from growth factors and mechanical strain are transduced into regulation of gene expression and protein synthesis. Involved is the sequential phosphorylation and activation of the cytoplasmic protein kinases MAP kinase kinase kinase (MEKK), MAP kinase kinase (MEK), and finally MAP kinase [115]. The MAP kinase cascade actually comprises three separate pathways which respond to different stimuli and instigate distinct cellular responses. Phosphorylation of one MAP kinase, which lies downstream of the MEKK Raf and is present in two isoforms termed extracellular-signal related kinase (ERK) 1 and 2, leads to the activation of regulatory proteins both in the cytoplasm and the nucleus [115].

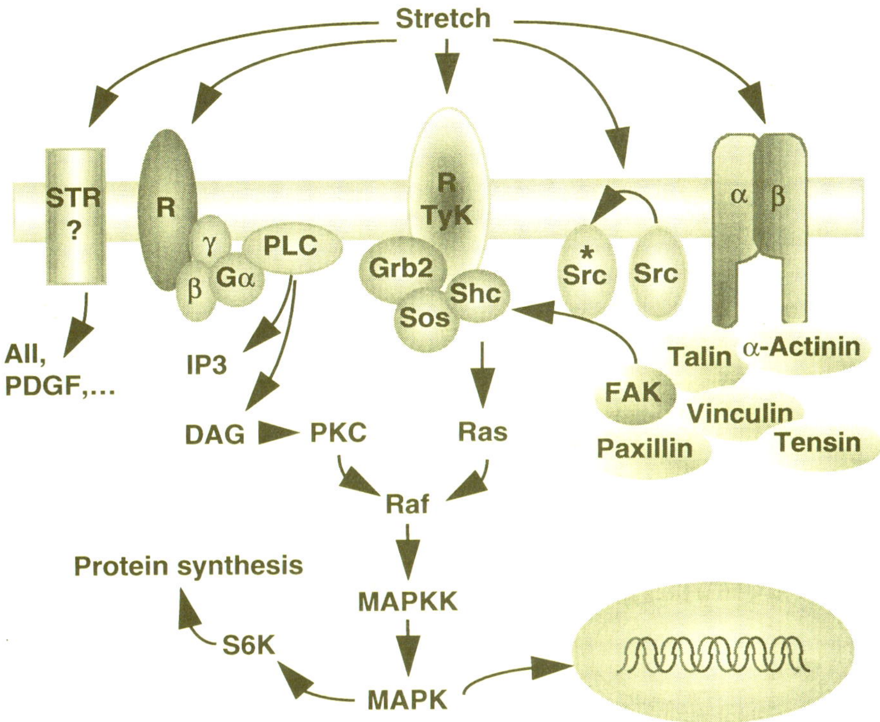

Figure 8. *Multiple pathways of transduction of mechanical stretch in vascular cells. Mechanical forces may act on α/β heterodimer integrins or activate non-receptor membrane tyrosine kinases including c-Src. Once stimulated, activated c-Src (Src*) translocates to focal contacts where it interacts with focal adhsion kinase (FAK) causing its autophosphorylation. Subsequently, the Shc-Grb2-Sos complex associates and downstream activation of mitogen-activated protein (MAP) kinase occurs. MAP kinase stimulates growth or protein synthesis via activation of S6 kinase (S6K). Alternatively, stretch acts on unknown stretch receptors (STR) that stimulate the renin-angiotensin system, leading to the production of angiotensin II (AII), which in turn acts on G-protein coupled receptors (R). Also, autocrine production of growth factors including platelet-derived growth factor (PDGF) may activate tyrosine kinase receptors (R TyK).*

A second branch of the MAPK family, termed stress-activated protein kinases (SAPK) since they are activated by such stimuli as UV light, heat shock, hypoxia or high osmolarity, includes kinases that phosphorylate the amino terminal of transcription factor c-jun (JNK) [116,117]. Finally, a third branch of the MAPK family comprises p38, also activated by osmotic stress [118]. In endothelial cells, physiological levels of shear stimulate ERK1/2 [104,119,120], whereas cyclic mechanical strain activates both ERK1/2 and JNK in SMC [79]. Furthermore, applying a high intraluminal pressure to aortas in organic culture induces a biphasic ERK 1/2 stimulation, characterized by an acute peak in activity, subsequent reversal, and a second more lengthy activation [121]. *In vivo*,

ERK 1/2 is transiently activated by acute hypertension [122] and by vessel wall injury with a balloon catheter [123,124].

Diverse pathways link mechanical strain to MAP kinase activation in vascular cells. Hence, G-protein and calcium-independent protein kinase C activation is involved in ERK 1/2 activation by shear stress [120], whereas in certain conditions, JNK may be more activated by shear than ERK 1/2, through sequential phosphorylation of Sos, Ras, and MEKK [119]. Furthermore, integrins are likely to be among the actors involved in the transmission of mechanical forces to the MAP kinase cascade, for several reasons. First, cellular response to stretch or shear stress *in vitro* varies widely depending on the nature of the substrate on which the cells are grown. For example, ERK 1/2 and JNK are both activated by cyclic mechanical strain in pronectin-grown neonatal SMC, whereas in their laminin-grown counterparts, only JNK is stimulated by cyclic strain [79]. Second, ERK1/2 activation by shear stress and by integrin-mediated adhesion to fibronectin occurs via a common herbimycin A-sensitive, PKC-dependent pathway in endothelial cells [104].

Src-family tyrosine kinases, which are inhibited by herbimycin A, have also been implicated in intraluminal pressure-induced ERK 1/2 activation in vascular organ culture, although a PKC-independent pathway was involved under these conditions [121]. Recent in vitro experiments report a role for c-Src in pressure-induced contraction of rat cerebral arteries [125]. Furthermore, both c-Src and Grb2 SH2 binding motifs have been involved in MAP kinase signaling pathways [113]. Accordingly, it was demonstrated that FAK overexpression enhances c-Src kinase activity and fibronectin-induced ERK 2 activity, whereas a Ras dominant negative, which blocked ERK activation, did not affect FAK phosphorylation or Src activity. Also, replacing the c-Src binding site on FAK prevented integrin signaling to ERK [113]. Finally, fibronectin-induced ERK 1/2 activation is Shc-dependent, and the Shc-ERK pathway is bridged by Ras [126]. Hence, these cumulative observations describe a pathway originating with integrin activation, focal adhesion assembly, activation of FAK by c-Src, association with Grb2 leading to Shc-dependent stimulation of Ras and subsequent activation of ERK 1/2 via the MAP kinase cascade (Figure 8). Accordingly, a dominant negative mutant of FAK was shown to attenuate shear-induced ERK2 and JNK activity in endothelial cells, as did a dominant negative mutant of Sos and an anti-vitronectin receptor antibody [103]. Likewise, both flow and β1 integrin activation stimulated ERK 1/2 and tyrosine phosphorylation of proteins. However, ERK 1/2 activation by β1 integrin activation occurred more slowly and to a lesser degree, and fewer proteins were tyrosine phosphorylated, than under flow conditions [105]. Multiple pathways are then likely to be recruited in the mechanotransduction of flow.

There is indeed evidence that integrin-mediated MAP kinase activation may in some cases bypass FAK, as demonstrated by the observation that a single chain tailless mutant of integrin α1 recruited Shc and activated ERK but not FAK, whereas an activating α6β1 antibody activated FAK but did not induce its association with Shc, and did not activate ERK [126]. Furthermore, the increase in adhesion-mediated ERK activation by shear was only partially affected by actin filament disruption [104] and JNK, but not ERK, remains activated by fibronectin despite the presence of cytochalasin D [111].

The MAP kinase cascade: downstream events

Events downstream to MAP kinase activation are numerous and varied. Once phosphorylated, ERK1/2 may translocate to the nucleus to phosphorylate transcription factors and thereby regulate cell cycle gene expression [127]. Both ERK 1/2 and JNK can lead to ternary complex formation at the serum response element, present on several gene promoters, and increased transcriptional activity [128]. Alternatively, phosphorylation of the translation regulator protein PHAS-I (phosphorylated heat- and acid-stable protein) promotes the dissociation of the PHAS-I - eukaryotic initiation factor (eIF) 4E complex, normally tightly bound when PHAS-I is relatively underphosphorylated, releasing eIF-4E which will facilitate initiation of translation in the nucleus [129,130]. Another downstream target of ERK in SMC is the 90 kDa ribosomal S-6 kinase RSK, which through activation of the transfer RNA-binding factor may provide a pathway essential for the initiation of translation [130]. Ultimately, ERK 1/2 activation coincides with enhanced c-fos and c-jun expression, and activation of the AP-1 transcription factor [122], and it is likely to play a significant role in regulating cell cycle progression of SMC [131] as well as protein synthesis [130].

The fact that ERK 1/2 can also induce cyclooxygenase-2 (COX-2) in SMC [132] may explain why activation of this MAP kinase does not necessarily result in increased cellular proliferation. Indeed, cytosolic phospholipase A_2 (PLA_2) is among the substrates of ERK 1/2 [133). PLA_2 catalyzes the release of arachidonic acid from phospholipids in the cell membrane [134], which will be transformed by the action of COX-2 into prostaglandins. The resulting elevated levels of prostaglandin E_2 and protein kinase A activation could counteract ERK 1/2-induced proliferation [132]. A further target of activated ERK 1/2 has been reported to be the contractile regulatory protein h-caldesmon, the high molecular weight form of caldesmon, indicating that ERK is involved in the regulation of contractile properties of the vascular wall [135]. Hence, in the end, the availability of downstream ligands may be a significant determinant of the biological outcome of ERK activation [132]. At length, ERK 1/2 activity is modulated by MAP kinase phosphatase (MKP-1), which dephosphorylates the enzyme [136]. Alternatively, the activation of ERK 1/2 may be terminated through a feedback loop implicating Ras/Raf mediated suppression of integrin activation [137].

ROLE OF THE RENIN ANGIOTENSIN SYSTEM AND GROWTH FACTORS

Induction of protein synthesis by stretch may occur in many cases via increased synthesis of growth factors or mitogenic agonists, among which angiotensin II (AngII) plays an important role. The pathways involved in the increased synthesis of these factors by mechanical strain are not yet clearly understood, although there is evidence that the AP-1 transcription factor downstream of ERK 1/2 activation may regulate growth factor expression [138]. Stimulation of protein and fibronectin synthesis by high intraluminal pressure in aortic organ culture was found not only to result from augmented angiotensin levels, but also to be further enhanced by addition of AngII to the culture medium [30]. In a similar fashion, the rise in transforming growth factor (TGFβ) mRNA expression brought about by AngII and stretch is additive [139], stretch induces parathyroid hormone related peptide mRNA and secretion synergistically with AngII [140], and both stretch- and AngII-induced DNA synthesis in collagen-plated SMC occurs in synergy [63]. Moreover, this latter effect is attenuated by platelet-derived growth factor (PDGF-AB) antibodies, whereas AngII and PDGF increase DNA synthesis in synergy [63], demonstrating that more than one factor may be implicated at once in the remodeling process. In addition, a role for extracellular matrix proteins cannot be excluded. Indeed, attachment of cells to fibronectin and antibody-induced aggregation of $\alpha 5\beta 1$ integrins enhances PDGF-induced increase in cytoplasmic pH [132], suggesting that integrins and growth factor receptors may act cooperatively.

In several circumstances, mechanical activation of vessels or vascular cells instigates the release of vasoactive factors that will be implicated in the ensuing changes in vessel structure and function. In organ culture, for example, angiotensin mediates the enhanced total protein and fibronectin synthesis induced by high intraluminal pressure [32]. Appropriately, angiotensin II is potentially involved in the stimulation of a number of intracellular pathways, leading in aortic SMC to hypertrophy, through enhanced protein synthesis, but not to hyperploidy [30,141]. Synthesis-promoting activities of AngII are transduced via the AT1 receptor [32] and the downstream signaling cascades include activation of PLC, phospholipase D, increased calcium and inhibition of adenylyl cyclase [142,143]. AngII may also induce protein synthesis in part via activation of the 70-kDa S6 kinase, by an ERK 1/2-independent pathway [144]. This is particularly interesting in light of the fact that intraluminal pressure induces both ERK 1/2 activation and protein synthesis, but only the latter effect is mediated by AngII [30,121]. Alternatively, AngII activates c-Src [145] which constitutes a probable pathway via which AngII phosphorylates both FAK and paxillin [142]. In fact, not only AngII but also epidermal growth factor (EGF) and thrombin activate tyrosine phosphorylation of paxillin, as demonstrated in rat aortic VSM [142]. Not surprisingly, growth factors (fibroblast growth factor, PDGF-BB, EGF) and integrins can activate the MAP kinase cascade in synergy, provided the integrins are both aggregated and occupied [146]. Like AngII, growth factors (insulin) may activate relatively downstream events in the signalizing cascade, such as 70-kDa S6 kinase stimulation and phosphorylation of PHAS-I, via MAP kinase independent pathways [129].

As it turns out, growth factors may bypass the MAP kinase cascade and function instead by activating the nuclear factor κ-B (NFκB) [147]. This family of transcription factors regulates the expression of genes encoding growth factors, inducible surface proteins and molecules involved in extracellular matrix remodeling [148]. NFκB is present in the cytosol in association with one of several inhibitors (generally identified as IkB), forming an inactive heteromeric complex. NFκB is released following phosphorylation and subsequent degradation of the IκB, allowing the active NFκB dimers to translocate to the nucleus and promote transactivation of target genes [107,149].

In parallel, recent reports propose a role for reactive oxygenated species in mechanical stress signal transduction. Indeed, stretch of SMC activates PKC, which presumably acts on NADPH oxidase, and thereby forms reactive oxygen species that then sequentially activate NFκB and DNA synthesis [150]. Alternatively, H_2O_2 in endothelial cells also induces f-actin reorganization, characterized by stress fiber formation and recruitment of vinculin to focal adhesions. These changes are modulated by activation of p38, followed by phosphorylation of heat shock protein HSP27 [151]. Furthermore, observations of shear stress induced oxygen free radical production and downstream HSP27 activation are combined in a single study describing sustained phosphorylation of HSP27 in endothelial cells subjected to shear stress, and consequent reorganization of cytoskeletal proteins and change of cell shape [66]. Finally, a recent study identified two members of the MAD protein family (for Mothers against decapentaplegic), Smad6 and Smad7, as unique among the MAD-related proteins, being expressed selectively in the endothelium *in vivo*, and activated by physiological levels of flow in endothelial cell cultures [152]. MADs traditionally act as second messengers distal to the TGFβ family of receptors [152].

CONCLUSION

The blood vessels possess autocrine and/or paracrine humoral mechanisms that enable them to react immediately to local hemodynamic changes: circumferential mechanical stress (which increases with pressure) and shear stress (which increases with blood flow). The immediate changes in local vasomotor tone provoked by these stress changes tend in most cases to normalize them. If the alterations in local geometry induced by changes in vasomotor tone are inadequate to compensate for the changes in mechanical parameters (stresses), the phenotype of the vascular cells may alter and give rise to local trophic changes. These changes, taking place over the longer term (from days to weeks), also tend to restore the mechanical stresses to their physiologie levels. Using purely local mechanisms, a blood vessel is capable of true self-regulation, enabling it to adapt to its mechanical environment. The transduction mechanisms between mechanical and biochemical or endocrine signals become to be elucidated. Located at the cell surface integrins are likely to be key mechanosensors. In parallel ion channels and other unknown stretch receptors presumably transduce the mechanical signal. As a result, several intracellular signaling pathways are activated including c-Src, FAK and MAP activated MAP cascade, as well as the renin angiotensin system. Understanding which signaling pathways are involved in the transduction of mechanical forces in the vascular wall should allow for a better approach of vascular remodeling. This may be of help in the development of novel therapeutic strategies for the treatment of cardiovascular diseases including hypertension, atherosclerosis, and restenosis following angioplasty.

REFERENCES

1. Glagov S. Intimal hyperplasia, vascular remodeling, and restenosis problem. Circulation. 1994;89:2888-2891.

2. Barbee KA, Macarak EJ, Thibault LE. Strain measurements in cultured vascular smooth muscle cells subjected to mechanical deformation. Ann Biomed Eng. 1994;22:14-22.

3. Tronc F, Wassef M, Esposito B, Henrion D, Glagov S, Tedgui A. Role of NO in flow-induced remodeling of the rabbit common carotid artery. Arterioscler Thromb Vasc Biol. 1996;16:1256-1262.

4. Glagov S. Hemodynamic risk factors: mechanical stress, mural architecture, medial nutrition and vulnerability of arteries to atherosclerosis, in The pathogenesis of atherosclerosis, Wissler RW and Geer JC, Editors. Williams & Wilkins CO: Baltimore. 1972 pp. 164-199.

5. Langille BL, Brownlee RD, Adamson SL. Perinatal aortic growth in lambs: relation to blood flow changes at birth. Am J Physiol. 1990;259:H1247-H1253.

6. Cho A, Courtman DW, Langille BL. Apoptosis (programmed cell death) in arteries of the neonatal lamb. Circ Res. 1995;76:168-175.

7. Leung DY, Glagov S, Mathews MB. Elastin and collagen accumulation in rabbit ascending aorta and pulmonary trunk during postnatal growth: correlation of cellular synthetic response with medial hypertension. Circ Res. 1977;41:316-323.

8. Bomberger RA, Zarins CK, Taylor KE, Glagov S. Effect of hypotension on atherogenesis and aortic wall composition. J Surg Res. 1980;28:401-409.

9. Wolinsky H. Response of the rat aortic media to hypertension: Morphological and chemical studies. Circ Res. 1970;26:507-522.

10. Wolinsky H, Glagov S. Comparison of abdominal and thoracic aortic medial structure in mammals: deviation from the usual pattern in man. Circ Res. 1969;25:677-685.

11. Owens GK. Control of hypertrophic versus hyperplastic growth of vascular smooth muscle cells. Am J Physiol. 1989;257:H1755-H1765.

12. Zarins CK, Bomberger RA, Glagov S. Local effects of stenoses: increased flow velocity inhibits atherogenesis. Circulation. 1981;64:221-227.

13. Tedgui A, Merval R, Esposito B. Albumin transport characteristics of rat aorta in early phase of hypertension. Circ Res. 1992;71:932-942.

14. Folkow B. Cardiovascular structural adaptation: its role in the initiation and maintenance of primary hypertension. The fourth Volhard lecture. Clin Sci Mol Med. 1978;55:3S-22S.

15. Levy BI, Michel JB, Salzmann JL, Azizi M, Poitevin P, Safar ME, Camilleri JP. Effects of chronic inhibition of converting enzyme on the mechanical and structural properties of arteries in rat renovascular hypertension. Circ Res. 1988;63:227-239.

16. Greenwald SE, Berry CL, Ramsey RE. The static mechanical properties and chemical composition of the rat aorta in spontaneously occuring and experimentally induced hypertension: the effect of an anti-hypertensive drug. Br J Exp Path. 1985;66:633-642.

17. Olivetti G, Anversa P, Melissari M, Loud AV. Morphometry of medial hypertrophy in the rat thoracic aorta. Lab Invest. 1980;42:559-565.

18. Berry CL, Greenwald SE. Effect of hypertension on the static mechanical properties and chemical composition of the rat aorta. Cardiovasc Res. 1976;10:437-451.

19. Levy BI, Benessiano J, Poitevin P, Lukin L, Safar M. Systemic arterial compliance in normotensive and hypertensive rats. J Cardiovasc Pharmacol. 1985;7:S28-S32.

20. Owens GK, Schwartz SM. Alterations in vascular smooth muscle mass in the spontaneously hypertensive rat. Role of cellular hypertrophy, hyperploidy, and hyperplasia. Circ Res. 1982;51:280-289.

21. Folkow B. Structural factor in primary and secondary hypertension. Hypertension. 1990;16:89-101.

22. Mulvany MJ. Vascular growth in hypertension. J Cardiovasc Pharmacol. 1992;20:S7-S11.

23. Laurent S. Arterial wall hypertrophy and stiffness in essential hypertensive patients. Hypertension. 1995;26:355-362.

24. Girerd X, Mourad JJ, Copie X, Moulin C, Acar C, Safar M, Laurent S. Noninvasive detection of an increased vascular mass in untreated hypertensive patients. Am J Hypertens. 1994;7:1076-1084.

25. Levy BI, Duriez M, Phillip M, Poitevin P, Michel JB. Effect of chronic dihydropyridine on the large arterial wall of spontaneously hypertensive rats. Circulation. 1994;90:3024-3033.

26. Mulvany MJ. Structure and function of small arteries in hypertension. J Hypertens. 1990;8:S225-S232.

27. Albaladejo P, Bouaziz H, Duriez M, Gohlke P, Levy BI, Safar ME, Benetos A. Angiotensin converting enzyme inhibition prevents the increase in aortic collagen in rats. Hypertension. 1994;23:74-82.

28. Qiu HY, Valtier B, Sruijker-Boudier HAJ, Levy BI. Mechanical and contractile properties of in situ localized mesenteric arteries in normotensive and spontaneously hypertensive rats. J Pharmacol Toxicol Method. 1995;33:159-170.

29. Struijker-Boudier HAJ, Van Eessen H, Fazzi G, DeMey JGR, Qiu HY, Levy BI. Disproportional arterial hypertrophy in hypertensive mREN-2 transgenic rats. Hypertension. 1996;28:779-784.

30. Bardy N, Karillon GJ, Merval R, Samuel J-L, Tedgui A. Differential effects of pressure and flow on DNA and protein synthesis, and on fibronectin expression by arteries in a novel organ culture system. Circ Res. 1995;77:684-694.

31. Bardy N, Merval R, Benessiano J, Samuel J-L, Tedgui A. Pressure and angiotensin II synergistically induce aortic fibronectin expression in organ culture model of rabbit aorta. Evidence for a pressure-induced tissue renin-angiotensin system. Circ Res. 1996;79:70-78.

32. Osol G. Mechanotransduction by vascular smooth muscle. J Vasc Res. 1995;32:275-292.

33. Burton AC. On the physical equilibrium of the small blood vessel walls. Am J Physiol. 1951;164:319-329.

34. Azuma T, Oka S. Mechanical equilibrium of blood vessel walls. Am J Physiol. 1971;221:1310-1318.

35. Virmani R, Avolio AP, Mergner WJ, Robinowitz M, Herderick EE, Cornhill JF, Guo SY, Liu TH, Ou DY, O'Rourke M. Effect of aging on aortic morphology in populations with high and low prevalence of hypertension and atherosclerosis. Comparison between occidental and Chinese communities. Am J Pathol. 1991;139:1119-1129.

36. Michel JB, Heudes D, Michel O, Poitevin P, Philippe M, Scalbert E, Corman B, Levy BI. Effect of chronic ang I-converting enzyme inhibition on aging processes .II. Large arteries. Am J Physiol. 1994;267:R124-R135.

37. Crouse JR, Goldbourt U, Evans G, Pinsky J, Sharrett AR, Sorlie P, Riley W, Heiss G. Arterial enlargement in the atherosclerosis risk in communities (ARIC) cohort. In vivo quantification of carotid arterial enlargement. The ARIC Investigators. Stroke. 1994;25:1354-1359.

38. Kakuta T, Currier JW, Haudenschild CC, Ryan TJ, Faxon DP. Differences in compensatory vessel enlargement, not intimal formation, account for restenosis after angioplasty in the hypercholesterolemic rabbit model. Circulation. 1994;89:2809-2815.

39. Steinke W, Els T, Hennerici M. Compensatory carotid artery dilatation in early atherosclerosis. Circulation. 1994;89:2578-2581.

40. Batellier J, Wassef M, Merval R, Duriez M, Tedgui A. Protection from atherosclerosis in vein grafts by a rigid external support. Arterioscler Thromb. 1993;13:379-384.

41. Hishikawa K, Nakaki T, Marumo T, Hayashi M, Suzuki H, Kato R, Saruta T. Pressure promotes DNA synthesis in rat cultured vascular smooth muscle cells. J Clin Invest. 1994;93:1975-1980.

42. Cheng GC, Libby P, Grodzinsky AJ, Lee RT. Induction of DNA synthesis by a single transient mechanical stimulus of human vascular smooth muscle cells: role of fibroblast growth factor-2. Circulation. 1996;93:99-105.

43. Mallat Z, Delcayre C, Tedgui A. Effects of stretch and pressure-induced crush of the arterial wall on the induction of immediate early protooncogenes. J Am Coll Cardiol. 1995;25:291A (abstract).

44. Abate C, Luk D, Gentz R, Rauscher FJ, Currant T. Expression and purification of the leucine zipper and DNA-binding domains of Fos and Jun: both Fos and Jun contact DNA directly. Proc Natl Acad Sci. 1990;87:1032-1036.

45. Schuermann M, Neuberg M, Hunter JB, Jenuwein T, Ryseck RP, Bravo R, Müller R. The leucine repeat motif in Fos protein mediates complex formation with Jun/AP-1 and is required for transformation. Cell. 1989;56:507-516.

46. Indolfi C, Esposito G, Dilorenzo E, Rapacciuolo A, Feliciello A, Porcellini A, Avvedimento VE, Condorelli M, Chiariello M. Smooth muscle cell proliferation is proportional to the degree of balloon injury in a rat model of angioplasty. Circulation. 1995;92:1230-1235.

47. Guyton JR, Hartley CJ. Flow restriction of one carotid artery juvenile rats inhibits growth of arterial diameter. Am J Physiol. 1985;248:H540-546.

48. Langille BL, Bendeck MP, Keeley FW. Adaptations of carotid arteries of young and mature rabbits to reduced carotid blood flow. Am J Physiol. 1989;256:H931-H939.

49. Kamiya A, Togawa T. Adaptive regulation of wall shear stress to flow change in the canine carotid artery. Am J Physiol. 1980;239:H14-H21.

50. Nadaud S, Philippe M, Arnal JF, Michel JB, Soubrier F. Sustained increase in aortic endothelial nitric oxide synthase expression in vivo in a model of chronic high blood flow. Circ Res. 1996;79:857-863.

51. Ben Driss A, Benessiano J, Poitevin P, Levy BI, Michel JB. Arterial expansive remodeling induced by high flow rates. Am J Physiol. 1997;272:H851-H858.

52. Jones GT, Stehbens WE. The ultrastructure of arteries proximal to chronic experimental carotid-jugular fistulae in rabbits. Pathology. 1995;27:36-42.

53. Greenhill NS, Stehbens WE. Scanning electron microscopic investigation of the afferent arteries of experimental arteriovenous fistulae in rabbits. Pathology. 1987;19:22-27.

54. Wong LCY, Langille BL. Developmental remodeling of the internal elastic lamina of rabbit arteries: effect of blood flow. Circ Res. 1996;78:799-805.

55. Hanemaaijer R, Sorsa T, Konttinen YT, Ding Y, Sutinen M, Visser H, van Hinsbergh VW, Helaakoski T, Kainulainen T, Ronka H, Tschesche H, Salo T. Matrix metalloproteinase-8 is expressed in rheumatoid synovial fibroblasts and endothelial cells. Regulation by tumor necrosis factor-alpha and doxycycline. J Biol Chem. 1997;272:31504-31509.

56. Owens MW, Milligan SA, Jourdíheuil D, Grisham MB. Effects of reactive metabolites of oxygen and nitrogen on gelatinase A activity. Am J Physiol. 1997;273:L445-L450.

57. Murell GAC, Jang D, Williams RJ. Nitric oxide activates metalloproteinase enzymes in articular cartilage. Biochem Biophys Res Commun. 1995;206:15-21.

58. Rajagopalan S, Meng XP, Ramasamy S, Harrison DG, Galis ZS. Reactive oxygen species produced by macrophage-derived foam cells regulate the activity of vascular matrix metalloproteinases in vitro. Implications for atherosclerotic plaque stability. J Clin Invest. 1996;98:2572-2579.

59. Brown LC, Messick FC, Kok MP, Hamiltom IG, Girard PR. Fluid flow stimulates metalloproteinase production and deposition into the extracellular matrix of endothelial cells. FASEB J. 1995;9:A617. Abstract.

60. Birukov KG, Bardy N, Lehoux S, Merval R, Shirinsky VP, Tedgui A. Intraluminal pressure is essential for the maintenance of smooth muscle caldesmon and filamin content in aortic organ culture. Arterioscler Thromb Vasc Biol. 1998;18:922-927.

61. Smith PG, Tokui T, Ikebe M. Mechanical strain increases contractile enzyme activity in cultured airway smooth muscle cells. Am J Physiol. 1995;268:L999-L1005.

62. Reusch P, Wagdy H, Reusch R, Wilson E, Ives HE. Mechanical strain increases smooth muscle and decreases nonmuscle myosin expression in rat vascular smooth muscle cells. Circ Res. 1996;79:1046-1053.

63. Sudhir K, Wilson E, Chatterjee K, Ives HE. Mechanical strain and collagen potentiate mitogenic activity of angiotensin-II in rat vascular smooth muscle cells. J Clin Invest. 1993;92:3003-3007.

64. Wilson E, Sudhir K, Ives HE. Mechanical strain of rat vascular smooth muscle cells is sensed by specific extracellular matrix/integrin interactions. J Clin Invest. 1995;96:2364-2372.

65. Thoumine O, Nerem RM, Girard PR. Oscillatory shear stress and hydrostatic pressure modulate cell-matrix attachment proteins in cultured endothelial cells. In Vitro Cell Dev Biol Animal. 1995;31:45-54.

66. Li S, Piotrowicz RS, Levin EG, Shyy YJ, Chien S. Fluid shear stress induces the phosphorylation of small heat shock proteins in vascular endothelial cells. Am J Physiol. 1996;40:C994-C1000.

67. Malek AM, Gibbons GH, Dzau VJ, Izumo S. Fluid shear stress differentially modulates expression of genes encoding basic fibroblast growth factor and platelet-derived growth factor-b chain in vascular endothelium. J Clin Invest. 1993;92:2013-2021.

68. Hsieh HJ, Li NQ, Frangos JA. Shear stress increases endothelial platelet-derived growth factor messenger RNA levels. Am J Physiol. 1991;260:H642-H646.

69. Furchgott RF, Vanhoutte PM. Endothelium-derived relaxing and contracting factors. FASEB J. 1989;3:2007-2018.

70. Bhagyalakshmi A, Frangos JA. Mechanism of shear-induced prostacyclin production in endothelial cells. Biochem Biophys Res Commun. 1989;158:31-37.

71. Kuchan MJ, Frangos JA. Shear stress regulates endothelin-1 release via protein kinase C and cGMP in cultured endothelial cells. Am J Physiol. 1993;264:H150-H156.

72. Golledge J, Turner RJ, Harley SL, Springall DR, Powell JT. Circumferential deformation and shear stress induce differential responses in saphenous vein endothelium exposed to arterial flow. J Clin Invest. 1997;99:2719-2726.

73. Zhao SM, Suciu A, Ziegler T, Moore JE, Burki E, Meister JJ, Brunner HR. Synergistic effects of fluid shear stress and cyclic circumferential stretch on vascular endothelial cell morphology and cytoskeleton. Arterioscler Thromb Vasc Biol. 1995;15:1781-1786.

74. Burridge K, Fath K, Kelly T, Nuckolls G, Turner C. Focal adhesions: transmembrane junctions between the extracellular matrix and the cytoskeleton. Annu Rev Cell Biol. 1988;4:487-525.

75. Haas TA, Plow EF. Integrin-ligand interactions: a year in review. Curr Opin Cell Biol. 1994;6:656-662.

76. Muller JM, Chilian WM, Davis MJ. Integrin signaling transduces shear stress-dependent vasodilation of coronary arterioles. Circ Res. 1997;80:320-326.

77. Hecker M, Mulsch A, Bassenge E, Busse R. Vasoconstriction and increased flow - two principal mechanisms of shear stress-dependent endothelial autacoid release. Am J Physiol. 1993;265:H828-H833.

78. Suzuki M, Naruse K, Asano Y, Okamoto T, Nishikimi N, Sakurai T, Nimura Y, Sokabe M. Up-regulation of integrin beta 3 expression by cyclic stretch in human umbilical endothelial cells. Biochem Biophys Res Commun. 1997;239:372-376.

79. Reusch HP, Chan G, Ives HE, Nemenoff RA. Activation of JNK/SAPK and ERK by mechanical strain in vascular smooth muscle cells depends on extracellular matrix composition. Biochem Biophys Res Commun. 1997;237:239-244.

80. Shrivastava A, Radziejewski C, Campbell E, Kovac L, McGlynn M, Ryan TE, Davis S, Glass DJ, Lemke G, Yancopoulos GD. An orphan receptor tyrosine kinase family whose members serve as nonintegrin collagen receptors. Molec Cell. 1997;1:25-34.

81. Abedi H, Dawes KE, Zachary I. Differential effects of platelet-derived growth factor BB on p125 focal adhesion kinase and paxillin tyrosine phosphorylation and on cell migration in rabbit aortic vascular smooth muscle cells and swiss 3T3 fibroblasts. J Biol Chem. 1995;270:11367-11376.

82. Schwartz MA, Lechene C. Adhesion is required for protein kinase-C-dependent activation of the Na^+/H^+ antiporter by platelet-derived growth factor. Proc Natl Acad Sci USA. 1992;89:6138-6141.

83. Sokabe M, Nunogaki K, Naruse K, Soga H. Mechanics of patch clamped and intact cell-membranes in relation to SA channel activation. Japn J Physiol. 1993;43:S197-S204.

84. Davis MJ, Donovitz JA, Hood JD. Stretch-activated single-channel and whole cell currents in vascular smooth muscle cells. Am J Physiol. 1992;262:C1083-C1088.

85. Lansman JB, Hallam TJ, Rink TJ. Single stretch-activated ion channels in vascular endothelial cells as mechanotransducers? Nature. 1987;325:811-813.

86. Matsumoto H, Baron CB, Coburn RF. Smooth muscle stretch-activated phospholipase C activity. Am J Physiol. 1995;37:C458-C465.

87. Lucchesi PA, Bell JM, Willis LS, Byron KL, Corson MA, Berk BC. Ca^{2+}-dependent mitogen-activated protein kinase activation in spontaneously hypertensive rat vascular smooth muscle defines a hypertensive signal transduction phenotype. Circ Res. 1996;78:962-970.

88. Olesen SP, Clapham DE, Davies PF. Haemodynamic shear stress activates a K^+ current in vascular endothelial cells. Nature. 1988;331:168-170.

89. Ohno M, Gibbons GH, Dzau VJ, Cooke JP. Shear stress elevated endothelial cGMP. Role of a potassium channel and G protein coupling. Circulation. 1993;88:193-197.

90. Kirber MT, Ordway RW, Clapp LH, Walsh JV, Singer JJ. Both membrane stretch and fatty acids directly activate large conductance Ca^{2+}-activated K^{+} channels in vascular smooth muscle cells. Febs Lett. 1992;297:24-28.

91. Takahashi T, Kawahara Y, Okuda M, Ueno H, Takeshita A, Yokoyama M. Angiotensin II stimulates mitogen-activated protein kinases and protein synthesis by a Ras-independent pathway in vascular smooth muscle cells. J Biol Chem. 1997;272:16018-16022.

92. Traub O, Berk BC. Shear stress-mediated stimulation of ERK 1/2 in endothelial cells is regulated by a Na^{+} channel. Circulation. 1997;96:I-50 (abstract).

93. Solowska J, Guan JL, Arcantonio EE, Trevithick JE, Buck CA, Hynes RO. Expression of normal and mutant avian integrin subunits in rodent cells. J Cell Biol. 1989;109:853-861.

94. Otey CA, Pavalko FM, Burridge K. An interaction between alpha-actinin and the beta 1 integrin subunit in vitro. J Cell Biol. 1990;111:721-729.

95. Horwitz A, Duggan K, Buck C, Beckerle MC, Burridge K. Interaction of plasma membrane fibronectin receptor with talin-a transmembrane linkage. Nature. 1986;320:531-533.

96. Bennett JP, Zaner KS, Stossel TP. Isolation and some properties of macrophage alpha-actinin: evidence that it is not an actin gelling protein. Biochemistry. 1984;23:5081-5086.

97. Burridge K, Mangeat P. An interaction between vinculin and talin. Nature. 1984;308:744-746.

98. Belkin AM, Koteliansky VE. Interaction of iodinated vinculin, metavinculin and alpha-actinin with cytoskeletal proteins. FEBS Lett. 1987;220:291-294.

99. Lo SH, Weisberg E, Chen LB. Tensin: a potential link between the cytoskeleton and signal transduction. Bioessays. 1994;16:817-823.

100. Kornberg L, Earp HS, Parsons JT, Schaller M, Juliano RL. Cell adhesion or integrin clustering increases phosphorylation of a focal adhesion-associated tyrosine kinase. J Biol Chem. 1992;267:23439-23442.

101. Burridge K, Turner CE, Romer LH. Tyrosine phosphorylation of paxillin and pp125FAK accompanies cell adhesion to extracellular matrix: a role in cytoskeletal assembly. J Cell Biol. 1992;119:893-903.

102. Huang MM, Lipfert L, Cunningham M, Brugge JS, Ginsberg MH, Shattil SJ. Adhesive ligand binding to integrin alpha IIb beta 3 stimulates tyrosine phosphorylation of novel protein substrates before phosphorylation of pp125FAK. J Cell Biol. 1993;122:473-483.

103. Li S, Kim M, Hu YL, Jalali S, Schlaepfer DD, Hunter T, Chien S, Shyy JY. Fluid shear stress activation of focal adhesion kinase. Linking to mitogen-activated protein kinases. J Biol Chem. 1997;272:30455-30462.

104. Takahashi M, Berk BC. Mitogen-activated protein kinase (ERK1/2) activation by shear stress and adhesion in endothelial cells - Essential role for a herbimycin-sensitive kinase. J Clin Invest. 1996;98:2623-2631.

105. Ishida T, Peterson TE, Kovach NL, Berk BC. MAP kinase activation by flow in endothelial cells Role of beta 1 integrins and tyrosine kinases. Circ Res. 1996;79:310-316.

106. Hildebrand JD, Schaller MD, Parsons JT. Paxillin, a tyrosine phosphorylated focal adhesion-associated protein binds to the carboxyl terminal domain of focal adhesion kinase. Mol Biol Cell. 1995;6:637-647.

107. Chen HC, Appeddu PA, Parsons JT, Hildebrand JD, Schaller MD, Guan JL. Interaction of focal adhesion kinase with cytoskeletal protein talin. J Biol Chem. 1995;270:16995-16999.

108. Schaller MD, Otey CA, Hildebrand JD, Parsons JT. Focal adhesion kinase and paxillin bind to peptides mimicking beta integrin cytoplasmic domains. J Cell Biol. 1995;130:1181-1187.

109. Gilmore AP, Romer LH. Inhibition of focal adhesion kinase (FAK) signaling in focal adhesions decreases cell motility and proliferation. Mol Biol Cell. 1996;7:1209-1224.

110. Lyman S, Gilmore A, Burridge K, Gidwitz S, White GC. Integrin-mediated activation of focal adhesion kinase is independent of focal adhesion formation or integrin activation Studies with activated and inhibitory beta(3) cytoplasmic domain mutants. J Biol Chem. 1997;272:22538-22547.

111. Miyamoto S, Teramoto H, Coso OA, Gutkind JS, Burbelo PD, Akiyama SK, Yamada KM. Integrin function: molecular hierarchies of cytoskeletal and signaling molecules. J Cell Biol. 1995;131:791-805.

112. Flinn HM, Ridley AJ. Rho stimulates tyrosine phosphorylation of focal adhesion kinase, p130 and paxillin. J Cell Sci. 1996;109:1133-1141.

113. Schlaepfer DD, Hunter T. Focal adhesion kinase overexpression enhances Ras-dependent integrin signaling to ERK2/mitogen-activated protein kinase through interactions with and activation of c-Src. J Biol Chem. 1997;272:13189-13195.

114. Chappel J, Ross FP, Abu-Ame Y, Shaw A, Teitelbaum SL. 1,25-dihydroxyvitamin D3 regulates pp60c-Src activity and expression of a pp60c-Src activating phosphatase. J Cell Biochem. 1997;67:432-438.

115. Seger R, Krebs EG. The MAPK signaling cascade. FASEB J. 1995;9:726-735.

116. Eppert K, Scherer SW, Ozcelik H, Pirone R, Hoodless P, Kim H, Tsui LC, Bapat B, Gallinger S, Andrulis IL, Thomsen GH, Wrana JL, Attisano L. MADR2 maps to 18q21 and encodes a TGFbeta-regulated MAD-related protein that is functionally mutated in colorectal carcinoma. Cell. 1996;86:543-552.

117. Zhang ZH, Vuori K, Wang HG, Reed JC, Ruoslahti E. Integrin activation by R-Ras. Cell. 1996;85:61-69.

118. Gimbrone MA, Nagel T, Topper JN. Biomechanical activation: An emerging paradigm in endothelial adhesion biology. J Clin Invest. 1997;99:1809-1813.

119. Li YS, Shyy JYJ, Li S, Lee JD, Su B, Karin M, Chien S. The Ras-JNK pathway is involved in shear-induced gene expression. Mol Cell Biol. 1996;16:5947-5954.

120. Tseng H, Peterson TE, Berk BC. Fluid shear stress stimulates mitogen-activated protein kinase in endothelial cells. Circ Res. 1995;77:869-878.

121. Birukov KG, Lehoux S, Birukova AA, Merval R, Tkachuk VA, Tedgui A. Increased pressure induces sustained PKC-independent herbimycin A-sensitive activation of extracellular signal-regulated kinase 1/2 in the rabbit aorta in organ culture. Circ Res. 1997;81:895-903.

122. Xu QB, Liu YS, Gorospe M, Udelsman R, Holbrook NJ. Acute hypertension activates mitogen-activated protein kinases arterial wall. J Clin Invest. 1996;97:508-514.

123. Lille S, Daum G, Clowes MM, Clowes AW. The regulation of p42/p44 mitogen-activated protein kinases in the injured rat carotid artery. J Surg Res. 1997;70:178-186.

124. Pyles JM, March KL, Franklin M, Mehdi K, Wilensky RL, Adam LP. Activation of MAP kinase in vivo follows balloon overstretch injury of porcine coronary and carotid arteries. Circ Res. 1997;81:904-910.

125. Masumoto N, Nakayama K, Oyabe A, Uchino M, Ishii K, Obara K, Tanabe Y. Specific attenuation of the pressure-induced contraction of rat cerebral artery by herbimycin A. Eur J Pharmacol. 1997;330:55-63.

126. Wary KK, Mainiero F, Isakoff SJ, Marcantonio EE, Giancotti FG. The adaptor protein Shc couples a class of integrins to the control of cell cycle progression. Cell. 1996;87:733-743.

127. Alvarez E, Northwood IC, Gonzalez FA, Latour DA, Seth A, Abate C, Curran T, Davis RJ. Pro-Leu-Ser/Thr-Pro is a consensus primary sequence for substrate protein phosphorylation. Characterization of the phosphorylation of c-myc and c-jun proteins by an epidermal growth factor receptor threonine 669 protein kinase. J Biol Chem. 1991;266:15277-15285.

128. Whitmarsh AJ, Shore P, Sharrocks AD, Davis RJ. Integration of MAP kinase signal transduction pathways at the serum response element. Science. 1995;269:403-407.

129. Brunn GJ, Hudson CC, Sekulic A, Williams JM, Hosoi H, Houghton PJ, Lawrence Jr JC, Abraham RT. Phosphorylation of the translational repressor PHAS-I by the mammalian target of rapamycin. Science. 1997;277:99-101.

130. Proud CG. Turned on by insulin. Nature. 1994;371:747-748.

131. Watson MH, Venance SL, Pang SC, Mak AS. Smooth muscle cell proliferation: expression and kinases activities of p34cdc2 and mitogen-activated protein kinases homologues. Circ Res. 1993;73:109-117.

132. Bornfeldt KE, Campbell JS, Koyama H, Argast GM, Leslie CC, Raines EW, Krebs EG, Ross R. The mitogen-activated protein kinase pathway can mediate growth inhibition and proliferation in smooth muscle cells. Dependence on the availability of downstream targets. J Clin Invest. 1997;100:875-885.

133. Davis RJ. The mitogen-activated protein kinase signal transduction pathway. J Biol Chem. 1993;268:14553-14556.

134. Dennis EA. Diversity of group types, regulation, and function of phospholipase A2. J Biol Chem. 1994;269:13057-13060.

135. Adam LP, Franklin MT, Raff GJ, Hathaway DR. Activation of mitogen-activated protein kinase in porcine carotid arteries. Circ Res. 1995;76:183-190.

136. Duff JL, Monia BP, Berk BC. Mitogen-activated protein (MAP) kinase is regulated by the map kinase phosphatase (MKP-1) in vascular smooth muscle cells. J Biol Chem. 1995;270:7161-7166.

137. Hughes PE, Renshaw MW, Pfaff M, Forsyth J, Keivens VM, Schwartz MA, Ginsberg MH. Suppression of integrin activation: A novel function of a Ras/Raf-initiated MAP kinase pathway. Cell. 1997;88:521-530.

138. Martin M, Vozenin MC, Gault N, Crechet F, Pfarr CM, Lefaix JL. Coactivation of AP-1 activity and TGF-beta1 gene expression in the stress response of normal skin cells to ionizing radiation. Oncogene. 1997;15:981-989.

139. Hirakata M, Kaname S, Chung UG, Joki N, Hori Y, Noda M, Takuwa Y, Okazaki T, Fujita T, Katoh T, Kurokawa K. Tyrosine kinase dependent expression of TGF-beta induced by stretch in mesangial cells. Kidney Int. 1997;51:1028-1036.

140. Noda M, Katoh T, Takuwa N, Kumada M, Kurokawa K, Takuwa Y. Synergistic stimulation of parathyroid hormone-related peptide gene expression by mechanical stretch and angiotensin II in rat aortic smooth muscle cells. J Biol Chem. 1994;269:17911-17917.

141. Geisterfer AAT, Peach MJ, Owens GK. Angiotensin II induces hypertrophy, not hyperplasia, of cultured aortic smooth muscle cells. Circ Res. 1988;62:749-756.

142. Leduc I, Meloche S. Angiotensin II stimulates tyrosine phosphorylation of the focal adhesion-associated protein paxillin in aortic smooth muscle cells. J Biol Chem. 1995;270:4401-4404.

143. Timmermans PBMWM, Wong PC, Chiu AT, Herblin WF, Benfield P, Carini DJ, Lee RJ, Wexler RR, Saye JAM, Smith RD. Angiotensin-II receptors and angiotensin-II receptor antagonists. Pharmacol Rev. 1993;45:205-251.

144. Giasson E, Meloche S. Role of p70 s6 protein kinase in angiotensin II-induced protein synthesis in vascular smooth muscle cells. J Biol Chem. 1995;270:5225-5231.

145. Ishida M, Marrero MB, Schieffer B, Ishida T, Bernstein KE, Berk BC. Angiotensin II activates pp60 (c-Src) in vascular smooth muscle cells. Circ Res. 1995;77:1053-1059.

146. Miyamoto S, Teramoto H, Gutkind JS, Yamada KM. Integrins can collaborate with growth factors for phosphorylation of receptor tyrosine kinases and MAP kinase activation: Roles of integrin aggregation and occupancy of receptors. J Cell Biol. 1996;135:1633-1642.

147. Bourcier T, Sukhova G, Libby P. The nuclear factor kappa-B signaling pathway participates in dysregulation of vascular smooth muscle cells in vitro and in human atherosclerosis. J Biol Chem. 1997;272:15817-15824.

148. Siebenlist U, Franzoso G, Brown K. Structure, regulation and function of NF-kappa B. Annu Rev Cell Biol. 1994;10:405-455.

149. Palombella VJ, Rando OJ, Goldberg AL, Maniatis T. The ubiquitin-proteasome pathway is required for processing the NF-kappa B1 precursor protein and the activation of NF-kappa B. Cell. 1994;78:773-785.

150. Hishikawa K, Oemar BS, Yang Z, Luscher TF. Pulsatile stretch stimulates superoxide production and activates nuclear factor-kappa B in human coronary smooth muscle. Circ Res. 1997;81:797-803.

151. Huot J, Houle F, Marceau F, Landry J. Oxidative stress-induced actin reorganization mediated by the p38 mitogen-activated protein kinase/heat shock protein 27 pathway in vascular endothelial cells. Circ Res. 1997;383-392.

152. Topper JN, Cai JX, Qiu YB, Anderson KR, Xu YY, Deeds JD, Feeley R, Gimeno CJ, Woolf EA, Tayber O, Mays GG, Sampson BA, Schoen FJ, Gimbrone MA, Falb D. Vascular MADs: Two novel MAD-related genes selectively inducible by flow in human vascular endothelium. Proc Natl Acad Sci USA. 1997;94:9314-9319.

6. ANGIOGENIC GROWTH FACTORS

Cedric J. Gaultier, Jean-Baptiste Michel
INSERM U 460 Remodelage cardiovasculaire Faculté de médecine X. Bichat Paris France

Hitherto, formation of new blood vessels was usually classified into two well-known embryological phenomenons: vasculogenesis and angiogenesis. Vasculogenesis is described as a primary in situ differentiation of endothelial cells from mesodermal precursors, leading to the formation of primary capillary poleaxe. Angiogenesis was thought to be exclusively the result of vessels' sprouting from pre-existing vessels. But recently, Asahara et al. have showed that circulating endothelial cell progenitors are also involved in the angiogenic process [1]. Concerning collateral vessel formation, a third concept, called "arteriogenesis" has been introduced by Schapper [2]. It consists of the remodeling of small capillaries into larger arterioles, with a muscular layer. Nevertheless, those three entities require:

- proliferation and migration with a spatial organisation of endothelial cells
- dissolution and regeneration of the vascular extracellular matrix, considered as vascular stake
- participation of mesenchymal perivascular cells (smooth muscle cells and pericytes).

During post-natal life, angiogenesis has been reported to be implicated in physiological situations, including tissue turnover, wound repair, ovulation and placenta growth, but also in pathological situations such as ischemia (cardiac and peripheral arterial diseases), rheumatoid arthritis and neoplasic tumors [3].

Throughout the past decade, converging works have permitted isolation of major growth factors, responsible for those different steps of angiogenesis: Vascular Endothelial Growth Factor (VEGF), Fibroblast Growth Factor (FGF), Platelet Derived Growth Factor (PDGF), Transforming Growth Factor (TGF), Hepatic Growth Factor (HGF) and Tumor Necrosis Factor (TNF). Since angiogenic growth factors are members of an exponential "growing" family, and that angiogenesis becomes a very complex process, there is a real need for a classification of growth factors. Moreover, growth factors have great diversity of cellular and molecular targets.

Before reviewing the best-known growth factors, it is first necessary to describe general mechanisms of growth factors in the angiogenic process and their modulators.

MECHANISMS OF GROWTH FACTORS IN ANGIOGENESIS

Among the different physiological or pathological situations that induce angiogenesis, hypoxia, shear-stress, metabolic and inflammatory mediators were seen to both exert a transcriptional or either a post-transcriptional regulation of the expression of angiogenic growth factors and their receptors [2]. Concerning coronary heart disease, two stimuli were intuitively found to be responsible for the initiation of the angiogenic process: hypoxia inside cardiac hypoperfused myocytes and increased shear stress in vessels of adjacent territories. Conversely, little is known about stimuli that induce the transition from the initial dormant phase to the exponential growing phase of solid tumors, called angiogenic switch [3]. There is also limited data available to explain the spontaneous angiogenic trend during embryogenesis.

Expression of growth factors have been reported for many cellular types, including cardiac, smooth muscle, brain, kidney, endocrine glands, inflammatory and endothelial cells as well. Those cytokines may be active in an intracrine, autocrine and principally paracrine pathways. Since the endothelial cell was considered the pivotal cell in the angiogenic process, researches was oriented towards isolation of endothelial growth factors and their specific tyrosine kinase receptors. Yet, recent works have emphazised the implication of circulating endothelial cell progenitors, perivascular and inflammatory cells during angiogenesis, yielding new "styles" of angiogenic factors.

Angiogenic potentials of growth factors depend on their ability to induce a proliferation of endothelial cells, a modification in cell adhesion (i.e. cell to cell and cell to extracellular matrix peptides interactions) and/ or a remodeling process of the extracellular matrix and the basement membrane. Main growth factors are direct-acting endothelial cell mitogens (FGF, VEGF, HGF). Other factors (PDGF, TGF-β and TNF-α) can be indirectly mitogenic, by promoting an overexpression of a direct-acting mitogen.

Cell adhesion rules out almost each phase of angiogenesis: proliferation, migration, alignment, polarity establishment of endothelial cells and perivascular cells as well [4]. α_v-β_3 integrin have been proposed to be one of the most relevant adhesion molecule, involved in the angiogenic process, since many evidences have been gathered [4]. However, other adhesion molecules (selectins, cadherins, immunoglobulins) need to be explored. In order to allow the invasion of endothelial cells in the perivascular space, growth factors induce the release of several proteases (serine proteases and metallo-proteases) that cleave the basement membrane and extra-cellular matrix. In addition, growth factors may also help in the resolution of angiogenesis, since some of them participate in the deposition of matrix peptides along the new formed vessels [3].

MODULATORS OF GROWTH FACTORS ACTIVITY

Co-factors: heparin-like molecules

Heparin-like molecules, found in the extracellular matrix and cell surfaces, have to be considered as majors "co-factors" of some angiogenic factors (FGF, VEGF, SH-HGF). Indeed, heparin-like molecules enhance bioavaibility of growth factors by protecting these factors against proteolysis and storage in the extracellular matrix, [5]. In addition, cell surface heparin facilitates the presentation of growth factors to their specific receptors. The trimolecular interaction of a growth factor, a high-affinity receptor and a low-affinity heparan sulfate receptor seems to be a prerequisite for proliferation of endothelial cells during angiogenesis [6]. This balance is impaired in presence of heparinase I and III that can inhibit FGF-2 induced neovascularization in vivo [7].

Inhibitors of angiogenesis

Endogenous negative regulators of angiogenesis interfere with expression or binding of growth factors on their receptors, and antagonize their downstream effectors. Beside the numerous well-known inhibitors of angiogenesis (Interferon α, Platelet Factor-4 (PF-4), TGF-β, Thrombospondin (TSP), Interleukin-12), several come from the proteolytic cleavage of endogenous regulators, involved in many physiological process, including angiogenesis [3]. Angiogenesis, by activating metalloproteases needed for extracellular matrix degradation, produces one of its own counter regulation factors: "endostatin" (fragment of collagen XVIII) [8]. Other degradation fragments were reported to be anti-mitogenic on endothelial cells: 38 kDa-fragment of plasminogen (angiostatin), 29 kDa-fragment of fibronectin, internal fragment of PF-4 and 16 kDa-N terminal prolactin [8].

Occasionally, antiproliferative activity is generated by a direct interaction between an inhibitor of angiogenesis and an angiogenic factor (PF-4 blocks FGF-2 dimerization and 16 kDa-prolactin inhibits VEGF) [9]. It is worth noting that malignant tumors produce angiogenic growth factors, but also some anti-angiogenic factors, such as TSP, endostatin and angiostatin [8].

PRINCIPAL ANGIOGENIC GROWTH FACTORS

Direct-acting endothelial cell mitogens

Vascular Endothelial Growth Factor (VEGF)

VEGF, initially called Vascular Permeability Factor (or VEGF-A), is the first isolated selective vascular endothelial cell growth factor [10]. Since 1989,

VEGF family also includes Placenta Growth Factor [11], VEGF-B [12], VEGF-C [13] and VEGF-D [14]. VEGF-A is a 45 kD homodimeric glycoprotein, that binds heparin. A signal sequence allows its secretion [10]. Alternative splicing of its mRNA provides 4 isoforms, with different biochemical properties: VEGF-121 is secreted and freely soluble; VEGF-165, the predominant isoform is secreted, but is almost bound to heparan sulfate; VEGF-189 and 206 are exclusively bound to the extracellular matrix [10]. VEGF expression have a wide distribution: brain, liver, pituitary cells, ovaries, kidneys, cardiomyocytes, several neoplasic cells, and smooth muscle cells in arteries [10]. Since there is no expression of VEGF in vascular endothelial cells in vivo and VEGF receptors are only present in endothelial cells, this suggests a purely paracrine action of VEGF. Hypoxia represents probably the main stimulus of VEGF expression, with a transcriptional and post-transcriptional regulation. Like erythopoietin, a "hypoxic responsive element" has been isolated in VEGF gene promoter [15], that is sensitive to a transcription factor, called Hypoxic Induced Factor. In addition, hypoxia increases VEGF mRNA stability [16]. With hypoxia, adenosine has an additive effect on VEGF expression, suggesting a possible independent regulatory role of tissue activity dependent paracrine metabolism [17, 18].

VEGF, being an effector of many cytokines (FGF-2 [19], PDGF [20], TGF-β [21] have probably a pivotal role in angiogenesis. Recently, VEGF was also found to be intricated with the coagulation cascade [22]; it can be up-regulated by the tissue factor or down-regulated by thrombospondin.

Three receptors of the tyrosine kinase family to VEGFs have been listed: Flt-1 [23] binds PlGF, VEGF-A and B. Flk-1 (KDR), responsible for mitogenic activity [24], binds VEGF-A, B, C and D. Flt-4 [25] binds VEGF-C and D. VEGF receptors are mainly found on endothelial cells, during angiogenesis [26, 27]. It is noteworthy that hypoxia can directly increase the synthesis of VEGF-receptors [28], permitting a tissue-targeting of VEGF, during ischemia induced angiogenesis. Beyond angiogenesis, the permanent expression of VEGF receptors by quiescent endothelium suggests a role of VEGF in the maintenance of endothelial integrity [29]. Recently, among CD34+ mononuclear blood cells, Asahara isolated some putative circulating endothelial cell progenitors, that are incorporated in post-natal angiogenic site [1]. Flk-1 is expressed by those monocytes, assimilated to angioblasts [1], and by early hematopoietic stem cells, but ceases to be expressed as soon as hematopoietic differentiation is initiated towards non-endothelial specificities. Flt-1, also present on monocytes seems to be involved in monocytes chemotaxis [30].

In vitro, VEGF (with evidence of synergism with FGF-2) stimulates the proliferation and migration of endothelial cells and forms tube-like structures [31], but it can also inhibit endothelial cell apoptosis [32]. Plasminogen activators and collagenases up-regulation by VEGF, induces extracellular matrix proteolysis [33, 34]. This major step allows the migration of endothelial cells, but also the releasing of growth factors, responsible for an auto-amplification. Finally, VEGF augments vessels permeability and may cause extravasation of plasma proteins, necessary for the formation of a new extracellular matrix [35]. Ziche have proposed that VEGF angiogenesis depends on nitric oxide production in endothelial cells, in opposition with FGF angiogenesis [36].

During embryogenesis, embryos lacking a single VEGF-A allele die at mid gestation and show blood island and angiogenesis impairment. The homozygote phenotype is more severe [37]. Flk-1 homozygous knock-out mice die at early stage and lack both endothelial and hematopoietic cells [27]. In contrast, Flt-1 homozygous knock-out mice have normal hematopoietic and endothelial progenitors, which cannot assemble into tubes [26]. Other members of VEGF family were principally studied in the embryo. VEGF-B is produced in the heart, skeletal muscle and pancreas [12]. VEGF-C, found in the heart, placenta, ovaries, is probably involved in lymphatic and venous vessels formation [13]. VEGF-D is most abundant in the heart, lung, skeletal muscle and small intestine. It is a mitogen for endothelial cells [14].

Administration of recombinant VEGF in animal model of ischemia can stimulate development of collateral arteries [38] and restore endothelium-dependent flow in arteries subserved by collateral vessels [39]. It can also accelerate reendothelialization, attenuate intimal hyperplasia in balloon-injured carotid artery [40]. Administration of antibodies directed against VEGF can suppress tumor growth [41].

Fibroblast Growth Factors (FGF)

FGF was initially considered as a mitogenic factor for brain's fibroblasts. Today, sixteen members, with widened targets and potentials have been isolated: FGF-1 to FGF-16 [42]. Two of them have been specially studied concerning their angiogenic power: acidic-FGF (or FGF-1) and basic-FGF (or FGF-2). Four isoforms of FGF-2 (18 to 24 kD) have been described, with a widespread expression. In the vascular compartment, FGF can be found in endothelial cells [43], smooth muscle cells [44] and macrophages [45]. By lacking an hydrophobic "signal sequence" [46], secretion of FGF-1 and FGF-2 cannot be done using the traditional exocytosis pathway. Other mechanisms have been speculated upon to explain their release: changing conformation of cytoskeleton, induced by shear-stress [47], cell necrosis or participation of a Na/K ATPase. Nevertheless, each isoform seems dedicated to different compartments with different biological effects [48]. For example, high molecular weight b-FGF, acts in an intracrine way: when synthesized in the cytoplasm, FGF is transferred into the nucleus to induce cell mitogenesis, with no participation of membranous receptors. At the opposite, small size FGF (18 kD) is secreted, and interacts with classic membranous receptors. This pathway tends to be reserved for cellular migration stimulation [48].

Binding of FGF-1 or FGF-2 to heparan sulphates enhances their autocrine or paracrine bioactivity. This link provides FGF a tissue storage [49], a protection against proteolysis [50] and improved presentation of FGF to its receptors [51]. FGF-receptors originate from the super-family of immunoglobulins, with 4 types identified: FGF-R1, R2, R3 and R4. With variable affinity, those tyrosin kinase-type receptors have been isolated on cardiomyocytes, endothelial and smooth muscle cells [52].

Unlike VEGF, FGF-2 knockout mice are morphologically normal and display decreased smooth muscle contractility and hypotension [53]. Despite its poor implication in embryonic angiogenesis, FGF is involved in the proliferation and maturation of cardiac myocytes [54].

In vitro, stimulation of endothelial cells by FGF-2 leads to the formation of capillary-like tubular structures, subsequent to proliferation and migration of endothelial cells [55].

The physiological involvement of FGF during ischemia-induced angiogenesis is not entirely understood. FGF's poor expression in cardiac or skeletal myocytes exposed to ischemia contrasts with its upregulation in adjacent well-perfused muscles, suggesting the participation of indirect mediators responsible for a paracrine induction of FGF [56]. However, monocytes exposed to ischemia shows a direct overexpression of FGF, that in turn can stimulate proliferation of endothelial cells [57]. On the other hand, increased shear stress, found upstream from an arterial occlusion, constitutes a stimulus for FGF expression. Shear stress can activate directly the endothelial cell's production of FGF [58], but it can also activate this production indirectly from the recruitment of monocytes [59].

A therapeutic use of FGF-2, in a model of chronic ischemia, has revealed an angiogenic response, represented by increased capillary density and blow flow [60]. Even if FGF-2 is also mitogenic for smooth muscle cells or fibroblasts in vitro, it does not increase neointimal accumulation at sites of desendothelialization, when administrated intravenously in vivo [61]. Furthermore, some studies conclude that exogenous FGF-2 seems to induce a better angiogenic response than an endothelial specific factor: VEGF [61]. In acute myocardial ischemia models, exogenous FGF-2 was able to significantly reduce the size of infarction [62]. According to Horrigan et al., FGF'-2 myocardial salvage seems independent of angiogenesis [63], but may be the result of improved endothelial cells or myocytes resistance to ischemia or the opening of pre-existing collateral vessels.

Scatter Factor/Hepatocyte Growth Factor (SH/HGF)

HGF, a 80 kDa heterodimeric cytokine, binds heparan sulfate as soon as it is secreted by mesenchymal cells [64]. Ischemia was found to be a positive stimulus of expression of both, HGF and its tyrosine kinase receptor [65]. In vitro, the formation of tube-like structures, subsequent to HGF's direct activation of endothelial cells proliferation and migration [66], is highly enhanced by an upregulation of VEGF-A, in smooth muscle cells [67]. In an ischemic hindlimb model, Van Belle has showed its potent angiogenic power, that is synergistic with VEGF [67]. On smooth muscle cells, HGF is not mitogenic, but seems involved in the relationship between endothelial cells and mesenchymal cells during angiogenesis.

Other mitogens

Angiopoietin-1 (Ang-1)

Angiopoietin-1, expressed by mesenchymal cells, is considered to be the principal ligand of TIE-2, a tyrosine-kinase receptor, expressed on vascular endothelial cells, and early hematopoietic cells [68]. Knock-out experiment of this receptor in mice induce vessels immaturity during embryogenesis, characterized by the lack of periendothelial supports and disorganization into large and small vessels. Mice with inactivation of Ang-1 gene display similar disorders in embryological vascular development. The fact that disorders seem less severe in Ang-1 deficient mice, acts as strong evidence for others TIE-2's ligands [68]. Angiopoietin-2 (ang-2) is a second ligand, that inhibits TIE-2 phosphorylation, and acts like a competitive inhibitor of Ang-1.

In opposition with VEGF, Ang-1 does not have any mitogenic or chemo-attractive effects on endothelial cells. According to Folkman's proposal [69], Ang-1 activates endothelium, which is releasing chemotactic factors (PDGF), in order to attract mesenchymal cells in contact with them. As a consequence of this contact, TGF-β is over expressed, and subsequently induces 3 major events. Firstly, mesenchymal cells differentiate into vascular smooth muscle cells or pericytes; secondly, proliferation of endothelial cells is inhibited and lastly, extra-cellular matrix deposition is stimulated. Hanahan's hypothesis [68] states that Ang-1 seems devoted to the maturation and stabilisation of vessels; related to endothelial cells-mesenchymal cells adhesion and matrix deposition; and Ang-2, from its induced disruption of endothelial cells-mesenchymal cells or cell-matrix links, plays an important role in the remodeling of vessels.

Platelet derived growth factor (PDGF)

From dimerization of two chains (A and B), platelets, macrophages, endothelial cells and vascular smooth muscle cells provide three kinds of PDGF: AA, BB or AB. Shear-stress [58] and hypoxia [57] are reported to respectively upregulates PDGF-BB expression in endothelial cells and in macrophages. Among the two types of tyrosine kinase receptors (α and β), shear-stress activates PDGF-β receptors expressed by endothelial cells, smooth muscle cells and pericytes [70].

PDGF is considered as a proangiogenic factor. Thus, it is not a direct mitogen for endothelial cells, but is able to enhance production of both FGF-2 and VEGF by smooth muscle cells [20, 21]. Recruitment of smooth muscle cells and pericytes represents probably the principal contribution of PDGF during angiogenesis, since those cells are first, responsible for extra-cellular matrix components deposition and second, help in the three dimensional organization of new vessels [69]. Lastly, chemoattraction of macrophages may accelerate extra-cellular matrix dissolution, needed during the first phase of angiogenesis.

Transforming Growth factor (TGF)

TGF-β is a 25 kDa homodimeric polypeptide. Its distribution includes kidneys, liver, heart, platelets and endothelial cells. TGF-β production is sensitive to the variation of shear stress exerted on endothelial cells [71]. Although in vitro studies found an inhibition of endothelial cells growth, either directly [72] or by antagonizing FGF-2 [73], TGF-β is clearly considered as a positive regulator of angiogenesis in vivo [74]. This may result from the overexpression of VEGF and FGF-2 by smooth muscle cells [21]. In addition, since this factor is able to inhibit endothelial cells proliferation, facilitate the differentiation of mesenchymal cells into smooth muscle cells, enhance the plasminogen activator inhibitor and promote fibronectin and collagen deposition, TGF-β has been proposed to be one of the effector that trigger the resolution of the angiogenic process [3].

CONCLUSION

During the last decade, significant progress has occurred in the understanding of angiogenesis. Since molecular mechanisms of several growth factors have been deciphered, a therapeutic use of angiogenic factors in some ischemic cardiovascular diseases is now envisaged. But, angiogenesis must not be summarized as the specific action of few factors on the proliferation and migration of endothelial cells. First, many other cells are involved in the process: inflammatory, mesenchymal cells and circulating progenitors of endothelial cells. Second, cell adhesion, apoptosis, remodeling of the extracellular matrix are other major events of angiogenesis. Third, activity of growth factors may be enhanced by cofactors or balanced by several inhibitors of angiogenesis. Fourth, some growth factors were found to be less active in adult life compared with embryological life, and vice versa. Finally, some dangerous side effects may emerge from the therapeutic use of growth factors.

REFERENCES

1. Asahara T, Murohara T, Sullivan A, Silver M, Van der Zee R, LI T, Witzenbichler B, Schatteman G, Isner J. Isolation of putative progenitor endothelial cells for angiogenesis. Science 1997;275:964-967.

2. Schaper W, Ito W. Molecular mechanisms of coronary collateral vessel growth. Circ Res 1996;79:911-919.

3. Pepper M. Manipulating angiogenesis: from basic science to the bedside. Arterioscler Thromb Vasc Biol 1997;17:605-619.

4. Bischoff J. Cell adhesion and angiogenesis (Perspectives series: cell adhesion in vascular biology). J Clin Invest 1997;100(11S)Suppl:37S-39S.

5. Folkman J, Klagsbrun M, Sasse J, Wadzinski M, Ingber D, Voldavsky I. Heparin-binding angiogenic protein-basic fibroblast growth factor is stored within basement membrane. Am J Pathol 1988;130:393-400.

6. Folkman J, Shing Y. Control of angiogenesis by heparin and other sulfated polysaccharides. Adv Exp Med Biol 1992;313:355-364.

7. Sasisekharan R, Moses M, Nugent M, Cooney C. Heparinase inhibits neovascularization. Proc Natl Acad Sci USA 1994;91:1524-1528.

8. O'Reilly M, Boehm T, Shing Y, Fukai N, Vasios G, Lane W, Flynn E, Birkhead J, Olsen B, Folkman J. Endostatin: an endogenous inhibitor of angiogenesis and tumor growth. Cell 1997;88:277-285.

9. D'Angelo G, Struman I, Martial J, Weiner RI. Activation of mitogen-activated protein kinases by vascular endothelial growth factor and basic fibroblast growth factor in capillary endothelial cells is inhibited by the antiangiogenic factor 16-kDa N-terminal fragment of prolactin. Proc Natl Acad Sci USA 1995;92:6374-6378.

10. Ferrara N. Vascular endothelial growth factor. Trend Cardiovasc Med 1993; 3:244-250.

11. Maglione D, Guerrerio V, Viglietto G, Delli-Bovi P, Persico M. Isolation of a human placenta cDNA coding for a protein related to vascular endothelial growth factor. Proc Natl Acad Sci USA 1991;88:9267-9271.

12. Olofsson B, Pajusola K, Kaipainen A, Voneuler G, Joukov V, Saksela O, Orpana A, Petersson RF, Alitalo K, Eriksson U. Vascular Endothelial Growth Factor B, a novel growth factor for endothelial cells. Proc Natl Acad Sci USA 1996;93:2576-2581.

13. Joukov V, Pajusola K, Kaipainen A, Chilov D, Lahtinen I, Kukk E, Saksela O, Kalkkinen N, Alitalo K. A novel Vascular Endothelial Growth Factor, Vegf-C, is a ligand for the Flt4 (Vegfr-3) and Kdr (Vegfr-2) receptor tyrosine kinases. EMBO Journal 1996;15:290-298.

14. Achen M, Jeltsch M, Kukk E, Mäkinen T, Vitali A, Wilks A, Alitalo K, Stacker S. Vascular endothelial growth factor D (VEGF-D) is a ligand for the tyrosine kinases VEGF receptor 2 (Flk1) and VEGF receptor 3 (Flt4). Proc Natl Acad Sci U S A 1998;95(2):548-553.

15. Levy AP, Levy NS, Wegner S, Goldberg MA. Transcriptional regulation of the rat vascular endothelial growth factor gene by hypoxia. J Biol Chem 1995;270:13333-13340.

16. Ikeda E, Achen MG, Breier G, Risau W. Hypoxia-induced transcriptional activation and increased mRNA stability of vascular endothelial growth factor in C6 glioma cells. J Biol Chem 1995;270:19761-19766.

17. Pueyo M, Chen Y, D'Angelo G, Michel JB. Regulation of vascular endothelial growth factor expression by AMPc in rat aortic smooth muscle cells. Exp Cell Res 1998;238:354-358.

18. Adair T, Gay W, Montani J. Growth regulation of the vascular system: evidence for a metabolic hypothesis. Am J Physiol 1990;259:R393-R404.

19. Stavri GT, Zachary IC, Baskerville PA, Martin JF, Erusalimsky JD. Basic fibroblast growth factor upregulates the expression of vascular endothelial growth factor in vascular smooth muscle cells. Synergistic interaction with hypoxia. Circulation 1995;92:11-14.

20. Stavri GT, Hong Y, Zachary IC, Breier G, Baskerville PA, Yla-Herttuala S, Risau W, Martin JF, Erusalimsky JD. Hypoxia and platelet-derived growth factor-BB synergistically upregulate the expression of vascular endothelial growth factor in vascular smooth muscle cells. FEBS Letters 1995;358:311-315.

21. Brogi E, Wu T, Namiki A, Isner JM. Indirect angiogenic cytokines upregulate VEGF and bFGF gene expression in vascular smooth muscle cells, whereas hypoxia upregulates VEGF expression only. Circulation 1994;90:649-652.

22. Shoji M, Abe K, Nawroth P, Rickles F. Molecular mechanisms linking thrombosis and angiogenesis in cancer. Trends Cardiovas Med 1997;7:52-59.

23. Shibuya M. Role of VEGF-flt receptor system in normal and tumor angiogenesis. [Review]. Adv Cancer Res 1995;67:281-316.

24. Waltenberger J, Claesson-Welsh L, Siegbahn A, Shibuya M, Heldin CH. Different signal transduction properties of KDR and Flt1, two receptors for vascular endothelial growth factor. J Biol Chem 1994;269:26988-26995.

25. Pajusola K, Aprelikova O, Korhonen J, Kaipainen A, Kerlovaara L, Alitalo R, Alitalo K. Flt-4 receptor tyrosine kinase contains seven immunoglobulin-like loops and is expressed in multiple human tissues and cell lines. Cancer Res 1992;52:5738-5742.

26. Fong GH, Rossant J, Gertsenstein M, Breitman ML. Role of the Flt-1 receptor tyrosine kinase in regulating the assembly of vascular endothelium. Nature 1995;376:66-70.

27. Shalaby F, Ho J, Stanford W, Fischer K, Schuh A, Schwartz L, Bernstein A, Rossant J. A requirement for Flk1 in primitive and definitive hematopoiesis and vasculogenesis. Cell 1997;89:981-990.

28. Tuder RM, Flook BE, Voelkel NF. Increased gene expression for VEGF and the VEGF receptors KDR/Flk and Flt in lungs exposed to acute or to chronic hypoxia. Modulation of gene expression by nitric oxide. J Clin Invest 1995;95:1798-1807.

29. Van der Zee R, Murohara T, Zhengyu L, Zollmann F, Passeri J, Lekutat C, Isner J. Vascular endothelial growth factor/vascular permeability factor augments nitric oxide release from quiescent rabbit and human vascular endothelium. Circulation 1997;95:1030-1037.

30. Barleon B, Sozzani S, Zhou D, Weich HA, Mantovani A, Marme D. Migration of human monocytes in response to vascular endothelial growth factor (VEGF) is mediated via the VEGF receptor flt-1. Blood 1996;87:3336-3343.

31. Pepper MS, Ferrara N, Orci L, Montesano R. Potent synergism between vascular endothelial growth factor and basic fibroblast growth factor in the induction of angiogenesis in vitro. Biochem Biophys Res Comm 1992;189:824-831.

32. Spyridopoulos I, Brogi E, Kearney M, Sullivan A, Cetrulo C, Isner J, Losordo D. Vascular endothelial growth factor inhibits endothelial cell apoptosis induced by tumor necrosis factor-alpha: balance between growth and death signals. J Mol Cell Cardiol 1997;29(5):1321-1330.

33. Pepper M, Ferrara N, Orci L, Montesano R. VEGF induces plasminogen activators and plasminogen activator inhibitor type 1 in microvascular endothelial cells. Biochem Biophys Res Commun 1991;181:902-908.

34. Unemori E, Ferrara N, Bauer E, Amento EP. Vascular endothelial growth factor induces interstitial collagenase expression in human endothelial cells. J Cell Physiol 1992;153:557-562.

35. Connoly D, Olander J, Heuvelman D, Nelson R, Monsell R, Siegel N, Haymore B, Leimgruber R, Feder J. Human vascular permeability factor. Isolation from U937 cells. J Biol Chem 1989;264:20017-20024.

36. Ziche M, Morbidelli L, Choudhuri R, Zhang H, Donnini S, Granger H. NO synthase lies downstream from VEGF- induced but not b-FGF-induced Angiogenesis. J Clin Invest 1997;99:2625-2634.

37. Carmeliet P, Ferreira V, Breier G, Pollefeyt S, Kieckens L, Gertsenstein M, Fahrig M, Vandenhoeck A, Harpal K, Eberhardt C, Declercq C, Pawling J, Moons L, Collen D, Risau W, Nagy A. Abnormal blood vessel development and lethality in embryos lacking a single VEGF allele. Nature 1996;380:435-439.

38. Takeshita S, Zheng L, Brogi E, Kearney M, Pu L, Bunting S, Ferrara N, Symes J, Isner J. Therapeutic angiogenesis. A single intraarterial bolus of vascular endothelial growth factor augments revascularization in a rabbit ischemic hind limb model. J Clin Invest 1994; 93:662-670.

39. Bauters C, Asahara T, Zheng LP, Takeshita S, Bunting S, Ferrara N, Symes JF, Isner JM. Recovery of disturbed endothelium-dependent flow in the collateral-perfused rabbit ischemic hindlimb after administration of vascular endothelial growth factor. Circulation 1995;91:2802-2809.

40. Asahara T, Bauters C, Pastore C, Kearney M, Rossow S, Bunting S, Ferrara N, Symes JF, Isner JM. Local delivery of vascular endothelial growth factor accelerates reendothelialization and attenuates intimal hyperplasia in balloon-injured rat carotid artery. Circulation 1995;91:2793-2801.

41. Kim KJ, Li B, Winer J, Armanini M, Gillett N, Phillips HS, Ferrara N. Inhibition of vascular endothelial growth factor-induced angiogenesis suppresses tumour growth in vivo. Nature 1993;362:841-844.

42. Miyake A, Konishi M, Martin F, Hernday N, Ozaki K, Yamamoto S, Mikami T, Arakawa T, Itoh N. Structure and expression of a novel member, FGF-16, on the fibroblast growth factor family. Biochem Biophys Res Commun 1998;243:1:148-152.

43. Vlodavsky I, Fridman R, Sullivan R, Sasse J, Klagbrun M. Aortic endothelial cells synthetize basic fibroblast growth factor which remains cell associated and platelet-derived factor-like protein which is secreted. J Cell Physiol 1987;131:402-408.

44. Speir E, Sasse J, Shrivastav S, Casscells W. Culture-induced increase in acidic and basic fibroblast growth factor activities and their association with the nuclei of vascular endothelial and smooth muscle cells. J Cell Physiol 1991;147:362-373.

45. Baird A, Mormede P, Bohlen P. Immunoreactive fibroblast growth factor in cells of peritoneal exsudate suggests its identity with macrophage-derived growth factor. Biochem biophys Res Commun 1985;126:358-364.

46. Abraham J, Mergia A, Whang J, Tumolo A, Friedman J, Hjerrild K, Gospodarowicz D, Fiddes J. Nucleotide sequence of a bovine clone encoding the angiogenic protein, basic fibroblast growth factor. Science 1986;233:545-548.

47. Sterpetti A, Cucina A, Fragale A, Lepidi S, Cavallaro A, Santoro d'Angelo L. Shear stress influences the release of platelet-derived growth factor and basic fibroblast growth factor by arteriel smooth muscle cells. Eur J Vasc Surg 1994;8:138-142.

48. Bikfalvi A, Klein S, Pintucci G, Rifkin D. Biological roles of fibroblast growth factor-2. Endocr Rev 1997;18:26-45.

49. Casscells W, Speir E, Sasse J, Klagsbrun M, Allen P, Lee M, Calvo B, Chiba M, Haggroth L, Folkman J, Epstein S. Isolation, characterization, and localization of heparin-binding growth factors in the heart. J Clin Invest 1990;85:433-441.

50. Gospodarowicz D, Cheng J. Heparin protects basic and acidic FGF from inactivation. J Physiol 1986;128:475-484.

51. Yayon A, Klagsbrun M, Esko J, Leder P, Ornitz D. Cell surface, heparin-like molecules are required for binding of basic fibroblast growth factor activity. Cell 1991;64:841-848.

52. Hughes S, Hall P. Overview of the fibroblast growth factor and receptor families: complexity functional diversity, and implications for future cardiovascular research. Cardiovasc Res 1993;27:1199-1203.

53. Zhou M, Sutliff R, Paul R, Lorenz J, Hoying J, Haudenschild C, Yin M, Coffin J, Kong L, Kranias E, Luo W, Boivin G, Duffy J, Pawlowski S, Doetschman T. Fibroblast growth factor 2 control of vascular tone. Nature Medicine 1998;4(2):201-207.

54. Engelmann G, Dionne C, Jaye M. Acidic-fibroblast growth factor and heart development: role in myocyte proliferation and capillary angiogenesis. Circ Res 1993;72:7-19.

55. Montesano R. Regulation of angiogenesis in vitro. Eur J Clin Invest 1992;22:504-515.

56. Walgenbach K, Gratas C, Shestak K, Becker D. Ischemia induced expression of b-FGF in normal skeletal muscle: a potential paracrine mechanism for mediating angiogenesis in ischaemic skeletal muscle. Nat Med 1995;1(5):453-459.

57. Kuwabara K, Ogawa S, Matsumoto M, Clauss M, Pinsky D, Lyn P, Leavy J, Witte L, Joseph-Silverstein J, Firie M, Torcia G, Cozzolino F, Kamada T, Stern D. Hypoxia-mediated induction of acidic/basic fibroblast growth factor and platelet-derived growth factor in mononuclear phagocytes stimulates growth of hypoxic endothelial cells. Proc Natl Acad Sci USA 1995;92:4606-4610.

58. Patrick C, McIntire L. Shear stress and cyclic strain modulation of gene expression in vascular endothelial cells. Blood Purif 1995;13:112-124.

59. Sampath R, Kukielka G, Smith C, Eskin S, McIntire L. Shear stress-mediated changes in the expression of leukocyte adhesion receptors on human umbilical vein endotheliel cells in vitro. Ann Biomed Eng 1995;23:247-256.

60. Unger E, Banai S, Shou M, Lazarous D, Jaklitsch M, Scheinowitz M, Correa R, Klingbeil C, Epstein S. Basic fibroblast growth factor enhances myocardial collateral flow in a canine model. Am J Physiol 1994;266:H1588-H1595.

61. Lazarous DF, Shou M, Scheinowitz M, Hodge E, Thirumurti V, Kitsiou AN, Stiber JA, Lobo AD, Hunsberger S, Guetta E, Epstein SE, Unger EF. Comparative effects of basic Fibroblast Growth Factor and Vascular Endothelial Growth Factor on coronary collateral development and the arterial response to injury. Circulation 1996;94:1074-1082.

62. Yanagisawa-Miwa K, Uschida Y, Nakamura F, T T, Kido H, Kamijo T, Sugimoto T, Kaji K, Utsumaya M, Kurashima C, Ito H. Salvage of infarcted myocardium by angiogenic action of basic Fibroblastic Growth Factor. Science 1992;257:1401-1403.

63. Horrigan M, MacIsaac A, Nicolini F, Vince D, Lee P, Ellis S, Topol E. Reduction in myocardial infarct size by basic-FGF after temporary cornary occluision in a canine model. Circulation 1996;94:1927-1933.

64. Nakamura T, Nishizawa T, Hagiya M, Scki T, Shimonishi A, Sugimura A, Tshiro K, Schimizu S. Molecular cloning and expression of human hepatocyte growth factor. Nature 1989;342:440-443.

65. Ono K, Matsumori A, Shioi T, Furukawa Y, Sasayama S. Enhanced expression of hepatocyte growth factor/c-Met by myocardial ischemia and reperfusion in a rat model. Circulation 1997;95:2552-2558.

66. Bussolino F, Direnzo M, Bocchietto E, Olivero M, Naldini L, Gaudino G, Tamagnone L. Hepatocyte growth factor is a potent angiogenic which stimulates endothelial cell mobility and growth. J Cell Biol 1992;119:629-641.

67. Van Belle E, Witzenbichler B, Chen D, Silver M, Chang L, Schwall R, Isner J. Potentiated angiogenic effect of scatter factor/hepatocyte growth factor via induction of vascular endothelial growth factor: the case for paracrine amplification of angiogenesis. Circulation 1998;97(4):381-390.

68. Hanahan D. Signaling vascular morphogenesis and maintenance. Science 1997;277:48-50.

69. Folkman J, D'Armore P. Blood vessel formation: what is its molecular basis? Cell 1996;87:1153-1155.

70. Risau W. Mechanisms of angiogenesis. Nature 1997;386:671-74.

71. Ohno M, Lopez F, Gibbons G, Cooke J, Dzau V. Shear stress induced TGF-b1 gene expression and generation of active TGF-b1 is mediated via a K channel. Circulation 1992;86 (supplI):I-87.

72. Hirschi K, D'Armore P. Pericytes in the microvasculature. Cardiov. Res 1996;32:687-698.

73. Baird A, Durkin T. Inhibition of endothelial cell proliferation by type B-transforming growth factor: interactions with acidic and basic fibroblast growth factors. Biochem Biophys Res Commun 1986;138:476-482.

74. Roberts A, Sporn M, Assoian R, Smith J, Roche N, Wakefield L, Heine U, Liotta L, Falanga V, Kehrl J, Fauci A. Transforming growth factor type-beta: rapid induction of fibrosis and angiogenesis in vivo and stimulation of collagen formation in vitro. Proc Natl Acad Sci USA 1986;83:4167-4171.

7. ANGIOGENESIS

Harry A.J. Struijker Boudier, Frank R.M. Stassen, Ferdinand A.C. le Noble
Department of Pharmacology, Cardiovascular Research Institute Maastricht (CARIM)
Maastricht, The Netherlands

The regulation of the architecture of the vascular system is an exiting and important area of research in developmental and vascular biology. The formation of blood vessels is a vital process in embryonic development and in physiological repair processes, such as wound healing and cyclical adaptations in the endometrium during the menstrual cycle. Disturbed vascular growth is now regarded as a key event in a variety of pathologies. Uninhibited blood vessel growth has been associated with various neoplastic and non-neoplastic diseases. This is widely accepted for tumor growth and diabetic retinopathy, but is now also believed to occur in psoriasis, atherosclerotic plaque growth and rheumatoid arthritis [1]. A lack of blood vessel growth is a key event in other pathologies. Certain developmental disorders, such as bowel atresia and unilateral facial atrophy, may arise from abnormal vascular development [1].

Ischemia-related diseases have an important basis in a dysfunction of normal vascular growth mechanisms [2]. This is particularly the case for ischemic heart diseases. In fact, following acute myocardial infarction, the lack of collateral vessel formation is a major negative prognostic indicator [3]. We have recently reviewed evidence for decreased microvascular growth as a pathogenic mechanism in hypertension [4]. Finally, deficiencies in microvascular growth appear an important step in peptic ulcer formation [5]. In this chapter we shall briefly review the physiological steps in vascular growth, the methods used for studying angiogenesis, the major molecular and cellular mechanisms of vascular growth and, finally, therapeutic aspects of angiogenesis.

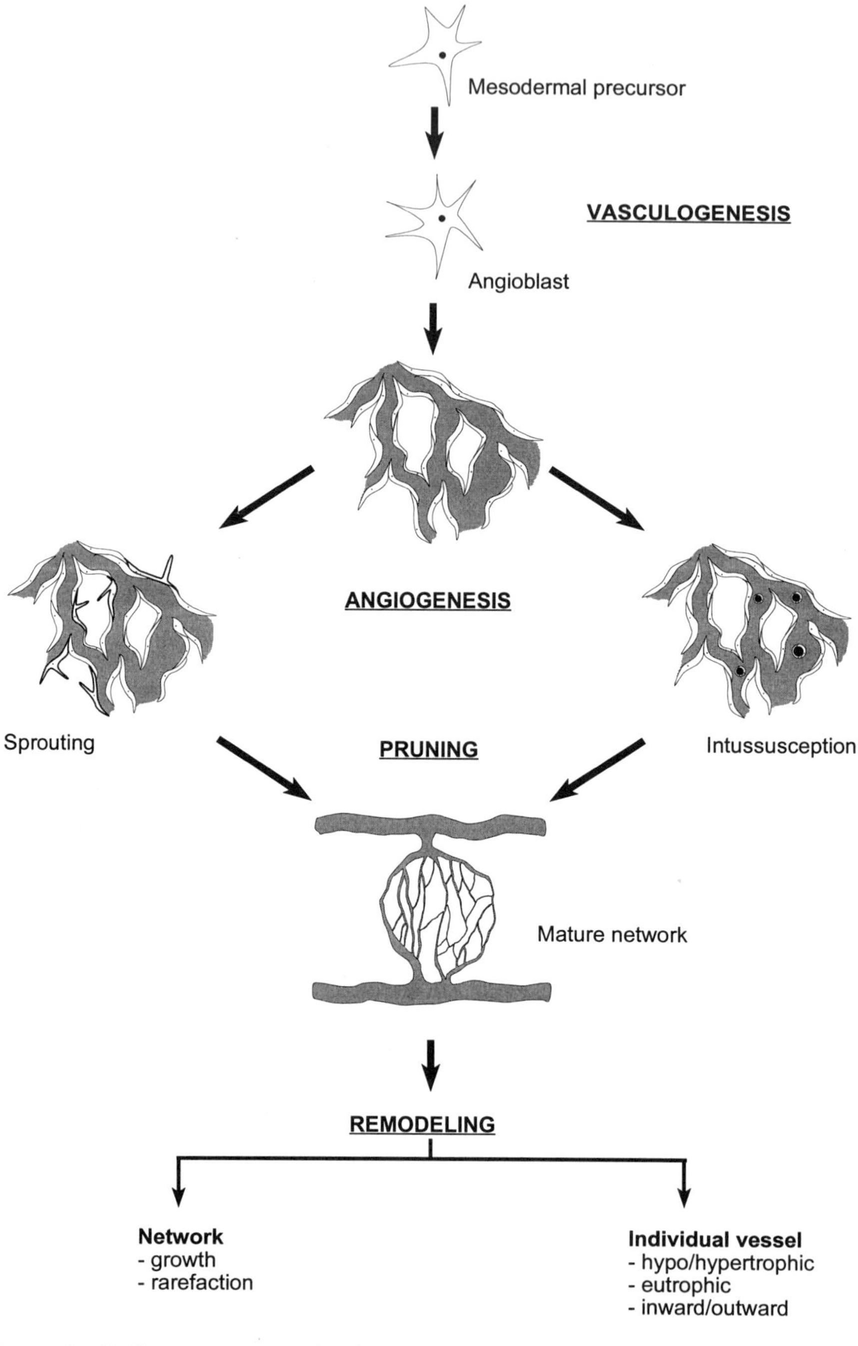

***Figure 1**. Different steps involved in embryonic and post-embryonic development of the vascular system*

ANGIOGENESIS: TERMINOLOGY AND STEPS INVOLVED

Growth of blood vessels involves several steps (figure 1). During embryonic development vasculogenesis is the de-novo development of blood vessels from differentiating endothelial cells. Endothelial precursor cells can be detected along with primitive blood cells, the angioblasts [6]. They are located in discrete zones referredto as blood islands. The fusion of these blood islands is the first evidence for a primitive vasculature. The subsequent development of different kinds of vessels (various branches of arterioles and venules) is a complex process, which varies per organ. In particular the origins of the endothelial cells and vascular smooth muscle cells constituting the vessel wall determine the ultimate structure of the vessel wall [6-8]. The development of vascular smooth muscle cell layers around the endothelial tube involves the migration and proliferation of different cell types, in particular pericytes, fibroblasts and vascular smooth muscle cells [9]. When the early endothelial tubes invade embryonic organs, they become surrounded by locally derived mesenchymal cells. The nature of these cells depends on the organ involved and the size of the vessel [9-11]. Whereas the endothelial cells are rather uniform, vascular smooth muscle cells seem much more heterogenous in origin and phenotypic behavior. Vascular smooth muscle cells may arise from fibroblasts or even endothelial cells [11,12]. They originate from the ventrolateral mesoderm and mesectodetermal-derived neural crest cells [13,14]. This heterogenous origin of vascular smooth muscle cells contributes to the later tissue-specific composition and function of the ultimate blood vessel.

The second process is angiogenesis *in sensu stricto* and involves the sprouting and growth of capillaries from existing vessels into a previous avascular part of a tissue. It is the main mode of vessel growth in adult organisms. Angiogenesis can be the result of sprouting of capillaries from pre-existing vessels or of intussusceptive growth by luminal division of existing capillaries. The first event in the sprouting mode of angiogenesis is disruption of the basement membrane that encompasses endothelial cells of existing capillaries. Subsequently, endothelial cells migrate into the extracellular space. This is followed by proliferation and formation of capillary sprouts which eventually connect to neighboring vessels, permitting flow [15]. The mechanisms involved in the non-sprouting form of angiogenesis are still less well understood. It is possible that intussusceptive growth is caused by the faster growth of the surrounding mesenchymal tissue than of the endothelium, thus leading to pressure exerted by the tissue to cause capillary collapse [15].

The third process in vascular development is the pruning and remodeling of existing capillaries, arteries and venules into a mature network. This process involves several aspects. Migration and proliferation of different cell types, in particular pericytes, fibroblasts and vascular smooth cells is needed to form additional cell layers around the endothelial tubes. The synthesis of matrix material and contractile proteins renders the vascular network its specific contractile and elastic properties. Finally, enlargement of the diameter and length of individual vessels by proliferation of both endothelial and smooth muscle cells is essential for the maturation of the network. These steps are - at least to a certain degree - reversible so that remodeling of the vascular network can also lead to vascular narrowing and even complete disappearance ("rarefaction") of small blood vessels. The role of endothelial and vascular smooth muscle cell apoptosis in this reversed remodeling is now a topic of intense research [16].

Although the word "angiogenesis" reflects only one aspect of vascular development, it is often used as a *pars pro toto* to include all above discussed forms of vascular development.

METHODS FOR STUDYING ANGIOGENESIS

Table 1 gives an overview of the major assays used to measure angiogenes. Such assays range in complexity from endothelial cell cultures to chronic in-vivo transparent chambers in various tissues in different mammalian species. Several recent review articles provide a critical evaluation of the potentials and limitations of these various assays [17-19].

Table 1*. Assays for angiogenesis studies*

In vitro
Endothelial cell cultures
Endothelial cell monolayers
Three-dimensional gel assays

In situ
Chick embryo CAM
Cornea micropocket assay

Matrix implants
Polymer/collagen gels

Exteriorized tissues
Subcutaneous/intradermal assays
Cremaster muscle
Mesentery

Chronic in-vivo transparent chambers
Hamster cheek pouch
Rabbit ear chamber
Cranial windows
Dorsal skin chamber

Endothelial cell cultures are particularly suited to study the influence of factors from the extracellular environment on cell growth. The source of the endothelial cell used for the culture is very important, because responses from cells of different origin vary considerably [19]. For example, endothelial cells derived from human dermal microvasculature form capillary-like tubes much faster than human umbilical vein cells [19]. Also the response to stretch may differ for endothelial cells derived from aorta when compared to vena cava [20].

Endothelial cell culture techniques have become increasingly sophisticated over the past years. Thus , the use of co-cultures with smooth muscle cells allows a more refined study of tube formation. The use of gels containing various substances, such as collagen, fibronectin, gelatin or basement membrane components influences the degree of endothelial cell migration, differentiation or proliferation. A major disadvantage of tissue culture is the potential phenotypic changes of cells brought into culture.

The anterior chamber of the eyes has been used for many decades as an in-situ angiogenesis assay. Initially, it was described for rabbit corneas, but has been expanded to include the corneas of guinea pigs, rats and mice. The assay involves the placement of an angiogenesis inducer into a corneal pocket to evoke vascular outgrowth from peripherally located limbal vasculature. Since the cornea is initially avascular, this assay measures vasculogenesis.

One of the most frequently used angiogenesis assay systems is the chick embryo chorio-allantois membrane (CAM). The CAM is an embryonic, developing vascular structure. Its major advantage, apart from low costs, is that it allows the study of the full spectrum of cellular events involved in vessel growth. On the other hand, it should be realized that it represents an embryonic model. Growth is most rapid around days 7-9 and it virtually stops around day 14 [21,22]. Quantification of vessel growth can be achieved on different levels of complexity, ranging from simple semi-quantitative density scales to sophisticated computer-analyzed network images [22].

In-vivo assays for angiogenesis have primarily followed three approaches: (1) vascularization of implanted polymer matrix gels, (2) microcirculatory preparations of exteriorized tissues, and (3) chronically implanted microcirculatory chambers in different tissues. The matrix implant methods have involved sterile polyester sponges or foam discs [23]. Vessel growth is assessed by histology of the vessels or, more indirectly, by blood flow measurements. The latter approach is sensitive to possible effects on vascular tone by the (anti-)angiogenic substances. Furthermore, a common problem associated with assays of vasculari-zation into matrix implants is the nonspecific host response to matrix implantation [18].

Intravital microscopy in either exteriorized or chronic window preparations is the most physiological but also the most complex way to study angiogenesis.

Already more than 100 years ago, Thoma [24] wrote an impressive monograph on vascular growth based on direct microscopic observations of developing vascular networks in the chick embryo. The early histological methods used by Thoma and his late 19th century contemporaries were refined in the 20th century. Two techniques have been particularly helpful in the study of the three-dimensional architecture and development of the microvasculature: light microscopy of dye injected tissue specimens and scanning electron microcopy (SEM) of corrosion casts. The use of India ink (and other dyes) has been particularly helpful in the quantification of capillary growth. In the past two decades, the introduction of in-situ immunohistochemistry has greatly facilitated the histomorphometric study of vascular development. Immunohistochemistry allows the recognition of a range of cellular and subcellular components of the vessel wall. Powerful new microscopy techniques (e.g. confocal microscopy, fluorescent imaging) enhance visualization of these cellular and subcellular components. An even more recent approach has been the selective manipulation of gene expression to alter the activity of growth factors and their receptors believed to play a role in vascular development. The reader is referred to a recent review article on the lessons learned from these genetic manipulations on vascular development [25].

The histological and molecular biology approaches discussed above are extremely useful in studying molecular mechanisms of angiogenesis. However, they require the removal of tissue from the intact organism and are, therefore, less suited for the study of the dynamic behavior of developing microvascular networks. Intravital microscopy, in particular in chronic window preparations, is the method of choice for studying the development of microvascular structures over time. We have recently reviewed the use of chronic window techniques elsewhere [26].

MEDIATORS OF ANGIOGENESIS: MECHANICAL FACTORS

There has been a recent explosion in research on factors involved in angiogenesis. A range of mechanical and chemical mediators have been implicated in the design of vascular architecture. The recognition of the importance of mechanical factors can be traced back to Thoma [24], who proposed that the rate of blood flow is an important determinant of both diameter of individual vessels and growth of vessels in a developing vascular network. Murray [27] subsequently proposed that vascular networks adapt their geometry on the basis of the principle of minimalization of work required to maintain adequate blood flow. Based on this concept, Glagov and co-workers [28] proposed that shear stress is the primary driving force for control of blood vessel architecture. In the mean time this "shear stress rule" for control of vascular structure has been confirmed in various models [29-31]. For single vessels, adaptation to shear stress implies that an acute reduction in flow is counteracted by a lumen reduction, whereas an increment in flow is counteracted by a lumen increase. Experimental studies have shown that long-term increments in tissue flow, such as during exercise or treatment with vasodilators, causes an increased number of arterioles and capillaries [30].

Computer modeling studies [32] indicate that adaptation to shear stress alone is not enough to explain the maintenance of a stable vascular network. Pressure and pressure-related circumferential wall stress are additional important mechanical factors in the control of microvascular network geometry [29-32]. Wall stress was known for a long time already as an important controlled parameter in relation to wall thickness and diameter of individual vessels. Besides these effects on the structure of single vessels, pressure may also affect microvascular network architecture. In fact, in chronic hypertension, accompanied by an increase in arteriolar pressure, capillary and small arteriolar rarefaction is a generally observed phenomenon [33].

An area of intensive present research is that of the molecular mechanisms mediating vascular growth responses to mechanical factors. Shear stress changes induce a range of molecular mechanisms in the vessel wall, including the altered expression of early growth genes and various peptide growth factors. Changes in wall tension also affect growth factors and matrix-integrin binding molecules. The reader is referred to a recent review by Davies [34] as well as the chapter by Tedgui et al. in this book for a critical discussion of the mechanical influences on vascular growth.

Table 2. *Molecular factors involved in angiogenesis*

Factor	Target
VEGF	Endothelial cell differentiation and growth
TGF-beta	Inhibition of EC growth Induction of SMC differentiation Stimulation of matrix growth
PDGF	Mural cell recruitment to capillary tubes
b-FGF	Hemangioblast differentiation
Tie-1 receptors	Endothelial cell layer organization
Angiopoietin/ Tie-2 receptors	Segmentation of early tubes to arterioles and venules
Angiotensin II	Mural cell recruitment to capillary tubes VSMC growth; matrix synthesis
Integrin $\alpha_v\beta_3$	Cellular attachment to capillary tubes

MEDIATORS OF ANGIOGENESIS: CHEMICAL FACTORS

Table 2 summarizes some of the major molecules involved in angiogenesis. The first is VEGF. Several isoforms of VEGF exist with different tissue expression patterns, affinity for cellular receptors and biological actions [35]. VEGF family members are bound to at least three receptor tyrosine kinases (Flt-1, Flt-4 and Flk-1) with different specificity and affinity. The Flt-1 and Flk-1 tyrosine kinases are expressed primarily by the endothelium. The major actions of VEGF are endothelial cell differentiation and growth. It thus plays a relatively early role in blood vessel formation. This early role was recently confirmed in a series of studies on targeted inactivation of various components of the VEGF/tyrosine kinase pathway. In these studies animals (mice) encountered embryonic lethality due to early vascular developmental abnormalities [25]. The expression of VEGF is strongly influenced by hypoxia, providing a physiological mechanism to promote blood vessel growth in situations of tissue ischemia.

TGF-beta comprises a family of peptides with pleiotropic effects on cellular growth and differentiation. A serine/threonine kinase has been identified as the cellular receptors of TGF-beta. However, TGF-beta may exert its biological effects in part through binding to other molecules, such as betaglycan and endoglin. The effects of TGF-beta on cells involved angiogenesis is rather controversial. On the one hand, it inhibits endothelial cell proliferation and impairs endothelial cell migration [36]. On the other hand, it is angiogenic in in-vivo assays. Most likely, TGF-beta stimulates the growth of other cell types involved in angiogenesis and influences integrin expression and matrix accumulation. Even its effects on endothelial cells may alter in an in-vivo environment in the presence of extracellular matrix and other growth factors. TGF-beta-1 deficient mouse embryos show vascular abnormalities. These do not occur at the level of endothelial cell proliferation and tube formation, but at later stages of maturation of the vascular network [37].

The PDGF family consists of at least three isoforms (PDGF-AA, -BB, and -AB). These isoforms have different effects on various cell types, the effect depending on the expression pattern of the PDGF receptors. PDGF may affect at least two aspects of vessel formation: stimulation of endothelial cell growth and recruitment of mural cell precursors during the process of tube formation and maturation of the vascular network. PDGF and PDGF-receptor deficient mice show abnormalities in vascular development. The most marked abnormality is the lack of recruitment of mural cells by the endothelial cells of capillary sprouts [36]. The abnormalities are confined to the microvasculature. Recent studies from D'Amore's laboratory suggest that PDGF is a critical actor in the differentiation of mural cells having been recruited to capillary tubes [36].

Basic FGF is a cationic polypeptide that is mitogenic for a number of cell types, including vascular endothelial cells. It increases the expression of the cellular proto-oncogene c-myc. It is an angiogenic factor in many angiogenesis assays. Gene disruption techniques have thus far failed to precisely define its role in vascular development. There is some evidence that b-FGF participates in the inductive processes that allow the early mesenchymal cells of the embryo to become hemangioblasts. Thus, b-FGF would be an essential mediator in the appearance of Flk-1 positive endothelial cells [7].

A recently described molecular mediator of angiogenesis is the Tie receptor family and its ligands, the angiopoietins. The Tie receptors comprise a class of receptor tyrosine kinases that appear specific for vascular endothelium. Angiopoietin-1 and angiopoietin-2 have recently been identified as ligands for the Tie-2 receptor [38,39]. Gene inactivation studies suggest that Tie-1 is not essential for early vasculogenesis. It appears to be required for the formation of an intact microvascular endothelial cell layer, which controls fluid exchange and hemodynamic stress regulation [25]. The angiopoietin/Tie-2 system seems involved in an even later stage of vascular development. Again on the basis of targeted gene inactivation studies, this system seems particularly active in determining the segmentation of an initially homogenous capillary network into larger and smaller caliber vessels [40,41].

Angiotensin II is the biologically active component of the renin-angiotensin system. In addition to its classical acute cardiovascular actions, recent research has concentrated on longer-term effects, including angiotensin II induced angiogenesis [22]. Its angiogenic activity has now been established in various angiogenesis assays. The nature of the receptor mediating angiotensin II induced angiogenesis is still a matter of controversy, with evidence to support the involvement of AT-1 receptors on the one hand, and a non-AT-1, non-AT-2 receptor on the other hand. The cellular target of angiotensin II in angiogenesis does not seem the endothelial cell, but rather the recruitment of mural cells to existing capillary tubes [22]. In particular, the process of arterialization of capillary tubes has been proposed as a target for angiotensin II [22].

Finally, cell adhesion molecules, such as integrins, play a critical role in the stabilization of newly formed vessels. Newly formed vessels express several integrins, including $\alpha_v\beta_3$. This integrin is localized in the tips of endothelial cells of sprouting vessels and its expression diminishes after vessel maturation [42]. Interference with the interaction of integrin $\alpha_v\beta_3$ and its natural substrates, such as fibrin and fibronectin induces apoptosis in the proliferating endothelial cells [43]. Immunological knock-out of integrin $\alpha_v\beta_3$ function caused an aberrant pattern of large artery formation [44].

THERAPEUTIC ASPECTS OF (ANTI-)ANGIOGENESIS

The impressive developments in the area of molecular angiogenesis research in the past decade has opened novel opportunities for therapeutic developments. On the one hand, anti-angiogenic drugs could be attractive for the treatment of tumors and other diseases characterized by uninhibited blood vessel growth, such as diabetic retinopathy and rheumatoid arthritis. On the other hand, "therapeutic angiogenesis" could be desirable in situations of insufficient (micro)vascular growth or need of collateral blood vessel growth, such as ischemia-related diseases.

Both pro- and anti-angiogenic therapies are actively investigated at the level of drug screening/development and clinical research. Research on anti-angiogenic therapies has the longest history and has been driven primarily by the desire to develop anti-tumor therapies [1].

Table 3. *Inhibitors of angiogenesis*

Polypeptides
Interferon-α and γ
Platelet factor IV
Tumor necrosis factor-α
Thrombospondin
Tissue inhibitors of metalloproteinase
Angiostatin/endostatin
Soluble cytokine receptors
Integrin $\alpha_v\beta_3$ antagonists and antibodies
VEGF neutralizing antibodies

Steroids
2-Methoxyoestradiol
Medroxyprogesterone acetate
Angiostatic steroids

Antibiotics
Fumagillin analogues
Minocycline
Herbimycin A

Others
Sulphated polysaccharides
Suramin and derivatives
Genistein
Thalidomide
Linomide

Table 3 contains a list of some of the molecules presently under investigation for this purpose. A number of these agents are directly derived from endogenous mediators of angiogenesis as discussed above. The anti-angiogenic potential of other drugs was discovered more accidentally and is now traced back to endogenous targets. It is beyond the scope of this chapter to review the potential clinical use of these agents for anti-angiogenic purposes. The reader is referred to two recent review articles for a critical discussion on anti-angiogenic therapy [45,46].

The development of pro-angiogenic therapies is of more recent date and has been inspired by the wish to develop novel therapies for wound repair and ischemia treatment. With respect to the latter, in particular the treatment of ischemia-related events in coronary artery disease and peripheral arterial occlusive disease has been a recent target for "therapeutic angiogenesis". In the treatment of coronary artery disease present therapeutic approaches aim at bypassing obstructions by means of coronary artery bypass grafting, restoring flow by angioplasty or reduction of myocardial oxygen demand by drug treatment. "Therapeutic angiogenesis" attempts to overcome some of the short-comings of these methods and provide a direct means of increasing blood flow. The following strategies have been considered: (1) stimulation of endogenous production of angiogenic factors, (2) gene transfer and (3) administration of peptidergic angiogenic factors. Among the angiogenic factors studied, VEGF and bFGF constitute the most widely studied compounds. In a gradual occlusion model in pigs, periadventitial bFGF or VEGF administration had beneficial effects on coronary flow, wall motion and endothelial cell receptor mediated function [47-49]. Similarly, in a rabbit model of chronic myocardial ischemia intrapericardial infusion of bFGF enhances new epicardial small vessel growth [50].

With respect to peripheral ischemia present therapeutic approaches include exercise, surgical revascularization, endovascular interventional surgical therapies and vasodilator drug treatment. The induction of growth of collateral vessels by means of "therapeutic angiogenesis" has now been demonstrated in animal models of hindlimb ischemia in the rabbit and rat. Again, like in the ischemic myocardium, VEGF and bFGF have been the most widely studied angiogenic molecules in these models [51-55]. Preliminary clinical studies show possible beneficial effects of VEGF gene transfer in patients with severe peripheral ischemia [56].

The therapeutic applications of agents influencing angiogenesis is promising, but still subject to critical analysis. In particular biopharmaceutical and pharmacokinetic issues (site and mode of administration, dose schedules) need to be resolved. Pharmacodynamic aspects, such as receptor interactions, tachyphylaxis and side-effects have not yet received systematic attention. These basic issues will undoubtedly generate a fascinating field of future research.

REFERENCES

1.Folkman J. Clinical applications of research on angiogenesis. New Engl J Med 1995;333:1757-1763.

2.Schaper W, Ito WD. Molecular mechanisms of coronary collateral vessel growth. Circ Res 1996;79:911-919.

3.Fujita M, Ohno A, Wada O, Miwa K, Nozawa T, Yamanishi K, Sasayama S. Collateral circulation as a marker of the presence of viable myocardium in patients with recent myocardial infarction. Am Heart J 1991;122:409-414.

4.Le Noble FAC, Stassen FRM, Hacking WJG, Struijker Boudier HAJ. Angiogenesis and hypertension. J Hypertens, 1998, in press.

5.Folkman J Szabo S, Stovroff M, McNeil P, Li W, Shing Y. Duodenal ulcer: discovery of a new mechanism and development of angiogenic therapy which accelerates healing. Ann Surg 1991;214:414-427

6.Risau W. Mechanisms of angiogenesis. Nature 1997;386:671-674.

7.Risau W, Flamme I. Vasculogenesis. Annu Rev Cell Dev Biol 1995;11:73-91.

8.Bergwerff M, De Ruiter MC, Poelmann RE, Gittenberger-de Groot AC. Onset of elastogenesis and downregulation of smooth muscle actin as distinguishing phenomena in artery differentiation in the chick embryo. Anat Embryol 1996;194:545-557.

9.Hirschi KK, D'Amore PA. Pericytes in the microvasculature. Cardiovasc Res 1996;32:687-699.

10.De Ruiter MC, Hogers B, Poelmann RE, Van Iperen L, Gittenberger-de Groot AC. The development of the vascular system in quail embryos: a combination of microvascular corrosion casts and immunohistochemical identification. Scanning Micr 1992;5:1082-1090.

11.Campbell GR, Campbell JH. Development of the vessel wall: overview. In: The vascular smooth muscle cell (Schwartz SM, Mecham RP, eds). San Diego: Academic Press, 1995;1-17

12.De Ruiter MC, Poelmann RE, Van Munsteren JC et al. Embryonic endothelial cells transdifferentiate into mesenchymal cells expressing smooth muscle actins in vivo and in vitro. Circ Res 1997;80:444-451.

13.Gittenberger-de Groot AC, Slomp J, De Ruiter MC, Poelmann RE. Smooth muscle cell differentiation during early development and during intimal thickening formation in the ductus arteriosus. In: The vascular smooth muscle cell (Schwartz SM, Mecham RP, eds). San Diego: Academic Press, 1995;17-37.

14.Katoh Y, Periasamy M. Growth and differentiation of smooth muscle cells during vascular development. Trends Cardiovasc Med 1996;6:100-106.

15.Hudlicka O, Brown M, Egginton S. Angiogenesis in skeletal and cardiac muscle. Physiol Rev 1992;72:369-417.

16.Moreau P, Tea B-S, Dam T-V, Hamet P. Altered balance between cell replication and apoptosis in hearts and kidneys of newborn SHR. Hypertension 1997;30(part 2):720-724.

17.Cockerill GW, Gamble JR, Vadas MA. Angiogenesis: models and modulators. Int Rev Cytol 1995;159:113-160.

18.Jain RK, Schlenger K, Höckel M, Yuan F. Quantitative angiogenesis assays: progress and problems. Nat Med 1997;3:1203-1208.

19.Hudlicka O, Brown MD, Egginton S. Angiogenesis: basic concepts and methodology. In: An introduction to vascular biology (Halliday A, Hunt BJ, Poston L, Schachter M, eds). Cambridge: University Press, 1998;3-20.

20.Iba T, Sain T, Sonoda T, Rosales O, Sumpio BE. Stimulation of endothelial secretion of tissue-type plasminogen activator by repetitive stretch. J Surg Res 1991;50:45-60.

21.Auerbach R, Auerbach W, Polakowski A. Assays for angiogenesis: a review. Pharmacol Ther 1991;51:1-11.

22.Le Noble FAC, Kessels-van Wylick LCGA, Hacking WJG, Slaaf DW, oude Egbrink MGA, Struijker Boudier HAJ. The role of angiotensin II and prostaglandins in arcade formation in a developing microvascular network. J Vasc Res 1996;33:480-488.

23.Andrade SP, Fan TPD, Lewis GP. Quantitative in-vivo studies on angiogenesis in a rat sponge model. Br J Exp Pathol 1987;68:755-766.

24.Thoma R. Untersuchungen über die Histogenese and Histomechanik des Gefässsystems. Stuttgart: F Enke, 1893.

25.Carmeliet P, Collen D. Genetic analysis of blood vessel formation. Role of endothelial versus smooth muscle cells. Trends Cardiovasc Med 1997;7:271-281.

26.Struijker Boudier HAJ, Crijns FRL, Stolte J, Van Essen H. Assessment of the microcirculation in cardiovascular disease. Clin Sci 1996;91:131-139.

27.Murray CD. The physiological principle of minimum work. 1. The vascular system and the cost of blood volume. Proc nat Acad Sci USA 1926;12:207-214.

28.Glagov S, Vito R, Giddens DP, Zarins CK. Micro-architecture and composition of artery walls: relationship to location, diameter and the distribution of mechanical stress. J Hypertens 1992;10(suppl 6):101-104.

29.Koller A, Kaley G. Shear stress dependent regulation of vascular resistance in health and disease: role of endothelium. Endothelium 1996;4:247-272.

30.Skalak TC, Price RJ. The role of mechanical stresses in microvascular remodeling. Microcirculation 1996;3:143-165.

31.Pries AR, Secomb TW, Gaehtgens P. Biophysical aspects of blood flow in the microvasculature. Cardiovasc Res 1996;32:654-667.

32.Hacking WJG, Van Bavel E, Spaan JAE. Shear stress is not sufficient to control growth of vascular networks: a model study. Am J Physiol 1996;270:H364-H375.

33.Struijker Boudier HAJ, Le Noble JLML, Messing MWJ et al. The microcirculation and hypertension. J Hypertens 1992;10(suppl 7):147-156.

34.Davies PF. Flow-mediated endothelial mechanotransduction. Physiol Rev 1995;75:519-560

35.Joukov V, Pajusola K, Kaipanen A et al. A novel vascular endothelial growth factor, VEGF-C, is a ligand for the flt-4 (VEGFR-3) and KDR (VEGFR-2) receptor tyrosine kinases. EMBO J 1996;15:290-298.

36.Beck,jr, L, D'Amore PA. Vascular development: cellular and molecular regulation. FASEB J 1997;11:365-373.

37.Dickson MC, Martin JS, Cousins FM, Kulkarni AB et al. Defective haematopoiesis and vasculogenesis in transforming growth factor-ß1 knock-out mice. Development 1995;121:1845-1854.

38.Davis S, Aldrich TH, Jones PF et al. Isolation of angiopoietin-1, a ligand for the angiogenic TIE-2 receptor by secretion-trap expression cloning. Cell 1996;87:1161-1169.

39.Maisonpierre PC, Suri C, Jones PF, Bartunkova S et al. Angiopoietin-2, a natural antagonist for TIE-2 that disrupts in vivo angiogenesis. Science 1997;277:55-60.

40.Suri C, Jones PF, Patan S et al. Requisite role of angiopoietin-1, a ligand for the TIE-2 receptor, during embryonic angiogenesis. Cell 1996;87:1171-1180.

41.Vikkula M, Boon LM, Carraway KL et al. Vascular dysmorphogenesis caused by an activating mutation in the receptor tyrosine kinase TIE-2. Cell 1996;87:1181-1190.

42.Clark RAF, Tonnesen MG, Gailit J, Cheresh DA. Transient functional expression of $\square_v\square_3$ on vascular cells during wound repair. Am J Pathol 1996;148:1407-1421.

43.Brooks PC, Montgomery AMP, Rosenfeld M et al. Integrin $\alpha_v\beta_3$ antagonists promote tumor regression by inducing apoptosis of angiogenic blood vessels. Cell 1994;79:1157-1164.

44.Drake CJ, Cheresh DA,Little CD. An antagonist of integrin $\alpha_v\beta_3$ prevents maturation of blood vessels during embryonic neovascularization. J Cell Sci 1995;108:2655-2661.

45.Pepper MS. Manipulating angiogenesis. From basic science to the bedside. Arterioscler Thromb Vasc Biol 1997;17:605-619.

46.Fan TD, Jaggar R, Bicknell R. Controlling the vasculature: angiogenesis, anti-angiogenesis and vascular targeting of gene therapy. Trends Pharmacol Sci 1995;16:57-66.

47.Sellke FW, Wang SY, Stamler A et al. Enhanced microvascular relaxations to VEGFand bFGF in chronically ischemic porcine myocardium. Am J Physiol 1996;271:H713-H720.

48.Pearlman JD, Hibberd MG, Chuang ML, Harada K, Lopez JJ et al. Magnetic resonance mapping demonstrates benefits of VEGF-induced myocardial angiogenesis. Nat Med 1995;1:996-997.

49.Harada K, Friedman M, Lopez JJ, Wang SY et al. Vascular endothelial growth factor administration in chronic myocardial ischemia. Am J Physiol 1996;270:H1791-H1802.

50.Landau C, Jacobs AK, Haudenschild CC. Intrapericardial basic fibroblast growth factor induces myocardial angiogenesis in a rabbit model of chronic ischemia. Am Heart J 1995;129:924-931.

51.Takeshita S, Zheng LP, Brogi E, Kearney M et al. Therapeutic angiogenesis: a single intra-arterial bolus of vascular endothelial growth factor augments revascularization in a rabbit ischemic hindlimb model. J Clin Invest 1994;93:662-670.

52.Takeshita S, Rossow ST, Kearney M et al. Time course of increased cellular proliferation in collateral arteries after administration of vascular endothelial growth factor in a rabbit model of lower limb vascular insufficiency. Am J Pathol 1995;147:1649-1660.

53.Bauters C, Asahara T, Zheng LP et al. Recovery of disturbed endothelium-dependent flow in the collateral-perfused rabbit ischemic hindlimb after administration of vascular endothelial growth factor. Circulation 1995;91:28-2-2809.

54.Yang HT, Deschenes MR, Ogilvie RW. Terjung RL. Basic fibroblast growth factor increases collateral blood flow in rats with femoral arterial ligation. Circ Res 1996;79:62-69.

55.Asahara T, Bauters C, Zheng LP et al. Synergistic effect of vascular endothelial growth factor and basic fibroblast growth factor on angiogenesis in vivo. Circulation 1995;92(suppl 2):365-371.

56. Baumgartner I, Pieczel A, Manor O et al. Constitutive expression of phVEGF165 after intramuscular gene transfer promotes collateral vessel development in patients with critical limb ischemia. Circulation 1998;97:1114-1123.

8. VASCULAR AGING

Joël Belmin
Departrment of Gerontology, René-Bigotini hospital, Sevran, France

Aging is responsible for important changes in vascular structure and function which in turn affect the function of the heart and of other organs. Aging is associated with large artery remodeling, characterized by a progressive increase in stiffness and changes in the function of the endothelium and of smooth muscle cells (SMC). The effects of aging on the arterial system should be differentiated from that of atherosclerosis. Even though atherosclerosis is highly prevalent among aged individuals, the two processes are distinct in many ways. Aging, a physiological process which affects the entire vascular system, starts after sexual maturation and leads to enlargement of the arterial lumen, whereas atherosclerosis is a disease that affects limited areas of arteries, which usually starts in infancy or in adolescence and tends to narrow the lumen. During the last decades, research advances in vascular biology have provided a better understanding of the process of vascular aging.

AGING AND ARTERIAL SYSTEM REMODELING

With advancing age, the morphology of the arterial system progressively changes. In a post-mortem study, the diameter of pressurized human aortas was found to be increased by 15-20% in subjects older than 65 years, compared with younger subjects [1]. Such age-related changes were also described in carotid [2,3] or coronary [4] arteries of normotensive humans and in laboratory animals [5] (Figure 1). These changes are predominant in elastic arteries, and in humans the diameter of femoral artery was not increased with age [2]. The length of the aorta and of other arteries was found to increase with age, contributing to the tortuosity of these vessels frequently observed in arteriograms from old patients. The mechanisms responsible for this arterial remodeling are not completely elucidated. The increase in blood pressure associated with aging is probably of minor importance. In WAG/Rij rats, aortic and carotid diameters were found to increase with age, although their arterial blood pressure remained stable throughout their life [5]. Moreover, chronic reduction of blood pressure by renin-angiotensin system blockade in these animals did not prevent the enlargement of arteries, also indicating that this process is pressure-independent [5].

Aging is associated with profound changes in the properties of elastin, a main component of the extracellular matrix. Elastic fibers from old animals [6, 7] and humans [8] appear to be disorganized, thinner and more fragmented, compared with those of younger individuals. Moreover, calcium bound to elastin was found to be increased in the aorta of old rats [5], which also indicates a qualitative change of elastin with aging. Elastase activity was found to be increased in the plasma and the arterial wall of old individuals, which might contribute to altered elastic fibers [9]. Age-related quantitative changes of elastin in the arterial wall are more controversial [5, 7, 10] and might depend on species and methods used. Elastic fiber properties play a major holding function in arteries and experimental alterations of elastin result in arterial dilatation and aneurysms [11]. It is thus likely that qualitative changes of elastin are the main determinant of age-related arterial enlargement.

Another important feature of age-related remodeling is the thickening of the intimal layer [12, 13] (Figure 1). In the human aorta, a 2 to 6-fold increase in the thickness of this layer was described between the ages of 20-40 and 65 or more [1]. The aged intima contains collagen, glycosaminoglycans, fragmented elastic fibers, SMC, and other mononucleated cells. These compounds are present in higher amounts than in the young intima [12, 13, 14, 15]. Age-related intimal thickening appears to be partly pressure-dependent, since it is postponed or limited by experimental blood pressure reduction in rats [5, 16]. Intimal thickening might be also the consequence of arterial enlargement and elastinolysis. The former increases stress applied to the arterial wall, which stimulates medial SMC growth and collagen synthesis; the latter facilitates SMC migration in the intima. The age-related changes in endothelial function (see under) might also contribute to intimal thickening. In humans, medial thickness was found to be unchanged with aging [1]. However, medial thickness increases with aging in rats [5, 6, 10] (Figure 2). It should be noted that this age-effect on medial thickness is much less pronounced than that of experimental hypertension [17]. Medial thickening might be related to changes in strains applied to the media as a consequence of increased lumen diameter with aging. This view is supported by the finding that chronic blood pressure reduction was found to postpone this age-related change [5].

AGING AND ARTERIAL STIFFNESS

Arterial stiffening is a hallmark of arterial aging. Evidence in humans was obtained by measuring pressure/volume relationships in aorta from subjects deceased at different ages [18]. In this study, increasing the pressure from 100 to 150 mmHg increased the volume by 70% in 20-24 year-old and 20-25% in 71-78 year-old subjects. Several studies using non invasive techniques confirmed that human arteries become more rigid with advancing age [2, 19, 20]. Interestingly, this age-related arterial stiffening was recorded in normotensive subjects, in different ethnic groups and in populations with varied life style habits [20].

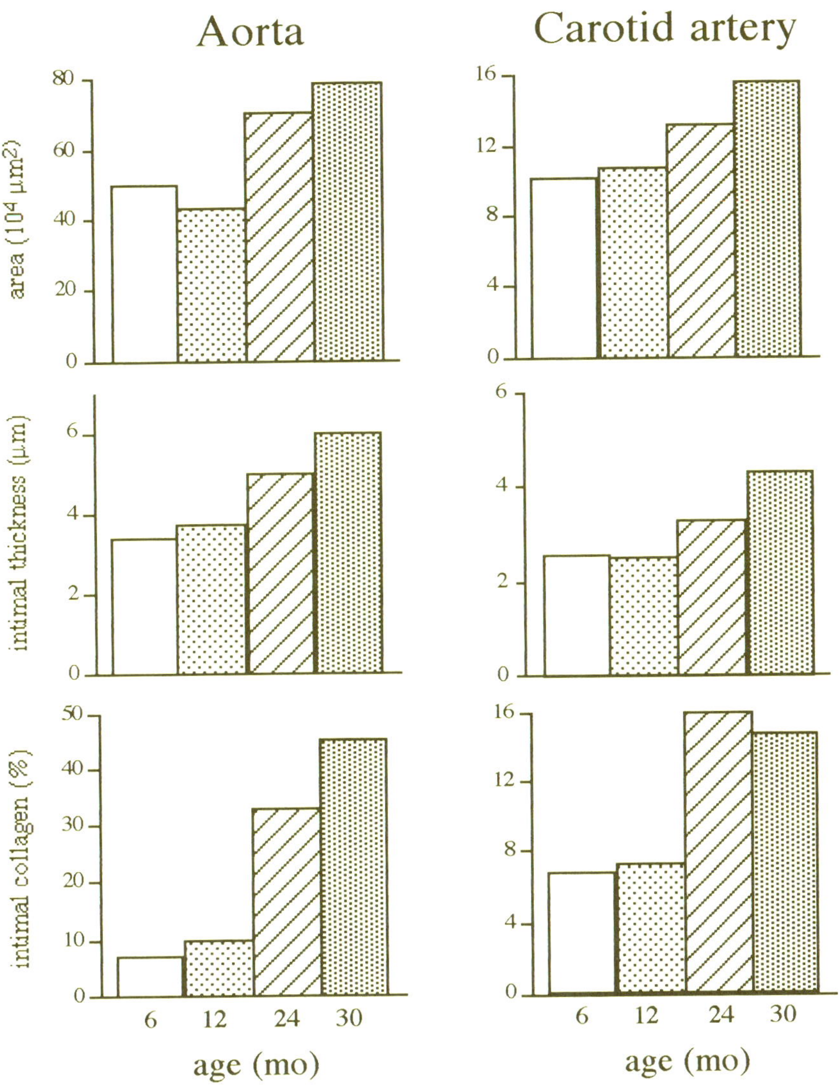

Figure 1. Cross-sectional area, intima thickness and intimal density of collagen in aorta (left panel) and carotid artery (right panel) of rats of different ages. Differences between age-groups were significant for each parameter and arterial site [5].

The in vivo arterial compliance depends both on extracellular matrix and on basal contractile tone of SMC. In old rats, SMC poisoning by potassium cyanate results in a much smaller change in compliance than in young animals, indicating that the contractile component in stiffening of aged arteries is minor [5]. In contrast, alterations of medial extracellular matrix play a major role in arterial stiffening. With aging, collagen density increases in the medial layer [5, 6, 10], whereas elastin amounts remain stable or decline. Therefore, the elastin/collagen ratio, which is a key determinant of viscoelastic properties of the wall, decreases with age [5, 6, 7, 10] (Figure 2). Moreover, the properties of collagen in aged arteries are modified by non-enzymatic glycation processes. Glucose has been shown to react with NH groups of proteins, leading to the formation of Schiff bases, Amadori products and advanced glycosilation end-products products (AGEs) [21, 22]. Since this process is very slow, long-living proteins like collagen and elastin are more concerned than proteins with a high turnover [23]. Increased glycation processes and AGE accumulation have been documented in the plasma and in the arteries of old individuals [24, 25]. AGEs exert several biological actions, one of which consists in cross-linking proteins of the extracellular matrix. These cross-links change the properties of collagen, which becomes more rigid, more resistant to proteolysis, and less soluble [23, 26].

Arterial stiffening has important implications for cardiovascular function in the aged. It alters the ability of large arteries to cushion pressure and flow during systole and to release pressure and flow towards distal arteries during diastole. As a consequence, for a given mean aortic blood pressure, brachial systolic blood pressure will be higher in old than in young subjects, and inversely diastolic blood pressure will be lower. This phenomenon contributes to the gradual increase in systolic blood pressure with age observed in human populations. The stiffening of arterioles also contributes to increased peripheral resistance and mean arterial blood pressure, accounting for the high prevalence of hypertension in the elderly [27]. As a result of stiffening, aortic impedance increases with age and ventricular-aortic coupling deteriorates [5, 28]. Moreover, arterial stiffness is responsible for an accelerated pulse wave propagation and for an earlier return of reflected waves [29]. Due to these changes, the cardiac afterload increases and the left ventricular mass gradually increases with age [30].

AGING AND ENDOTHELIAL DYSFUNCTION

Aging is associated with multiple alterations in endothelial function. The turnover of endothelial cells and their DNA synthesis increases in aged animals in association with higher protein synthesis [31]. The number of circulating endothelial cells, an indicator of their turnover, increases with age in rats [32].

The barrier function of the endothelium was found to be altered with age. In the aorta of 30 month-old rats, we found a two-fold increase in endothelial permeability to albumin compared with that of 10 month-old rats [33] (Figure 3). In this study, the high endothelial permeability contrasted with a decrease in the distribution volume of albumin in the media of old rats, a factor reducing the transport of the macromolecules across the media (Figure 3). The altered transport might favor the entry of plasma macromolecules across the endothelium and their trapping in the intima [34], which might contribute to age-related modifications of this layer. Interestingly, both increased endothelial cell turnover [35] and interaction between AGEs and endothelial cells [36] can alter the barrier function of the endothelium.

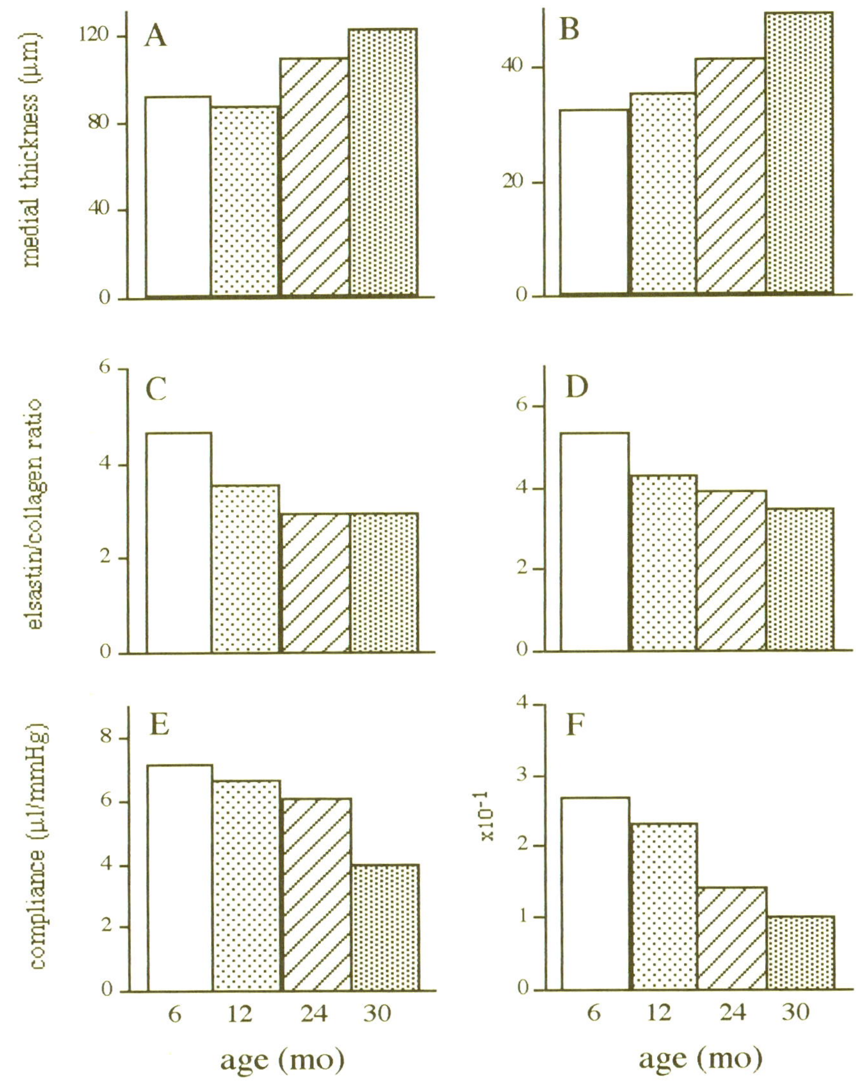

Figure 2. *Changes of medial thickness and elastin/collagen ratio in the aorta (A and C) and the carotid artery (B and D) of rats of different ages. Differences between age-groups were significant for each parameter and arterial site. Systemic arterial compliance (E) and in situ carotid compliance at 100 mmHg (F) significantly declined with age [5].*

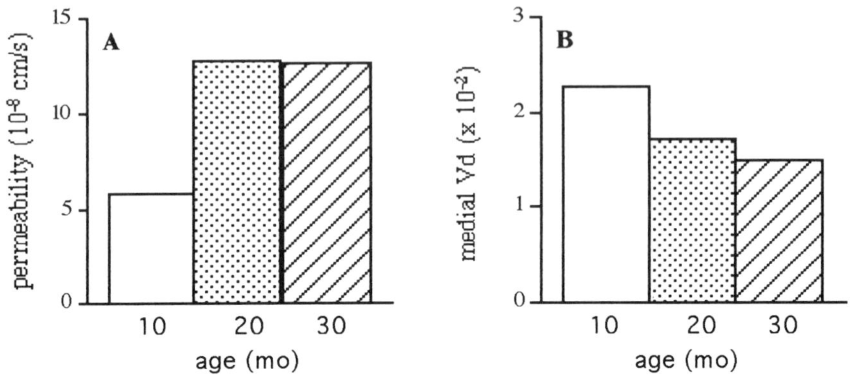

Figure 3. *Endothelial permeability (A) and apparent distribution volume in the media (B) of albumin in the aorta of 10, 20 and 30 month-old rats. Differences between age groups were significant for the two parameters [33].*

The release of vasoactive substances, including prostacyclin (PGI_2), nitric oxide (NO) and endothelin, is a major function of the endothelium. PGI_2 has potent vasodilatatory effects and inhibits platelet aggregation. PGI_2 released by human endothelial cells from aged donors was found to be reduced compared with those of young donor [37]. Moreover, a 3 to 4 fold decline in plasma levels of PGI_2 was documented in 60-70 year old subjects, compared with 20-30 year old subjects [38]. Aging is associated with a marked impairment of endothelium-dependent relaxation, a response essentially mediated by NO. In old animals, acetylcholine-induced relaxation of precontracted vessels is reduced compared with younger animals [39-42] (Figure 4). In humans with angiographically normal coronary arteries, the changes in coronary blood flow and diameter following acetylcholine injection were found to be reduced with age [43, 44].

The vasomotor response of forearm arteries to acetylcholine was also found to be impaired with age in humans [45, 46]. The effect of age on blood flow responses indicates that this impaired control of relaxation is not limited to proximal arteries but also affects distal circulation. In basal conditions, the endothelium exerts a relaxing effect on arteries. In vivo administration of L-NAME inhibits the basal production of NO and provokes vasocontriction and blood pressure increase. These responses are unchanged [47] or even increased [48] in old rats, suggesting that the basal tone of NO is unchanged or increased, but not reduced with advancing age.

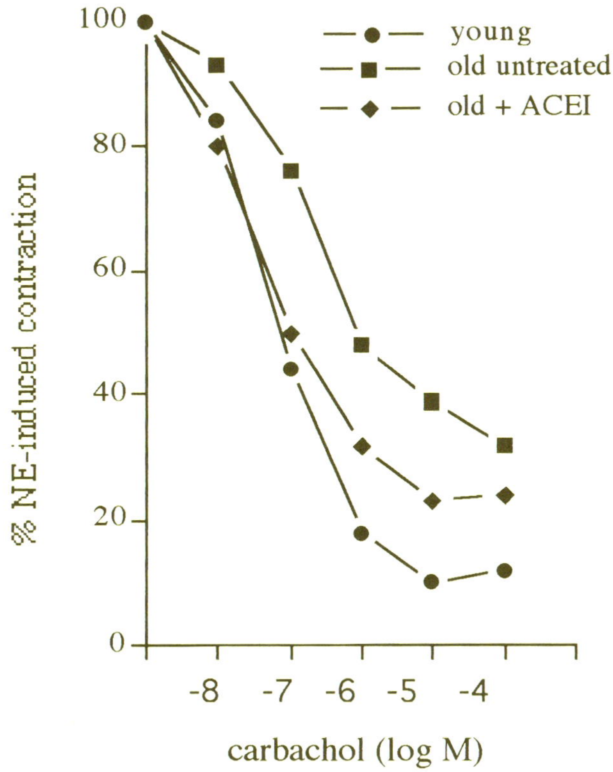

Figure 4. *Endothelium-dependent relaxation of arterial rings pre-contracted by norepinephrine, in reponse to increasing doses of muscarinic agonist in 6 and 30 month-old rats. Treatment by an angiotensin I-converting inhibitor (ACEI) from 6 to 30 months of age partially prevented the age-related impairment in endothelium-dependent relaxation [66].*

The mechanisms accounting for the impaired endothelium-mediated relaxation in aging are not completely elucidated. In laboratory animals, this impairment was observed in various arteries and species [49]. Endothelium-dependent relaxation in aged arteries remained altered when preparations were stimulated by ionophores or various agonists instead of acetylcholine, indicating that age-dependent endothelial dysfunction was not due to a defect in the stimulation or in the response of muscarinic receptors. Paterno et al. [50] exposed aortic rings of young rats to the effluent from pressurized and perfused carotid arteries of old or young animals, stimulated by acetylcholine. This experimental model provided the opportunity to investigate whether the age-related defect resulted from an impaired NO release or from an impaired response to NO. In this model, the effluent from aged arteries produced a relaxation similar as that from young arteries, suggesting that NO production and release was not impaired with age, and implying that the effects of NO might be diminished in aged arteries. Since the direct effects of NO donors on arteries or on SMC are not reduced by aging [44, 51], these results suggested that NO had a limited access to the SMC of the media. Interestingly, AGEs which are present in intima and inner media of aged arteries were shown to quench NO [52] and thus might contribute to impairment of the endothelium-dependent relaxation. However, other experiments conclude that aging is associated with a decrease in NO production and release. Endothelial NO synthase expression was found to be reduced in aorta from old rats compared with that in young rats [31, 53, 54], in association with a decrease in wall content of cyclic guanosine monophosphate (cGMP), the intracellular messenger of NO [31]. On the contrary, another study performed in rat aorta reported a 7-fold increase in the percentage of endothelial cells expressing NO synthase with aging [55]. Recently, the in vitro NO production by rat vessel rings, directly measured by a microsensor technique, was found to be reduced with aging in the aorta, but unchanged in the pulmonary artery [56].

The endothelium produces endothelin-1, a substance which exerts a contractile effect on SMC and which potentiates the effects of other vasocontrictors. Endothelin production and the activity of endothelin converting enzyme in rat aorta were found to be increased with age [53]. However, the contractile effects of endothelin-1 were modified in a complex manner in aged arteries. Aging was found to be associated with a reduced ability of the endothelium to counteract contractile responses to endothelin and with a reduced responsiveness of SMC to this vasoconstrictor [51, 57].

Multiple endothelial dysfunctions may have important consequences in aged individuals. Alteration of endothelial control of vasomotor tone might be responsible for the impaired vascular adaptations to changes in flow, like exercise and ischemia. Moreover, alterations in macromolecular transport and in PGI_2 synthesis might facilitate the development of atherosclerosis and thrombosis, the main determinants of cardiovascular diseases in elderly populations.

AGING AND VASOMOTOR RESPONSES OF ARTERIES

A large number of studies have investigated the effect of aging on vasomotor responses of arteries. The effect of aging on vasodilatatory responses depends on the agent. Vasodilatatory effects of papaverine [39, 43], sodium nitroprusside [40] or other nitrovasodilatators [47, 58], are similar in young and old arteries. However, relaxation induced by isoproterenol or ß-adrenergic agonists was found to be reduced with aging in arteries of laboratory animals [59] and in human veins [60]. This impaired response seems to be due to the effect of aging on ß-adrenergic receptors, mostly on ß2-subtype receptors, resulting in an altered coupling with intracellular messenger response [61].

The effects of age on vasoconstrictive responses of arteries are more complex. They depend on the pharmacological agents studied, but also on the vascular bed, on the species and on the experimental conditions (see review in 48]. In intact arteries, the endothelium exerts a modulatory influence on vasocontrictive responses to some agents. Thus, since endothelial function is impaired with aging, the presence/absence of the intact endothelium should be taken into account when investigating the responses of SMC. Finally, some age-related changes in calcium-effector coupling might also intervene to modify the contractile responses of aged arteries [62]. The vasoconstrictive responses to 5-hydroxytryptamine [40, 63], histamine [64] and angiotensin II [65] were not modified in aged arteries in most, but not all, studies [48], and that to KCl was found be be unchanged [57] or slightly increased [65, 66]. The response to endothelin-1 was found to be decreased in rat arteries, even when the endothelium was removed [50]. This reduced responsiveness of SMC to endothelin-1 might be due to down-regulation of receptors because of the age-related increase in endothelin-1 production. The response of aged arteries to norepinephrine (NE) is a controversial point and was described to be unchanged [57, 60], decreased [66] or increased [53] with age [48]. The endothelium is a key modulator of the effects of age on vasomotor responses to NE [65, 66]. When the endothelium was removed [66] or in presence of L-NAME [65], NE-induced vasocontriction was increased in 30 month-old rats compared with 6 or 12 month-old rats, indicating an increased sensitivity of aged SMC to NE. In contrast, the same experiments performed with intact endothelial function showed that NE-induced vasocontriction declines with age [65, 66]. This modulation might be related to diffential actions of NE on vascular cells, inducing direct contraction of SMC, but also stimulating the release of relaxing and contracting factors by endothelial cells. This complex equilibrium seems to be modified in aged arteries.

PROLIFERATION AND CYTOKINES IN THE AGED ARTERIAL WALL

Aged arteries exhibit greater proliferative and synthetic responses to injury than young arteries. The neointima formed following balloon catheter-injury was found to be much thicker in aged rats than in young [67]. Similar results were obtained with a catheter-induced de-endothelization, a more subtle injury which does not alter the medial layer [68]. After injury, a high proliferative activity was found in aged arteries [67]. The enhanced proliferative response of the aged arterial wall was not related to circulating factors but to intrinsic properties of the wall. Using an experimental model of aorta cross-transplantations between young and old rats, Hariri et al. [68] showed that the increased thickening of neointima in response to injury depended on the age of the artery and not on that of the host. SMC from old rats in culture also exhibit a higher replicative rate that SMC from young animals, as shown by a higher percentage of cells in the S cell cycle phase and a larger ^{3}H-thymidine uptake [69]. By contrast, fibroblasts from aged arteries did not exhibit a higher replicative rate in culture than fibroblasts from young arteries [70].

The high proliferative response of aged SMC might be related to an elevated production of growth factors. Lysates of aged SMC stimulated the growth of target cells more than lysates of young SMC [71]. In this model, the stimulating effect of aged SMC was found to be reduced by antibodies against platelet derived growth factor (PDGF) to a level similar to that of SMC lysates from young animals [71]. Moreover, aging was found to be associated with an overexpression of PDGF receptors in the rat aorta [72]. In the human aorta, the number of SMC expressing heparin-binding epidermal growth factor (EGF)-like growth factor, another potent mitogen, tends to increase with age [73]. Using an organ culture model, we found that interleukin 6 (IL-6) was produced in larger amounts by the aorta of 30 month-old rats compared with 10 month-old rats [74]. When aortas were stimulated by lipopolysaccharide, the differences in IL-6 production between young and aged rats remained large and significant. Since IL-6 has a potent stimulating effect on SMC proliferation [75], this alteration might contribute to the age-related changes in growth control of SMC. In the same experiment, we also found that under stimulated conditions tumor necrosis factor α (TNF) was produced in a larger amount by aorta from old rats than by aorta from young rats [74]. TNF participates in the cascade activation of proinflamatory cytokines and is able to induce IL-1 release, to promote extracellular protein synthesis, to activate endothelial cells and to promote transendothelial leukocyte migration [76].

Another possible mechanism accounting for the enhanced proliferative response of aged SMC is an impaired growth control by antiproliferative factors. McCaffrey et al. [77] studied the effect of aging on the production and action of TGF-ß1, a strong antiproliferative factor for SMC. Gene expression, synthesis and activity of TGF-ß1 were similar in the aortas of young and old rats, but the responses to TGF-ß1 were markedly reduced in aged arteries. Increasing concentrations of TGF-ß1 did not suppress DNA synthesis of aged SMC in culture, whereas this synthesis was markedly reduced in cells from young animals. This defective action was associated with a diminished binding of TGF-ß1 to its receptor on SMC and to its increased degradation in old rats [77].

SUSCEPTIBILITY OF THE AGED ARTERIAL WALL TO ATHEROSCLEROSIS

Atherosclerosis is strongly linked to advance in age. In population studies, the prevalence of atherosclerosis, its extent and severity are statistically correlated with age [78]. Age is the strongest risk factor for the clinical complications of atherosclerosis, which may be due to a longer exposure to risk factors in old individuals. However, age remains a strong risk factor even in the absence of other risk factors [79]. Experiments conducted in rabbits [80] and monkeys [81] documented a marked susceptibility of the aged arterial wall to atherosclerosis. In old rabbits fed a cholesterol-rich diet for 16 months, a dramatic increase in the area of atherosclerotic lesions was observed compared with young rabbits fed the atherogenic diet for the same duration [80]. Several age-related modifications of vascular cells or matrix might favor plaque formation and progression. The age-related changes in macromolecular transport within the wall might increase the LDL concentration in the intima. AGE-modifed collagen and sulfated glycosaminoglycans, which are both present in the aged intima, can bind and trap LDL [5, 82]. The presence of mononuclear cells in the aged intima and the proliferative behavior of aged SMC are factors involved in plaque formation.

Moreover, AGEs interact with specific receptors on the cell membrane of macrophages resulting in their activation and the release of TNF, IL-1 and PDGF [83, 84], other factors involved in plaque progression. AGEs also interact with endothelial cells leading to oxidative stress of these cells, to expression of vascular cell adhesion molecule-1 [85] and to increased recruitment of monocytes in the intima. In old individuals, impairment in PGI_2 production and other endothelial functions may also contribute to complications of atherosclerotic lesions. Hence, aging interacts by several mechanisms with factors involved in atherosclerosis, and this may account for the strong links between the two processes.

MODULATION OF VASCULAR AGING: TOWARDS ANTI-AGING INTERVENTIONS

Even if aging is known to be an ineluctable process, recent evidences show that some age-related changes might be postponed or prevented by therapeutic interventions.

Dietary restriction was shown to increase life span of laboratory animals, and its effect on metabolism, renal function and the immune system has been widely studied. In mice, dietary restriction was shown to lessen non enzymatic glycation of collagen and reduce collagen crosslinks in the aorta [86]. However, the effect of dietary restriction on vascular aging has not been extensively studied. Protein restriction was shown to prevent renal aging and nephropathies in rodents. In old rats fed a 12% protein diet ad libidum during adult life, the endothelium-dependent relaxation is not altered, contrary to that of old rats fed a 23% protein diet [87].

Chronic blood pressure reduction has also been studied as an anti-aging intervention on the cardiovascular system. Haudenschild and Chobanian [16] observed that the blood pressure of normotensive Wistar-Kyoto rats submitted to a combination of antihypertensive drugs from 3 to 14 months of age was greatly decreased. The intima layer of these rats was thinner than that of control rats and the number of subendothelial cells was diminished. However, in this study, blood pressure significantly increased with age in control rats, making it difficult to distinguish between the effects of hypotension and those of prevention of increased blood pressure. Another experiment was conducted on WAG/Rij rats. In this strain, blood pressure remains stable with advancing age [5], giving the opportunity to study the effects of aging independently of those of blood pressure elevation. Chronic blood pressure reduction was achieved by prolonged blockade of the renin-angiotensin system by chronic treatment with an angiotensin converting enzyme inhibitor from 6 to 30 months of age [5]. This study showed that some aspects of vascular aging were not modified by chronic blood pressure reduction and might be considered as pressure-independent processes. This was the case for aorta enlargement, medial elastic/collagen composition and arterial compliance. In contrast, chronic blood pressure reduction was found to prevent or postpone some other aspects of vascular aging which might be considered as pressure-dependent processes, especially intimal and medial thickening [5] as well as endothelium-dependent relaxation [66] (Figure 4).

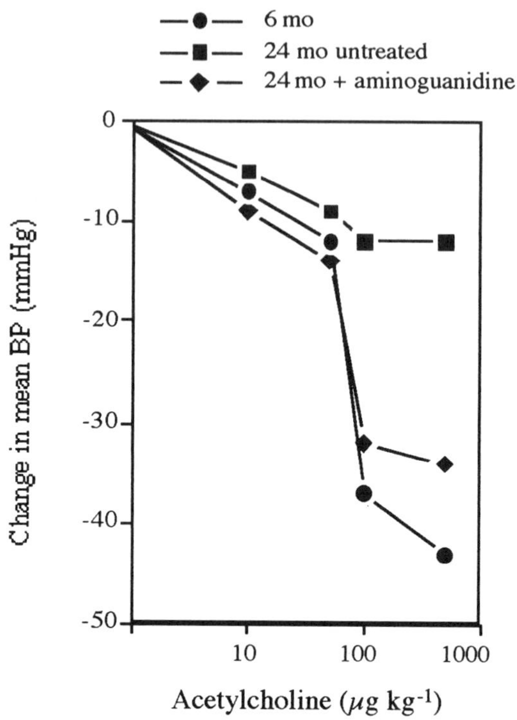

Figure 5. *Impaired vasodilatatory effect of acetylcholine in vivo, measured by the decrease in mean blood pressure, in 24 month-old rats as compared with 6 month-old rats. Treatment with aminoguanidine, which reduces tissue levels of AGEs, from 6 to 24 months of age prevented this age-related alteration in response [88].*

Since non enzymatic glycation is of major importance in vascular aging, pharmacological blockade of this process is a promising approach. Aminoguanidine (AG), an AGE inhibitor, was found to reduce tissue accumulation of AGEs in aging [88] and diabetes [89]. In rats given AG from 6 to 24 months of age, the endothelium-dependent relaxation was found to be preserved contrary to untreated control rats [88] (Figure 5). In another study [90], AG treatment in old rats from 24 to 30 months of age was found to prevent the age-related decrease in carotid distensibility and the increase in aortic input impedance and in total peripheral resistance (Figure 6). Moreover, the age-related enlargement of the carotid artery was completely prevented by AG treatment [90]. These studies demontrated that inhibition of glycation processes postpones or prevents the major features of vascular aging, including remodeling and stiffening of large arteries and impairment of endothelium-dependent relaxation. Interestingly, AG treatment completely prevented the increase in cardiac mass associated with aging [90] despite the absence of a significant effect on blood pressure or cardiac output (Figure 6).

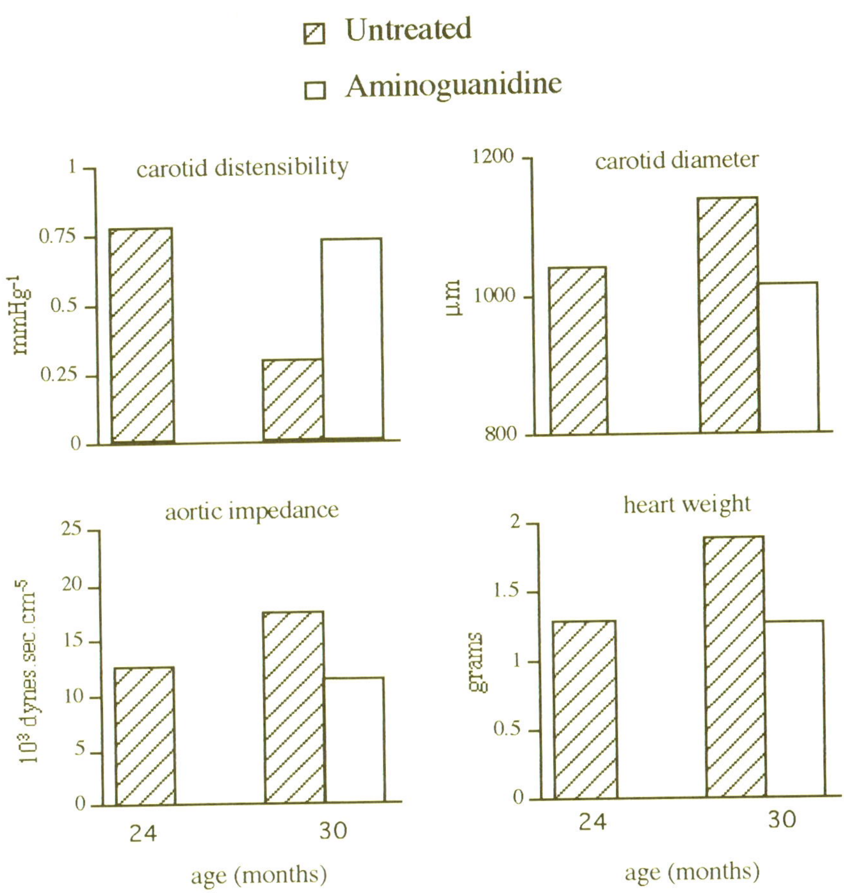

Figure 6. *Carotid artery distensibility and diameter, aortic characteristic input impedance and heart weight in 24 and 30 month-old rats. The differences between the age groups were significant. Treament with aminoguanidine from 24 to 30 months of age prevented the age-related changes of the four parameters [90].*

Estradiol was found to improve acetylcholine-induced relaxation in animal and human arteries. In post-menopausal women, administration of estrogen improves the endothelium-dependent function [91]. Moreover, estrogen exerts an endothelium-independent relaxating effect on the smooth muscle of human coronary arteries [92]. This effect on the arterial wall is believed to contribute to the protection of middle-aged women and post-menopausal women receiving substitutive hormonal therapy against cardiovascular diseases.

REFERENCES

1. Virmani R, Avolio AP, Mergner WJ, et al. Effect of aging on aortic morphology in populations with high and low prevalence of hypertension and atherosclerosis. Am J Pathol 1991;139:1119-29.

2. Benetos A, Laurent S, Hoeks AP, Boutoutrie PH, Safar ME. Arterial alterations with aging and high blood pressure. Arterioscl er Thromb 1993;13:90-97.

3. Bonithon-Kopp C, Touboul PJ, Berr C, Magne C, Ducimetière P. Factors of carotid enlargement in a population aged 59 to 71 years. The EVA study. Stroke 1996;27:654-60.

4. Keohane SG, Adams CW, Poston RN. Coronary arterial dimensions and cell populations in ageing man. Atherosclerosis 1988;69:103-8.

5. Michel JB, Heudes D, Michel O, et al. Effect of chronic ANG I-converting enzyme inhibition on aging processes. II. Large arteries. Am J Physiol 1994; 267: R124-35.

6. Cliff WJ. The aortic tunica media in aging rats. Exp. Mol Path 1970;13:172-89.

7. Fornieri C, Quaglino D, Mori G. Role of the extracellular matrix in age-related modifications of the rat aorta. Ultrastructural, morphometric and enzymatic evaluations. Arterioscler Thromb 1992;12:1008-16.

8. O'Rourke MF, Avolio AP, Lauren PD, Yong J. Age-related changes of elastic lamellae in the human thorcic aorta. J Am Coll Cardiol 1987;29:A53.

9. Robert L, Jacob MP, Frances C, Godeau G, Hornebeck W. Interaction between elastin and elastases and its role in aging of arterial wall, skin, and other connective tissues. A review. Mech Aging Dev 1984;28:155-66.

10. Fischer GM. Effects of spontaneous hypertension and age on arterial connective tissue in the rat. Exp. Geront 1976;11:209-15.

11. Anidjar S, Salzmann JL, Gentric D, Lagneau P, Camilleri JP, Michel JB. Elastase-induced experimental aneurysms in rats. Circulation 1990;82:973-81.

12. Gerrity RG, Cliff WF. The aortic tunica intima in young and aging rats. Exp Mol Pathol 1972;16:382-402.

13. Guyton JR, Lindsay KL, Dao DT. Comparison of aortic intima and inner media in young adult versus aging rats. Am J Pathol 1983;111:234-46.

14. Haudenschild CC, Prescott MF, Chobanian AV. Aortic endothelial and subendothelial cells in experimental hypertension and aging. Hypertension 1981;3(suppl I):148-153.

15. Richardson M, Hatton MWC, Moore S. Proteoglycan distribution in the intima and media of the aortas of young and aging rabbits: an ultrastructural study. Atherosclerosis 1988;71:243-56.

16. Haudenschild CC, Chobanian AV. Blood pressure lowering diminishes age-related changes in the rat aortic intima. Hypertension, 1984;6 (suppl I):62-68.

17. Wolinsky H. Long-term effects of hypertension on the rat aortic wall and their relation to concurrent aging changes. Morphological and chemical studies. Circ Res 1972;30:301-9.

18. Hallock P, Benson IC. Studies on the elastic properties of human isolated aorta. J Clin Invest 1973;16:595-602.

19. Avolio AP, Chen S, Wang R, Zhang CL, Li MF, O'Rourke MF. Effects of aging on changing arterial compliance and left ventricular load in a northen Chinese urban community. Circulation 1983;68:50-8.

20. Avolio AP, Deng FG, Li WQ, et al. Effects of aging on arterial distensibility in populations with high and low prevalence of hypertension: comparison between urban and rural communities in China. Circulation 1985;71:202-10.

21. Brownlee M, Cerami A, Vlassara H. Advanced glycosylation end products in tissue and biochemical basis of diabetic complications. N Engl J Med 1988;307:205-11.

22. Brownlee M, Vlassara H, Cerami A. Nonenzymatic glycosylation and the pathogenesis of diabetic complications. Ann Intern Med 1984;101:527-37.

23. Sell DR, Monnier VM. Structure elucidation of a senescence cross-link from human extracellular matrix. J Biol Chem 1989; 264:21597-602.

24. Reiser KM. Non enzymatic glycation of collagen in aging and diabetes. Proc Soc Exp Biol Med 1991;196:14-29.

25. Schnider SL, Kohn RR. Glucosylation of human collagen in aging and diabetes mellitus. J Clin Invest 1980;66:1179-81.

26. Monnier VM, Kohn RR, Cerami A. Accelerated age-related browning of human collagen in diabetes mellitus. Proc Natl Acad Sci 1984;81:883-87.

27. Safar M. Ageing and its effects on the cardiovascular system. Drugs 1990;39(Suppl 1): 1-8.

28. Mérillon JP, Motte G, Masquet G, Azancot I, Guiomard A, Gourgon R. Relationship between physical properties of the arterial system and left ventricular performance in the course of aging and in permanent arterial hypertension. Eur Heart J 1982;3(suppl A):95-102.

29. Nichols WW, Avolio AP, Kelly RP, et al. Effects of age and hypertension on wave travel and reflections. In: O'Rourke MF, Safar M and Dzau V, eds. Arterial vasodilatation: mechanisms and therapy. London, Edwards Arnold, 1993:23-40.

30. Gerstenblith G, Frederiksen J, Yin FC, Fortuin NJ, Lakatta EG, Weisfeldt ML. Echocardiographic assessment in a normal adult aging population. Circulation 1977;56:273-8.

31. Schwartz MS, Benditt EP. Aortic endothelial cell replication. I. Effects of age and hypertension in the rat. Circ Res 1977;41:248-55.

32. Challah M, Nadaud S, Philippe M, et al. Circulating and cellular markers of endothelial dysfunction with aging in rats. Am J Physiol 1997;273:H1941-48.

33. Belmin J, Corman B, Merval R, Tedgui A. Age-related changes in albumin endothelial permeability and distribution volume in the rat aorta. Am J Physiol 1993;264:H679-H685.

34. Fry DL. Mass transport, atherogenesis and risk. Arteriosclerosis 1987;7:88-100.

35. Lin SJ, Jan KM, Schuessler G, Weinbaum S, Chien S. Enhanced macromolecular permeability of aortic endothelial cells in association with mitosis. Arteriosclerosis 1988;73:223-32.

36. Esposito C, Gerlag H, Brett J, Stern D, Vlassara H. Endothelial receptor-mediated binding of glucose-modified albumin is associated with increased monolayer permeability and modulation of cell surface coagulant properties. J Exp Med 1989;170:1387-407.

37. Tokunaga O, Yamada T, Fan J, Watanabe T. Age-related decline in prostacyclin synthesis by human aortic endothelial cells. Am J Pathol 1991;138:941-9.

38. Masotti G, Poggesi L, Galanti, et al. Prostacyclin production in man. In: Lewis PJ, O'Grady J, eds. Clinical pharmacology of prostacyclin. New York: Raven Press, 1981:9-20.

39. Hongo K, Nakagomi T, Kassell NF et al. Effects of aging and hypertension on endothelium-dependent vascular relaxation in rat carotid artery. Stroke 1988; 19:892-7.

40. Koga T, Takata Y, Kobayashi K, Takishita S, Yamashita Y, Fujishima M. Ageing supresses endothelium-dependent relaxation and generates contraction mediated by the muscarinic receptors in vascular smooth muscle of normotensive Wistar-Kyoto and spontaneously hypertensive rats. J Hypertens 1988; 6(suppl 4):243-245.

41. Soltis EE. Effet of age on blood pressure and membrane-dependent vascular responses in the rat. Circ Res 1987;10:889-897.

42. Atkinson J, Tatchum-Talom R, Capdeville-Atkinson C. Reduction of endothelial function with age in the mesenteric arterial bed of the normotensive rat. Br J Pharmacol 1994 ;111:1184-8.

43. Egashira K, Inou T, Hirooka Y, et al. Effects of age on endothelium-dependent vasodilation of resistance coronary artery by acetylcholine in humans. Circulation 1993;88:77-81.

44. Yasue H, Matsuyama K, Matsuyama K, Okumura K, Morikami Y, Ogawa H. Responses of angiographically normal human coronary arteries to intracoronary injection of acetylcholine by age and segment. Circulation 1990;81:482-90.

45. Gerhard M, Roddy MA, Creager SJ, Creager MA. Aging progressively impairs endothelium dependent vasodilation in forearm resistance vessels of humans. Hypertension 1996;27:849-53.

46. Celermajer DS, Sorensen KE, Spiegelhalter DJ, Georgakopoulos D, Robinson J, Deanfield JE. Aging is associated with endothelial dysfunction in healthy men years before age-related decline in women. J Am Coll Cardiol 1994;24:471-6.

47. Tominaga M, Fujii K, Abe I, Takata Y, Kobayashi K, Fujishima M. Hypertension and aging impair acetylcholine-induced vasodilatation in rats. J Hypertens 1994;12:259-68.

48. Reckelhoff JF, Manning RD. The role of endothelium-derived nitric oxide in control of renal vasculature in aging male rats. Am J Physiol 1993;265:R1126-31.

49. Dohi Y, Kojima M, Sato K, Lüscher TF. Age-related changes in vascular smooth muscle and endothelium. Drugs Aging 1995;7: 278-91.

50. Paterno R, Faraci FM, Heistad DD. Age-related changes in release of endothelium-derived relaxing factor from the carotid artery. Stroke 1994;25:2457-60.

51. Dohi Y, Lüscher TF. Aging differentially affects direct and indirect actions of endothelin-1 in perfused mesenteric arteries of the rat. Br J Pharmacol 1990;100:889-93.

52. Bucala R, Tracey K, Cerami A. Advanced glycosylation end-products quench nitric oxide and mediate defective endothelium-dependent vasodilatation in experimental diabetes. J Clin Invest 1991;87:432-38.

53. Barton M, Cosentino F, Brandes RP, Moreau P, Shaw S, Lüsher TF. Anatomic heterogenity of vascular aging: role of nitric oxide and endothelin. Hypertension 1997;30:817-24.

54. Chou TC, Yen MH, Li CY, Ding YA. Alterations of nitric oxide synthase expression with aging and hypertension. Hypertension 1998;31:643-48.

55. Aliev G, Miah S, Turmaine M, Burnstock G. An ultrastructural and immunocytochemical study of thoracic aortic endothelium in aged Sprague-Dawley rats. J Submicrosc Cytol Pathol 1995;27:477-90.

56. Tschudi MR, Barton M, Bersinger NA, et al. Effect of age on kinetics of nitric oxide release in rat aorta and pulmonary artery. J Clin Invest 1996;1598:899-905.

57. Ishihata A, Katano Y, Morinobu S, Endoh M. Influence of aging to the contractile response to endothelin of rat thoracic aorta. Eur J Pharmacol 1991;200:199-201.

58. Hajdu MA, McElmurry RT, Heistad DD, Baumbach GL. Effects of aging on cerebral vascular responses to serotonin in rats. Am J Physiol 1993;264:H2136-40.

59. Fleish JH, Maling HM, Brodie BB. Beta-receptor activity in aorta: variations with age and species. Circ Res 1970;26:151-62.

60. Feldman RD. A low-sodium diet corrects the defect in ß-adrenergic response in older subjects. Circulation 1992;85:612-18.

61. O'Donnel SR, Wanstall JC. Beta-1 and beta-2 adrenoreceptor-mediated responses in preparations of pulmonary artery and aorta from young and aged rats. J Pharmacol Exp Ther 1984;228:733-8.

62. Capdeville-Atkinson C, Oster L, Thorin-Trescases N, Robert A, Corman B, Atkinson J. Effect of chronic ANG I-converting enzyme inhibition on aging processes. V. Intracellular calcium-vasoreactivity coupling. Am J Physiol 1995;268:R1394-1400.

63. Blauw GJ, Van Blummelen P, Chang PC, Vermeij P, van Zwieten PA. Arterial dilatation and venous constriction induced by serotonin in elderly in the elderly. Drugs 1988;36(suppl 1):74-7.

64. Owen TL. Effects of age on blood pressure and small vessels reactivity in male rabbits. Blood Vessels 1986;23:271-78.

65. Lang MG, Noll G, Lüscher TF. Effect of aging and hypertension on contractility of resistance arteries: modulation by endothelial factors. Am J Physiol 1995;269:H837-44.

66. Atkinson J, Tatchum-Talom R, Corman B. Effect of chronic ANG I-converting enzyme inhibition on aging processes. III. Endothelial function of mesenteric arterial bed of rat. Am J Physiol 1994;264:R136-43.

67. Stemerman MB, Weinstein R, Rowe JW, Maciag T, Fuhro R, Gardner R. Vascular smooth muscle cell growth kinetics in vivo in aged rats. Proc Natl Acad Sci 1982;79:3863-6.

68. Hariri RJ, Alonso DR, Hajjar DP, Coletti D, Weksler ME. Aging and arteriosclerosis. Development of myointimal hyperplasia after endothelial injury. J Exp Med 1986;164:1171-8.

69. Hariri RJ, Hajjar DP, Coletti D, Alonso DR, Weksler ME, Rabellino E. Aging and arteriosclerosis. Cell cycle kinetics of young and old arterial smooth muscle cells. Am J Pathol 1988;131:132-6.

70. Bochaton-Pialat ML, Gabbiani F, Ropraz P, Gabbiani G. Age influences the replicative activity and the differenciation features of cultured rat aortic smooth muscle cell population and clones. Arterioscl er Thromb 1993;13:1449-55.

71. McCaffrey TA, Nicholson AC, Szabo PE, Weksler ME, Weksler BB. Aging and arteriosclerosis. The increased proliferation of arterial smooth muscle cells isolated from old rats is associated with increased Platelet-Derived Growth Factor-like activity. J Exp Med 1988;167:163-74.

72. Sarzani R, Arnaldi G, Takasaki I, Brecher P, Chobanian AV. Effect of hypertension and aging on PDGF and PDGF-receptor expression in rat aorta and heart. Hypertension 1991;18(suppl3):93-99.

73. Miyagawa J, Higashiyama S, Kawata S, et al. Localisation of heparin-binding EGF-like growth factor in the smooth muscle cells and macrophages of human atherosclerotic plaques. J Clin Invest 1995;95:404-11.

74. Belmin J, Bernard C, Corman B, Merval R, Esposito B, Tedgui A. Increased production of tumor necrosis factor and interleukin-6 by arterial wall of aged rats. Am J Physiol 1995;268:H2288-93.

75. Ikeda U, Ikeda M, Oohara T, et al. Interleukin 6 stimulates growth of vascular smooth cells in a PDGF-dependent manner. Am J Physiol 1991;269:H1713-17.

76. Pober JS, Cotran RS. Cytokines and endothelial cell biology. Physiol Rev 1990;70:427-51.

77. McCaffrey TA, Falcone DJ. Evidence for an age-related dysfunction in the antiproliferative response to transforming growth factor-ß in vascular smooth muscle cells. Mol Biol Cell 1993;4:315-22.

78. Weber G, Bianciardi G, Bussani R, et al. Atherosclerosis and aging: a morphometric study on arterial lesions of elderly and very elderly necropsy subjects. Arch Pathol Lab Med 1988; 112:1066-70.

79. Kannel WB, Gordon T. Evaluation of cardiovascular risk in the elderly: the Framingham study. Bull N Y Acad Med 1978;54:573-91.

80. Spagnoli LG, Orlandi A, Mauriello A, De Angelis C, Ramaci MT. Age-dependent increase of rabbit aortic atherosclerosis. A morphometric approach. Path Res Pract 1992;188:637-42.

81. Bullock BC, Clarkson TB, Lehner NDM, Lofland HB Jr, St. Clair RW. Atherosclerosis in Cebus albifrons monkeys: clinical and pathologic studies. Exp Mol Pathol 1969;10:39-62.

82. Brownlee M, Vlassara H, Cerami A. Nonenzymatic glycosylation products on collagen covalently trap low density lipoprotein. Diabetes 1985;34:938-41.

83. Kirkstein M, Aston C, Hintz R, et al. Receptor-specific induction of insulin-like growth factor I in human monocytes by advanced glycosylation end-products modified proteins. J Clin Invest 1992;90:439-46.

84. Vlassara H, Brownlee M, Manogue R, Dinarello CA, Pasaglan A. Cachectin/TNF and IL-1 induced by glucose-modified proteins: role in normal tissue remodeling. Science 1988;240:1546-48.

85. Schmidt AM, Hori O, Chen JX, et al. Advanced glycation end-products interacting with their endothelial receptor induce expression of vascular cell adhesion molecule-1 (VCAM-1) in cultured human endothelial cells and in mice: a potential mechanism for accelerated vasculopathy of diabetes. J Clin Invest 1995;96:1395-1403.

86. Reiser KM. Influence of age and long-term dietary restriction on enzymatically mediated crosslinks and nonenzymatic glycation of collagen in mice. J Gerontol Biol Sci 1994;49:71-79.

87. Sonaka I, Futami Y, Kobayashi T, Umezawa T, Maki T. Effects of dietary protein restriction on nitrogen balance and cardiovascular functions in aged rats. J Gerontol Biol Sci 1993;48:145-50.

88. Li YM, Steffes M, Donnely T et al. Prevention of cardiovascular and renal pathology of aging by the advanced glycation inhibitor aminoguanidine. Proc Natl Acad Sci 1996;93:3902-7.

89. Hammes HP, Martin S, Federlin K, Geisen K, Brownlee M. Aminoguanidine treatment inhibits the development of experimental diabetic retinopathy. Proc Natl Acad Sci USA 1991;88:11555-58.

90. Corman B, Duriez M, Poitevin P, et al. Aminoguanidine prevents age-related arterial stiffening and cardiac hypertrophy. Proc Natl Acad Sci USA 1998;95:1301-6.

91. Reis SE, Gloth ST, Blumenthal RS, et al. Ethinyl estradiol acutely attenuates abnormal coronary vasomotor responses to acetylcholine in postmenopausal women. Circulation 1994;84:52-60.

92. Mügge A, Riedel M, Barton M, Kuhn M, Lichtlen P. Endothelium independent relaxation of human coronary arteries by 17ß-oestradiol in vitro. Cardiovasc Res 1993;27:1939-42.

II. The Vascular Wall under Pathological Conditions

9. VESSEL AND INFLAMMATION

Catherine Bernard
Department of Anesthesiology, Lariboisière Hospiral, Paris, France

Abbreviations and alternative names
EC: endothelial cell
VSMC: vascular smooth muscle cell
LPS: lipopolysaccharide

Cytokines:
TNF: Tumor Necrosis Factor
IL-1 to-18: Interleukin 1 to- 18
IL-1ra: IL-1 receptor antagonist
IFN-γ: Interferon-gamma
CSF: Colony Stimulating Factor
GM-CSF: Granulocyte-Macrophage CSF
G-CSF: Granulocyte-CSF
M-CSF: Macrophage-CSF
TGFβ: Transforming growth factor-β
OSM: Oncostatin-M
LIF: Leukemia Inhibitory Factor

Selectins:
L-selectin: lymphocyte-selectin: LECAM-1: CD62L
E-selectin: endothelial selectin: ELAM-1/CD62E
P-selectin: platelet-selectin/GMP-140/CD62P

Immunoglobulin superfamily:
ICAM-1: intercellular adhesion molecule-1/CD54
VCAM-1: vascular cell adhesion molecule-1
PECAM-1: platelet-endothelial cell adhesion molecule/CD31

Integrins:
LFA-1: CD11a-CD18/α1β2
MAC-1: CD11b-CD18/α2β2
VLA-4: Very late antigen/α4β1

Metalloproteases and their inhibitors
MMP: matrix metalloproteinase
MT-MMP: membrane-type metalloproteinase
TIMP: tissue inhibitor of metalloproteinases
PAI: plasminogen activator inhibitor
uPA: urokinase plasminogen activator

Chemokines
GRO-α:growth-related oncogene
IP-10: interferon-inducible protein-10
MCP: monocyte chemotactic protein, MCP-1, -2, -3, -4, -5; MCP-1/ MCAF: monocyte chemotactic and activating factor
MIP: macrophage inflammatory protein, MIP-1α, -1β
RANTES: regulated upon activation of normal T-cell expressed and secreted

Growth factors
aFGF: acidic fibroblast growth factor/FGF-1
bFGF: basic fibroblast growth factor/FGF-2
VEGF: vascular endothelial growth factor, or vascular permeability factor
PDGF: platelet derived growth factor
TGFα: transforming growth factor α
EGF: epithelial growth factor

THE LANGUAGE OF CYTOKINES WITHIN THE VESSEL WALL

Research in vascular biology has progressed remarkably in the last decade, resulting in a better understanding of vascular cell responses to inflammatory stimuli. The role of vascular cells during inflammatory processes is critical, being both target and sources of various proinflammatory cytokines. The secondary mediators involved in the regulation of the complex interactions between vascular cells and blood cells include adhesion molecules, chemokines, metalloproteinases and growth factors.

Cytokines

Cytokines are regulatory peptides that participate in host defense and repair and coordinate cellular responses. Cytokines include therefore lymphocyte-derived factors known initially as lymphokines, monocyte-derived factors called monokines, haematopoietic "colony stimulating factors" and connective tissue "growth factors". The development of tissue culture techniques enabled to detect the presence of soluble factors in tissue culture supernatants with effects on morphology, motility, proliferation, differentiation and functional specialized capabilities of different cell types. Discovery of the cytokines derived from lymphocytes and monocytes revolutionized the conceptual basis of cell-mediated immunity. It is now well accepted that biological activities of every cell type are regulated by cytokines. This is also the case for vascular cells.

Whereas endocrine hormones are secreted by specialized tissue and present in the circulation to maintain homeostasis, cytokines usually act over short distances as autocrine and paracrine intercellular signals in local tissues. Cytokines are often not produced constitutively but are generated in pulses and produce their actions by binding to specific high affinity cell surface receptors and by changing gene expression in the target cells.

Cytokines interact in a network with three essential features.
1) The biological redundancy: individual cytokines have multiple overlapping synergistic activities.
2) The biological amplification: cytokines induce each other as well as their own production in a cytokine cascade.
3) And finally, targets are often sources, adding in complexity of the concurrent messages.

TNF, IL-1 and IL-6 are cytokines mainly produced by macrophages, but various other types of cells are found to be sources, especially vascular cells. The vessel now has to be considered as a critical partner in the immune and endocrine system. Vascular cells in culture are able to secrete copious amounts of cytokines (TNF, IL-1 and IL-6) with different secretory abilities between endothelial and smooth muscle cells [1-3]. LPS has been shown to activate aortic tissue in culture to produce cytokines, and IL-1 is able to induce its own expression in an autocrine secretory action [4, 5]. Moreover, the actions of cytokines on vascular cells introduce us to the novel concepts of activation of non leukocyte cell types.

Cytokines expressed by vascular cells

During inflammation, vascular cells contribute to the regulation of the immune response by autocrine production of cytokines (Table 1). TNF, IL-1 and IL-6 are expressed by vascular cells. Human vascular SMC are both sources and targets of TNF [6]. IL-6 is constitutively released by vascular cells in culture, its production can increase up to represent around 4% of newly synthetized proteins by activated vascular cells [3].

SMC and EC express both IL-1α and IL-1β as cell-associated, but SMC neither contain mature IL-1β, nor processed recombinant 33kD IL-1β precursor into its mature 17kD form. Despite this failure, SMC express IL-1 converting enzyme (ICE) but possess in its cell membrane compartment an inhibitory factor of IL-1β processing [7]. ICE is a protease that cleaves IL-1β precursor to yield the active cytokine but is also the proximal protease (caspase 1) involved in the apoptotic pathway, since additional substrates for ICE have been shown to be CPP-32 (caspase-3). Vascular cells have thus complex properties to regulate inflammation.

EC are also important sources of haematopoietic growth factors including stem cell factor, IL-3, GM-CSF, G-CSF and M-CSF [8]. IL-15 is a recently identified cytokine, implied in T cell migration, and has been shown to be produced by ECs in response to IFNγ [9].

The inflammatory response and its temporal patterns may result from a dynamic balance between proinflammatory cytokines and antiinflammatory cytokines. The proinflammatory repertoire of vascular cells includes TNF, IL-1, IL-8, while the antiinflammatory repertoire expressed by vascular cells is represented predominabtly by TGFβ. IL-6 which is produced in large amounts by SMC is a good marker of the inflammatory response but seems to possess antiinflammatory rather than proinflammatoary properties [10]. EC also present a low expression of IL-1ra. Vascular cells may not be able to express the antiinflammatory cytokines IL-10, IL-4 or IL-13. However, EC respond to IL-4 and IL-13, but their actions are ambiguous, favoring inflammation for some of them (sustained expression of adhesion molecules: VCAM-1 and P-selectin) or opposing inflammation (downregulation of ICAM-1, E-selectin and RANTES chemokines). Information on the interaction of IL-10 with vascular cells are still fragmentary, but IL-10 has been shown to inhibit antigen presentation by human dermal microvascular ECs [11]. In addition, vascular derived growth factors such as VEGFs and FGFs may serve as deactivating factors for vascular cells.

The effects of a combination of cytokines are, in part, function of the receptor repertoire of the cell [12]. There are three major families of cytokine receptors: the haematopoietic growth factor receptor family, the TNF receptor family, and members of the immunoglobulin supergene family.

Table 1. Cytokines expressed by vascular cells

Endothelial cell	Smooth muscle cell
TNF	TNF
IL-1 IL-6, IL-8, IL-15	IL-1 IL-6, no IL-8
TGFβ	TGFβ
LIF,OSM?	LIF, OSM?
IL-1ra	IL-1ra
CD40L	CD40L

Cytokine receptors expressed by vascular cells *(Table 2)*

ECs express c-kit, the β-chain common to the receptors for GM-CSF, IL-3 and IL-5, and the α-chain for IL-3 and GM-CSF receptors [8]. ECs also express both p55 and p75 TNF receptors, the latter being the most abundant on the cell membrane while the p55, the most abundant overall, being detectable mainly in the cytoplasm [13]. TNF activates ECs predominantly via p55. The transmembrane form of TNF is the prime ligand for p75, being implicated in juxtacrine interactions between Ecs and monocytes. ECs express the type I IL-1 receptor and the IL-1 receptor antagonist, but not the IL-1 type II decoy receptor.

ECs express gp130, the common signal transducing component of the functional receptor complexes for the IL-6 cytokine family (IL-6, IL-11, LIF, OSM, ciliary neurotrophic factor, cardiotrophin-1) but do not possess the specific IL-6 receptor subunit [12]. Several in vitro studies did not find any effect of IL-6 on ECs despite indications from in vivo studies for the role of IL-6 in angiogenesis. Interestingly, soluble IL-6 receptors complexed with IL-6 potently activate ECs, inducing the synthesis of adhesion molecules (E-selectin, ICAM-1, VCAM-1), of cytokines (IL-6 itself), and of chemokines (IL-8) [8]. Moreover, at the initiation of the inflammatory response, cooperations between ECs and leukocytes are such that neutrophils and monocytes express IL-6 receptors and shed them upon stimulation in close vicinity of Ecs. The formation of a complex between IL-6, which is constitutively expressed by EC, and soluble IL-6 receptors lead to activation of ECs [14]. Receptors for other members of the IL-6 family contain the signalling gp130 subunit and cytokine-specific subunit: OSM activates ECs via the binding to OSM receptor II combined to gp130.

IL-4 and IL-13 have similar activities on Ecs because both can bind the non γ_c subunit of the IL-4 receptor, IL-4 receptor α. ECs do not express the γ_c subunit common to IL-4 and IL-13 receptors.

Table 2. Cytokine receptors expressed by vascular cells

Endothelial cell	Smooth muscle cell
TNFR I (p55) and II (p75)	TNFRs?
IL-1R I and II	IL-1Rs?
gp130, no IL-6R	gp130, no IL-6R
IL-8R	no IL-8R
TGFβR I and II	TGFβ R I and II
OSMR II	OSMR II
IL-3R	IL-3R?
IL-4/IL-13R (non-γc)	IL-4/IL-13R (non-γc)
IFNγR?	IFNγR?
GM-CSFR	GM-CSFR?
EPOR	EPOR?
CD40	CD40

Interaction of cytokines with vascular matrix

As proposed by Nathan and Sporn [15], "if we consider cytokines as specialized symbols in the language of intercellular communication, the response of cells to cytokines can be markedly affected by the extracellular matrix in which most cells are normally imbedded." Indeed, matrix components present binding sites for cytokines: type IV collagen, fibronectin, thrombospondin bind TGFβ, while heparan sulfate proteoglycans bind both TNFα and IFNγ.

Cytokine inducers

LPS is a potent activator of various cell types, via its specific recognition. One protein has been identified as an LPS receptor is the CD14 molecule. CD14 is a cell surface molecule anchored in the membrane by phosphatidylinositol linkage and specifically expressed by monocytes , but is also present in a soluble form in the serum, as it is shedded from activated cells. CD14 is a polyspecific "sentinel receptor" of the innate immunity, capable of recognizing various conserved bacterial molecules, but was recently described also as a ligand for apoptotic cells to be phagocytized. ECs and VSMC do not express CD14, but are able to produce LBP. Vascular cells actually utilized LPS-soluble CD14 complexes formed in plasma and catalyzed by LBP to become activated by LPS [16]. Moreover, as to what ECs see in vivo is whole blood and not only plasma, the same authors demonstrate that blood cells dramatically increased endothelial cell activation in response to LPS by a factor of 1000 and that this effect was due to monocyte-derived TNFα and IL-1β [17]. The monocyte thus acts as an intravascular amplificator of LPS effects by secreting TNFα and IL-1β that stimulate secondarily ECs. Vascular cells may become activated by Gram positive products such as lipoteichoic acid via CD14.

Shear stress increases the release of IL-1 and IL-6 in EC [18], while mechanical deformation promotes secretion of IL-1α and IL-1ra by EC [19].

Cytokine inhibitors

Two special vascular cell-derived factors play an autoregulatory role on the vascular response to inflammation. NO constitutively produced by endothelial cells inhibits adhesion molecules expression and cytokine production, via NFκB/IκB signaling inhibition [20, 21]. Growth factors not only modulate vascular cell survival and growth, but also may act as downregulators of inflammatory mediators. TGFβ is a potent anti-inflammatory cytokine targeting vascular cells as it was first shown for monocytes/macrophages. FGFs can inhibit the expression by ECs of Tissue Factor and other inflammatory genes (tissue plasminogen activator, plasminogen activator-2 and IL-8) [22], while PDGF inhibit the expression of iNOS in VSMC. Interestingly, this regulation can be exerted in an autocrine/juxtacrine way.

CD40-CD40 ligand system

CD40 ligand is a transmembrane protein structurally related to the cytokine TNF and its expression was, until recently, restricted to activated CD4+ T cells, while CD40, a member of the TNF-receptor superfamily is found on B cells, on monocytes/macrophages. However, CD40 ligand is coexpressed constitutively with CD40 in vascular cells and that stimulation with IL-1β, TNFα and IFNγ increases surface level and de novo synthesis of CD40L in both EC and VSMC [23]. Crosslinking of the CD40 molecules on vascular cells induces functional changes that contribute to cell-mediated inflammatory responses including upregulation of cytokines, costimulatory and adhesion molecules. The endpoint of these changes has been shown to be the recruitment of leukocytes.

The expression of CD40 by endothelial cells has been found in many tissues such as skin, spleen, lung, and can be upregulated in vitro by IFNγ, TNF and IL-1 [24]. Stimulation of human endothelial cells and vascular smooth muscle with CD40L induces the synthesis of pro-IL-1β followed by mature IL-1 through the activation of IL-1β-converting enzyme (caspase-1) [25]. CD40 ligation on endothelial cells induces the expression of E-selectin, VCAM-1, and ICAM-1, and can occur during adhesion of activated T cells or that of activated platelets [24].

Indeed, it has been recently shown that platelets have preformed CD40L and express it on their surface within seconds, as a part of the "basic platelet activation", in concert with CD63, P-selectin and several other proteins [26]. The interaction between CD40L on platelets and CD40 on endothelial cells results in the induction of endothelial chemokines (IL-8, MCP-1) and adhesion molecules (E-selectin, ICAM-1, VCAM-1), thereby generating signals for the recruitment of the leucocytes at the site of injury [26]. Platelets may thus also directly initiate an inflammatory response of the vessel wall and provide a link between haemostasis and inflammation. The generation of inflammatory signals by platelets within the vessel wall may be involved in the pathogenesis of both atherosclerosis [27] and sepsis.

Early activation signals in vascular cells

Few studies report the role of tyrosine phosphorylation in the cytokine activation of vascular cells. Genistein-sensitive step is involved in the induction of plasminogen activator, VCAM-1 and E-selectin, IL-6 by proinflammatory cytokines such as TNF, IL-1 and OSM in both EC and VSMC. In EC, TNF has been shown to activate the treee MAPkinases: p42ERK, p38 and p54JNK downstream in the MAPkinase cascade. A role of ceramide has been suggested in the activation of EC by TNF and IL-1.

Cytokines modulate gene transcription via the activation of STAT (signal transducers and activators of transcription) by various members of the JAK (Janus kinase) family. There are four known JAK: JAK1, JAK2, JAK3 and Tyk2. JAK3 seems to be restricted to some cells in contrast to the ubiquitous of the others. JAKs are unique in containing 2 kinases domains. STAT proteins are usually present in a quiescent form in the cytoplasm of cells and contain a single tyrosine residue that is phosphorylated in response to cellular activation by cytokines. The STATs diverge by their carboxyl terminii but bind a common consensus sequence. Therefore the specificity of signal transduction is not strictly attribuable to JAK/STAT interactions.

The JAK-STAT pathway is only beginning to be explored in both EC and VSMC. IL-4 and IL-13 activate JAK2 phosphorylation and the STAT6 transcription factor, leading to the induction of VCAM-1 [28]. OSM activates JAK2-STAT3 in EC and Kaposi sarcoma cells and JAK1/STAT1 in human VSMC (Figure 1A-B) [29]. GM-CSF activates JAK2 which induces in EC a functional program related to angiogenesis. In addition, IL-6/IL-6R complexes may activate EC via STAT3 phosphorylation.

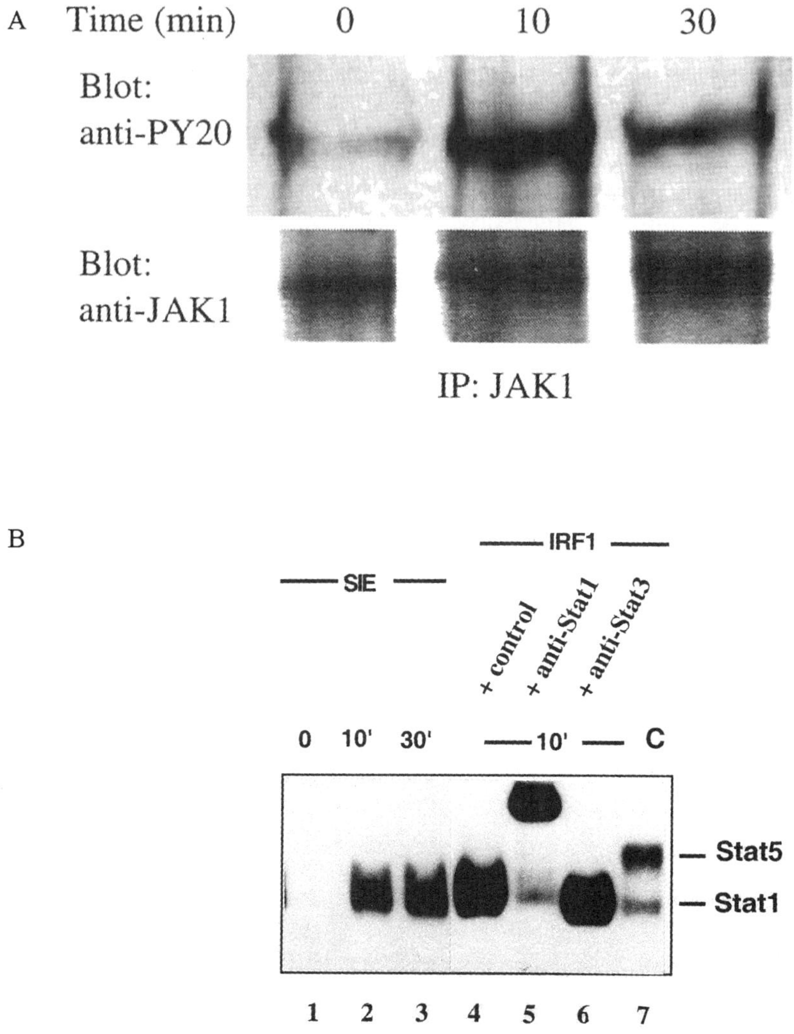

Figure 1. *A. OSM-induced tyrosine phosphorylation of JAK1. Human aortic SMC were stimulated with 1 ng/ml OSM for the indicated times. Immunoprecipitation was performed in the presence of anti-JAK1 antibodies and western-blotted with anti-PY20 antibodies. The same blot was probed with anti-JAK1 antibodies to determine equal loading. IP indicates immunoprecipation.*

B. Specific activation of STAT1 in human aortic SMC in response to OSM. The presence of activated STAT factors in nuclear extracts from OSM-activated SMC was evaluated by EMSA using two different STAT-responsive elements as probes (m67 SIE or IRF1-GAS), as described in Materials and Methods. To identify the STAT factors activated, nuclear extracts were next incubated with the labeled probes in the presence of specific anti-STAT antibodies (supershift assays).

Cytokines of the IL-6 family utilize gp130 as a common signal-transducing factor, since biological functions of these cytokines in ECs are inhibited by the addition of anti-gp130 monoclonal antibodies. gp130-mediated signaling involves two distinct pathways, STAT-dependent and ras-dependent [30]. Although gp130 does not possess tyrosine kinase domain, ligang binding gp130 leads to activation of associated tyrosine kinases, JAK1, JAK2, Tyk2. Phosphorylated JAK kinases in turn phosphorylate the latent cytoplasmic transcription factor STAT3, allowing binding of the target genes to an IL-6 response element. The ras/MAP kinase cascade is also activated following gp130 stimulation and leads to the critical phosphorylation of NF-IL6.

Cytokine activation of vascular cells may also result in the recruitment of proteins of the death pathway. Activation of TNFRI (p55) results in the binding of TNFRI-associated death domain protein (TRADD), TNFR-associated protein 2 (TRAF2) and RIP, with subsequent signaling pathways leading to the activation of transcription factor NF-KB or the MAP kinase, JNK. Alternatively, TRADD can recruit Fas-associated protein with death domain (FADD), which leads to apoptosis. Ceramide is the product of hydrolysis of sphingomyelin and participate to the signalling but does not seem to be sufficient to induce a biological response. Instead, ceramide may have a cooperative role in the MAP kinase activation and in the canonical NF-κB activation.

Activation of NF-κB in vascular cells (Table 3)

The expression of inducible genes leading to the synthesis of cytokines, chemokines, adhesion molecules and autacoids relies on transcription factors. Among the primary transcription factors, NF-κB has gained since a few years a considerable attention. NF-κB plays a central role in the regulation of inflammatory mediators [31, 32].

Usually, NF-κB is a dimer composed of 2 DNA-binding proteins (p50 and p65) and is held in an inactive form in the cytoplasm by binding to the inhibitory protein Ik-B-a which prevents the nuclear uptake of NF-κB. Activation of NF-κB involves the release of I kappa B-a and the translocation of p50/p65 subunits to the nucleus. Binding of p50/p65 subunits to the κB consensus motif in the 5' region of the promotor of specific target genes results in their expression. Events leading to the activation of NF-κB relays upon the phosphorylation of I kappa B followed by its ubiquitination and its proteolytic degradation into the proteasome. The mechanisms of activation of I-κB kinase that leads to the phosphorylation of I-κB are unclear, but it seems that the generation of intracellular oxygen-derived free radicals plays a pivotal role. In EC, TNF induces Ser phosphorylation and subsequent proteolytic degradation of I-κB-α and two kinases have been recently identified that specifically phosphorylate I-κB-α subunit: p36 and p41.

Two major mechanisms are likely to play a role in the activation of NF-κB: phosphorylation of the inhibitory molecule I-κB and free radical-dependent oxidation. Stimuli for the activation of NF-κB are therefore LPS, lipoteichoic acid from cocci Gram + and bacterial exotoxins, proinflammatory cytokines (TNF, IL-1β), and also oxidants (hydrogen peroxide) and viruses (HIV-1, herpes viruses). The most amazing stimulation of NF-κB expression is however the mechanical one, which is quite specific for vascular cells. Indeed, shear stress activates NF-κB.

The intracellular redox status of the cell is extremely important in the regulation of NF-κB /I-κB by preventing the activation of I-κB kinase. Anti-oxidants, aspirin, N-acetylcysteine may inhibit the activation of NF-κB. In EC, TNF induces Ser phosphorylation and subsequent proteolytic degradation of I kappa B-a and two kinases have been recently identified that specifically phosphorylate I-κB -α subunit: p36 and p41.

Some factors may inhibit NF-κB activity. Glucocorticoids enhance the formation of I-κB, and the zinc finger transcription factor A20, an immediate-early gene in TNF-activated EC, has recently been shown to inhibit NF-κB activity in EC. NO also inhibits NF-κB activity in EC by the induction and stabilization of I-κB-α. The activation of NF-κB also can be attenuated by inhibiting the proteolytic degradation of I kappa B in the proteasome. Several serine protease inhibitors such as calpain inhibitor 1 have been reported to inhibit the proteolytic cleavage of proteins in the proteasome.

The proteins regulated by NF-κB within vascular cells include: TNFα, IL-1, IL-6, IL-8, G-CSF, M-CSF, GM-CSF, chemokines, adhesion molecules (Selectins, ICAM-1, VCAM-1), inducible enzymes (iNOS, COX-2 and phospholipase A2). Moreover, NF-κB itself induces the synthesis of Ik-B-α to regenerate an inactive form of NF-κB and insure the transiency of NF-κB activation.

Table 3. Regulation of cytokines by NF-κB

Stimuli that activate NF-κB

- Bacterial products: LPS
- Cytokines: TNFα, IL-1β
- PAF
- Oxidants: Hydrogen peroxide
- Shear stress

→

Proteins regulated by NF-κB

- Cytokines:
  TNFα, IL-1 β, IL-6
- Haematopoietic growth factors:
  GM-CSF, M-CSF, G-CSF
- Chemokines:
  IL-8, MCP-1, MIP, GRO
- Inflammatory enzymes:
  iNOS, COX2, PLA2
- Adhesion molecules:
  ICAM-1, VCAM-1, E-Selectin

Nuclear receptors and vascular cells

The peroxisome-proliferator-activated receptors (PPAR) are members of the nuclear hormone receptor superfamily of transcription factors that mediate ligand-dependent transcriptional activation and repression. The PPAR receptor class consists of PPARα, PPARγ and PPARδ. Like other nuclear receptors, PPAR contain a ligand binding domain and a central DNA binding domain, which interacts with PPAR response elements in the promotor of target genes. It has been shown that PPARγ agonists, including several prostanoids, suppress monocyte proinflammatory cytokine production, suggesting a role for PPARγ agonists to modulate the inflammatory response [33, 34]. In addition, stimulation of PPARγ with PGJ2 inhibits matrix metalloprotease "MMP9" gelatinolytic activity and secretion [35].

PPARα deficient mice show a prolonged response to inflammatory stimuli. Interestingly, PPARα, but not PPARγ, is expressed in aortic SMC and fibrates ; the PPARα synthetic agonists inhibit IL-1-induced IL-6 production and COX-2 expression in these cells [36]. The inhibition of COX-2 induction occurs at a transcriptional level as a result of PPARα repression of NF-κB signaling.

Adhesion molecules

A system of adhesion molecules orchestrate cellular interactions between vascular cells and circulating blood cells, with activation of constitutive adhesion molecules but also induction of new receptors. These molecules, as cytokines and chemokines, partially overlap adding in safety to the defense system. The cellular adhesion molecules are classified into four major families: the selectins, the immunoglobulin superfamily, the integrins and the cadherins.

Selectins

The term selectin was originally proposed to highlight at the same time the presence of the lectin domain and the selective nature of the expression and function of these molecules. The selectins expressed by endothelial cells, but also by vascular smooth muscle cells, platelets and some leukocytes, interact with carbohydrate ligands on leukocytes. In this regard, the selectins are unique: in constrast to other cellular adhesion molecules which function via protein-protein interactions, selectins function as carbohydrate-binding proteins. There are three known selectins:

- L-selectin is found mainly on lymphocytes and mediates their homing to lymph nodes. In part of this role, L-selectin constituvely expressed on their leucocyte surface may participate in the adhesion of neutrophils, monocytes and T lymphocytes to activated endothelium. Interestingly, leucocytes rapidly shed this selectin following activation.

- E-selectin is induced in endothelial cells by proinflammatory cytokines: TNF and IL-1, with peaking expression at 4-6 hours and a return to basal level in about 24 hours. E-selectin supports the binding of neutrophils, monocytes, memory T lymphocytes, eosinophils and basophils. Human E-selectin gene reveals in its promoter region NF kappa B and AP-1 binding sites. In vivo, E-selectin is mostly found on the endothelium of postcapillary venules at sites of active inflammation and, sometimes in case of systemic inflammatory response, on capillary endothelium. This expression is associated predominantly to mononuclear infiltrate, but also to neutrophil inflitrate.

- P-selectin is stored in platelets (α-granules) and endothelial cells (Weibel-Palade bodies) and is rapidly released when these cells are stimulated. New P-selectin synthesis can also be induced by TNF and IL-1 as for E-selectin, and some other cytokines such as IL-4 and OSM induce a sustained expression of P-selectin in endothelial cells for 48 hours [37]. These selectins play a part in leukocyte rolling, by binding leukocyte with low affinity allowing them to quickly decelerate. While rolling, leukocytes may be activated by chemokines by increasing the affinity of their integrins. The P-selectin expression is associated predominantly to neutrophil infiltrate.

Selectins work in conjunction with other cell-associated molecules: for example, E-selectin binding to neutrophils activates their CD11/CD18 molecules which can in turn bind to ICAM-1. The generation via gene targeting of animals deficient for selectins has provided valuable insight into their functions in vivo. P-selectin knockout mice display severe defects in leucocyte rolling and extravasation in the early phases of inflammation [38] while E-selectin knockout mice [39] are free of such defects. However, the gene target disruption of both selectins induces profound alterations of leucocyte recruitment and enhances susceptibility to infection.

Immunoglobulin superfamily

Members of the immunoglobulin superfamily contain immunoglobulin-like structures and include ICAM-1, VCAM-1 and PECAM-1. ICAM-1 is constitutively expressed on the endothelium and its level of expression is increased by proinflammatory cytokines and LPS. ICAM-1 can bind integrins, such as CD11a/CD18 and CD11b/CD18. VCAM-1 is induced by proinflammatory cytokines and LPS on endothelial and smooth muscle cells and serves as ligand for some integrins present on T and B lymphocytes and basophils. PECAM-1 or CD31 is constitutively expressed at high level on endothelium (concentrated at the intercellular junctions), at low level on monocytes and certain lymphocytes and is involved in the transendothelial migration.

Integrins

The term integrin suggested originally the presumed role of these proteins to integrate cells within the extracellular matrix. The integrins are membrane heterodimeric glycoproteins present on leucocytes with 2 subunits designated α and β. There are at least 11 α subunits and 6 β subunits, and the variety of their combinations form more than 16 integrins. The target aminoacid sequence for many integrins is the tripeptide: Arg-Gly-Asp (RGD) sequence, which endows integrins to bind to extracellular matrix components containing this recognition signal: such as fibronectin, fibrinogen, laminin, various collagens, vitronectin, thrombospondin, Von Willebrand Factor. The 3 major members of the $\beta 2$ integrin group are CD11a/CD18, CD11b/CD18, CD11c/CD18. CD11b/CD18 is responsible for the firm adhesion of neutrophils to endothelium preceeding transmigration. The counterreceptors for LFA-1 (CD11a/CD18 or $\alpha_L\beta_2$) are the intercellular adhesion proteins ICAM-1 and ICAM-2, while the countereceptor for the $\alpha_4\beta_2$ integrin is VCAM-1. $\alpha_M\beta_2$ (MAC-1 or CD11b/CD18) binds to fibrinogen. Leucocyte integrins need to be activated before to mediate leucocyte binding. In addition, there is increasing evidence that the cell movements that take place during tissue repair such as wound healing depend on integrin-mediated interactions.

Endothelial cell-leucocyte adhesion

The endothelial cell-leucocyte adhesion is a cascade involving successive adhesion molecules and includes (Figure 2):

- an initial phase named rolling where leucocytes are subjected to deceleration on the luminal surface of endothelium by low affinity binding;
- a secondary phase where leucocytes bind to endothelium wih firm adhesion and are activated at the same time;
- the emigration phase, which does not necessarily follow the leucocyte adhesion, constitutes the transendothelial cell migration and requires a chemokine gradient.

The rolling is mediated by members of the selectin family. P-selectin stored in Weibel-Palade bodies is quickly moved to the endothelial cell surface in 5 to 10 minutes by inflammatory stimuli such as thrombin, leucotriene C4 and free radicals. E-selectin is more slowly upregulated by proinflammatory cytokines such as TNF and IL-1 or LPS via a process requiring de novo protein synthesis and 4-6 hours. Ligands of these selectins are oligosaccharides related to sialyl Lewis x expressed constitutively on the cell surface of leucocytes. The binding between leucocyyte and endothelium is of low affinity. The rolling is a prerequisite to firm adhesion and allows leucocyte to deccelerate in the conditions of flow.

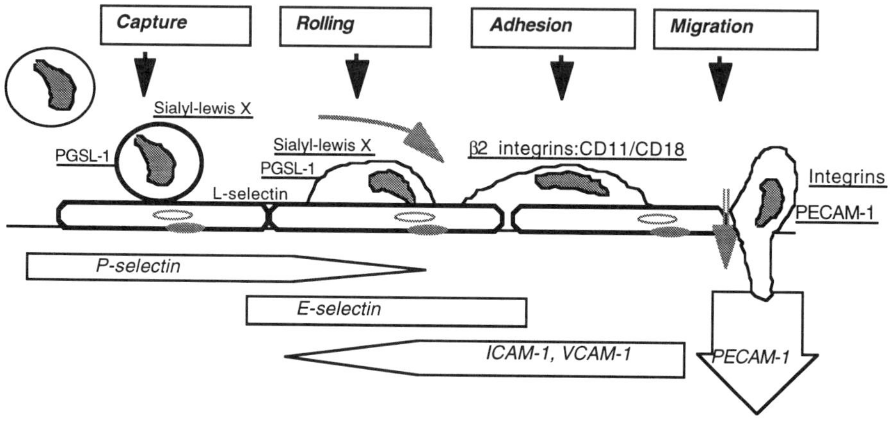

Figure 2. *Different steps involved in endothelial cell-leucocyte adhesion*

The firm adhesion does not occur after rolling unless another set of adhesion molecules is engaged. The interaction b2 integrin molecules expressed by leucocytes and immunoglobulin superfamily members (ICAM-1, VCAM-1) expressed by endothelial cells mediates this firm adhesion. An important characteristic of the leucocyte integrins is that they exist under basal conditions in an inactive conformation and that they thus need a conformational change under activation to confer adhesive property. The increase in expression of ICAM-1 by proinflammatory cytokines (TNF, IL-1, IFNγ) and LPS peaks about 8 h after stimulation.

The transmigration of leucocytes across endothelial monolayer requires a chemotactic gradient. These chemotactic factors includes leucotrienes, complement fractions, and chemokines. Another member of immunoglobulin superfamily expressed at high levels by endothelium named PECAM-1 or CD31 is implicated in the phase of transmigration. Its particular localization has first suggested this role in that it is concentrated at the junctions between endothelial cells. It seems that leucocytes proceed through a molecular zipper made of CD31 molecules on opposing ECs. Further progression of leucocytes through the endothelial junctions towards the subendothelial matrix is dependent on CD31-mediated transactivation of b1 integrin. This adhesion cascade is finely regulated being balanced by both adhesion and de-adhesion events to allow leucocyte to acquire a vectorial motility.

Chemokines

The term chemokine is a contraction of "chemotactic cytokine." Chemokines are essential to induce directed locomotion of leukocytes into tissues subjected to an inflammatory insult. Of the cytokines, members of the chemokine family have 3 main properties: the activation of leukocyte integrins, the chemoattraction and the activation of leukocyte effector functions (for reviews see [40-43]). More than 40 chemokines have been identified in the past few years, detected initially in some inflammatory diseases with no known biological activity, such as IP-10 in viral hepatitis. Chemokines are low molecular weight peptides (8-10 kd) defined by a common structural motif consisting of 2 intramolecular disulfide bridges.

Chemokines are subdivided into two main families (Table 4) : the α- and β- families for which the position of cysteine residues is distinctive: in the α-chemokines, one aminoacid is inserted between 2 cysteines (CXC), whereas in the β-chemokines, the first 2 cysteine residues are adjacent (CC). In the α-chemokines, some are chemotactic for neutrophils and others are for lymphocytes. This family includes IL-8, IP-10, GROα. The β-chemokines do not act on neutrophils, but attract monocytes, lymphocytes, basophils and eosinophils. This family includes MCP-1 to -5, MIP-1a and b, RANTES.

Table 4. *Chemokine families*

C-X-C subfamily	C-C subfamily
GRO	MCP1-5
IL-8	MIP1α & β
IP10	RANTES
PF-4	

Chemokines induce cell attraction by binding to specific G-protein-coupled cell-surface receptors on target cells, namely CXC-R (CXCR 1 to 4) or CCR (CCR1 to 8) [44, 45]. CXCR1 is restricted to neutrophils, whereas some CR are widely expressed. Moreover, the expression of chemokine receptors may be constitutive for some of them, inducible for others, and sometimes downregulated. T lymphocytes may express CXCR3 when differentiated into T helper 1, or CCR3 in the case of T helper 2 phenotype. Some chemokine receptors are also expressed by non haematopoietic cells such as endothelial cells, suggesting a role different from leukocyte chemotaxis. Chemokines also interact with two non-signalling molecules expressed by endothelial cells or vascular matrix: the Duffy antigen and a group of heparan sulfate proteoglycans. These molecules may serve as capture sites, clearing chemokines from the circulation but the major role of proteoglycans is to immobilize chemokines and to create a gradient allowing leukocyte directed locomotion.

Tissue inflammation is associated with a dramatic increase in the secretion of chemokines which have been identified in many organs and many different cell types [40]. The majority of CC and CXC chemokines are produced by activated monocytes/macrophages as well as endothelial [46] and epithelial cells. Smooth muscle cells are also able to express chemokines [47].

In general, chemokines are not stored within cells and their production are induced by a variety of stimuli, such as proinflammatory cytokines (TNFα and IL-1), bacterial products (LPS) and virus. An exception is made by the chemokine PF4 which is stored within a-granules of platelets. Inhibition of chemokine production may be brought about by TGFβ, IL-4, IL-10, depending on chemokine and cell types [47]. Neutralization of chemokine activity may result from the binding to antibodies or to extracellular matrix components.

Chemokine receptor activation leads to a cascade of cellular events first mediated by G proteins. Rho proteins are then activated and induce cell motility through actin-dependent processes. Chemokines are both chemotactic factors and adhesive molecules, providing the signals to convert leukocyte low affinity fixation to the endothelium to high affinity fixation, by a conformational change in the integrins [48]. For example, IL-8 induces binding between LFA-1 on neutrophils and ICAM-1 on the endothelium and MIP-1β stimulate T-cell adhesion via interactions between LFA-1 and ICAM-1 but also between VLA-4 and VCAM-1. In the same time, leukocytes are promoted by chemokines to an activated phenotype. Subsequently, a cytoskeleton reorganisation allows leukocytes to cross the endothelium (diapedesis) to migrate into the tissue. The ability of chemokines to bind to proteoglycans of the extracellular matrix of the vascular wall provides a gradient of immobilized chemokines along which leukocytes move and their effector functions are activated.

α-chemokines (CXC) either inhibit or promote angiogenesis, depending on a structural aminoacid motif. IL-8 and GRO-α promote angiogenesis, whereas IP-10 and PF-4 inhibit angiogenesis [49].
In atherosclerosis, MCP-1 is highly expressed and deletion of its receptor CCR2 has been recently reported to prevent atherosclerosis in apoE-/- mice [50].

The chemokine receptors serve as coreceptors for two important human pathogens: plasmodium and HIV, allowing microbe entry into the cells. Some genetic polymorphisms for CCR5 modulate human sensitivity to HIV1 infection. Kaposi's sarcoma-associated herpes virus 8 encodes an active chemokine receptor that stimulates the proliferation of KS cells derived from normal ECs [51].

Matrix metalloproteinases and their inhibitors

Matrix proteins are expressed by inflammatory cells and vascular cells and modulate leucocyte functions and cytokine activity. Cytokines are able to regulate matrix component synthesis: TNF, IL-1 and TGFβ stimulate synthesis of type I and III collagens by VSMC, while IFNg inhibits their synthesis. The tissue composition of the extracellular matrix depends on its turnover, according to synthesis and degradation.

MMPs (review in [52]) are a family of Zn^{2+}- and Ca^{2+}-dependent enzymes of at least 12 members, devoided to the resorption of extracellular matrices and divided into 3 main groups based broadly on substrate preferences: collagenases, gelatinases and stromyelysins. The collagenases degrade type I to III collagens, gelatinases degrade collagen type IV, stromyelysins have a broad substrate specificity (proteoglycans, laminin, fibronectin, gelatin and basement membrane collagens). Another family of MMPs are bound to cell membrane and are involved in cell migration.

A number of cytokines and growth factors have been shown to induce and to stimulate the expression of MMPs, including TNFα, IL-1, PDGF and TGFβ. Interestingly, MMPs are able to cleave TNFα propeptide to an active form, suggesting their important role in the course of inflammation. Stimulated T lymphocytes induce the expression of MMP-1 (intertitial collagenase), MMP-2 (gelatinase A), MMP-3 (stromyelysin), MMP-9 (gelatinse B) in human VSMC by cell contact via CD40 ligation [53]. The interaction of macrophages with lymphocytes using CD40 and its ligand also upregulates MMPs. Mechanical forces also regulate MMPs activity: cyclic strains inhibit PDGF- or TNF-induced synthesis of MMP-1 by VSMC [54]. Macrophages in human atheroma contain PPARg and this expression may down-regulate the activity of MMP9 in the plaque [35].

The activation of the latent proenzymes is the second level to control MMPs. Plasmin is one of the most potent activators promoting cleavage of the propeptides to the active molecules.

The MMPs are inhibited by a family of specific inhibitors, TIMPs. TIMPs are secreted multifunctional proteins which bind to and neutralize the active enzymes MMPs. TIMP1 is synthetized by most types of connective tissue cells, including macrophages and vascular cells, acts against all MMPs, and is highly inducible by cytokines and hormones. TIMP-1 et -3 are upregulated in VSMC by PDGF-AA and TGFβ [55]. TIMP2 expression is mostly constitutive following the pattern of expression of gelatinase A, with which it interacts specifically. Several cell lines produce MMPs, including leucocytes especially macrophages and basophils, and vascular cells mostly smooth muscle cells. The same cell type also produces TIMPs. MMP9, also referred to as gelatinase B, is the predominant MMP secreted by monocytes/macrophages in vitro.

MMPs and TIMPs are involved in numerous processes occurring during inflammation: matrix degradation, cell migration and leucocyte emigration, cytokine activation, cell apoptosis and angiogenesis.

RESPONSES OF VASCULAR CELLS TO INFLAMMATORY AGENTS

The vascular cell responses to inflammatory stimuli can be distinguished schematically into 5 different states: vascular activation, vascular cytostasis and cell energetic depletion, vascular cell death, vascular protection, and vascular growth.

Vascular activation

The activation of vascular cells, as described for leukocytes, includes both the loss of constitutive functions and the induction of new properties [56, 57].

Endothelial cell activation

Morphological changes and increase in endothelial permeability

Endothelial morphological changes occur in cytokines-treated endothelial cells. TNF increases the permeability to macromolecules of endothelial cells monolayers, via an action on membrane-G-proteins. Endothelial injury can also account for this leakiness. In vivo endothelial activation and endothelial injury often coexist. Harlan et al. showed a direct morphological effect of LPS on bovine endothelial cells, including dilation of the intercellular junctions, cell contraction and ruffling of the surface membrane [58]. Young et al. described, using electron microscopy, endothelial injury in vessels isolated from septic animals, with separation of the endothelium from the internal elastic lamina media and increase in interendothelial gaps [59]. Many factors may favor the cytotoxicity of LPS : complement system, cytokines, free radicals and leukocyte products. Indeed, cytokine-treated endothelial cells are more susceptible to injury. Oxygen free radicals released by both endothelial and blood cells are known to be cytotoxic and interaction between oxygen and nitrogen free radicals might lead to the formation of peroxynitrite which is a higher toxic radical. In addition, adhesive leukocytes may mediate LPS endothelium injury through the release of proteases.

Endothelial NOS dysfunction

LPS and several proinflammatory cytokines impair ecNOS expression or activity. IL-1 inhibits the acetylcholine-induced vasodilation of pre-contracted rabbit aorta [60]. TNF impairs coronary responses to receptor-mediated and endothelium-dependent vasodilators such as acetylcholine and ADP. Conversely, relaxant responses to A 23187 (that releases EDRF independently of endothelial cells receptor) and thoses induced by endothelium-independent vasodilators ($NaNO_2$) are not affected. Several mechanisms underly this impairment: cytokines may induce free oxygen radicals release from endothelial cells which inactivate NO; In vitro the expression of the constitutive endothelial NOS decreases with time while the inducible isoform might be expressed upon exposition to cytokines. Moreover, a negative autoregulatory role of NO resulting from iNOS activity has been described leading to eNOS mRNA downregulation [61]. TGFβ preserves endothelium-dependent relaxation and may be an intrinsic cytoprotective cytokine, since it can antagonize effects of TNF, inhibits free radical generation and adhesion of neutrophils to the endothelium [62].

Proadhesive state

The recruitment of leukocytes at sites of inflammation is a multistep process, already described in the chapter "adhesion molecules", and involves rolling, firm adhesion and emigration of leukocytes from the circulation to the subendothelial space. Rolling of neutrophil is mediated by the selectins : L-selectin binding to endothelial GlyCAM, P- and E-selectins binding to sialyl Lewisx (sLex)-related determinants on the surface of the neutrophil. IL-8 is critical to allow firm adhesion, which is mediated by the binding of the integrins (CD11a and CD11b) to ICAM-1. Diapedesis of the neutrophil is partially mediated by PECAM-1. For monocyte, the initial rolling depends mainly on interaction between L-selectin and an inducible L-seletin ligand expressed on activated endothelium. MCP-1 seems critical to allow monocyte adhesion by triggering β_1- and β_2-integrin activation. Stable arrest can thus be observed via binding to ICAM-1 and VCAM-1. Ultimately, monocyte transmigration involves PECAM-1 as well as β_1- and β_2-integrins. This entire process occurs within minutes. The kinetics of expression of these adhesion molecules suggests that earlier induced ELAM1 participates in neutrophils adhesion at the early phase of inflammation, IL-8 (neutrophil activating factor) secreted by cytokine activated endothelial cells promoting neutrophils diapedesis through the endothelium. Maximal expression of ICAM-1 occurs after that of ELAM-1, while INCAM-110/VCAM-1 begins to be induced. At a later time, IL-1 and TNF cause endothelial cells to secrete MCP-1 to recruite monocytes to inflammatory sites. These time-dependent modifications of adhesive molecules correlate with changes in leukocyte subtype population involved in the inflammatory response. However, disparities in the kinetics for adhesion molecules have been observed between in vitro and in vivo endothelial activation. Indeed, P-selectin expression is sustained in vivo up to 24h resulting mainly from de novo synthesis while ICAM-1 expression is more transient in vivo than in vitro [63].

Apart from humoral stimuli, endothelial cells are constantly exposed to a spectrum of hemodynamic forces generated by pulsatile blood flow and including: hydrostatic pressures, cyclic strains and wall shear stresses [64]. Biomechanical forces induce endothelial structural changes and modulate gene regulation. When endothelial cells are exposed to physiologically laminar shear stress, ICAM-1 expression is rapidly induced and lasts at least 48h, while E-selectin and VCAM-1 remain unchanged [65]. It has been identified positive and negative shear stress responsive elements in the promoters of various biomechanically responsive genes. Transient upregulated genes: Tissue factor, MCP-1, TGFβ, PDGFs, ICAM-1. Sustained upregulated genes: manganese superoxide dismutase, cyclooxygenase-2, ecNOS. Downregulated genes: VCAM-1, ICAM-1, endothelin-1.

In a manner similar to leukocytes, platelets roll on the endothelium of venules, this process depending mainly on endothelial P-selectin. Endothelial cells represent a major target of TNF which orchestrates interactions between microvascular endothelial cells and platelets. These events have been described in an experimental model of cerebral malaria [66]. TNF induces adhesion and fusion of platelets to brain microvascular endothelial cells, depending on LFA-1 expressed on the platelet surface and ICAM-1 on EC. The platelet-endothelial fusion has been reported to support the integrity of endothelium in physiological conditions. In pathological conditions, platelet fusion is accompanied by the transfer of cytoplasm and surface molecules from platelets to endothelial cells and results in increased leukocyte adherence to EC and the enhancement of EC injury [66]. These microvascular endothelial cells alterations induced by TNF in the presence of platelets may be essential in the development of microvascular hemorrhages and leukocyte sequestration in vivo.

Prothrombotic state

A crucial physiological function of the endothelium is to provide an antithrombotic surface, thereby inhibiting platelet adhesion and clotting. This function depends on a multistep control of thrombin generation and actions. Indeed, the extracellular matrix of the endothelium promotes the activity of antithrombin III, ECs prevent thrombin formation via tissue factor pathway inhibitor and help to contain thrombin activity through the expression of thrombomodulin and this latter interaction both activates the anticoagulant protein C in conjunction with that of protein S synthetized by ECs, and promotes ECs fibrinolytic activity.

The pivotal step in the shift from an anticoagulant to a procoagulant surface during inflammation is the induction of Tissue Factor (TF). TF dramatically promotes factor VIIa-dependent activation of factors X and IX. Synthesis of TF is induced in vitro by various stimuli such as endotoxin, TNF and IL-1, hypoxia, shear stress, but also by exposure of anionic phospholipids that may occurs during apoptosis or platelet activation.

Smooth muscle cell activation

In vitro, cytokines have been identified as potent inhibitors of vascular contraction via an endothelium-independent mechanism. TNF and IL-1 have been reported to inhibit the phenylephrine-induced contraction of vessels [67]. Indomethacin did not prevent the inhibitory effect of IL-1. Contractions caused by potassium depolarization also were depressed, indicating that the effects of cytokines are not limited to receptor-dependent vasoconstriction. The depressive effect of cytokines on vascular contractility requires protein synthesis, for it is prevented by pretreatment with cycloheximide. Vessels excised from endoxin-treated animals [68], or exposed *in vitro* for few hours to bacterial fragments, *E. coli* LPS [67], or lipoteichoic acid from *Aureus staphyloccocus* [69] show depressed contractility in response to vasoconstrictor agonists. The endotoxin-induced hypocontractility might be mediated by autocrine production of cytokines TNF and IL-1 by the vascular cells [70]. The distal inflammatory mediator that impress contractile dysfunction has been shown to be NO formed from the activity of iNOS expressed by smooth muscle cells [71-73]. Indeed, both EC and VSMC are able to express inducible NOS under inflammatory stimulation.

The two main stimuli described to induce iNOS are IFNγ and LPS. The mouse iNOS promoter-enhancer has a complex structure, suggesting that numerous transcriptional factors from various agents may be active. LPS inducibility depends on the unique NF-κB sequence. IFNγ can synergize with LPS to induce transcription of iNOS requirng interferon regulatory factor-1 (IRF-1), but may also stabilize iNOS mRNA. The transcription induction is not limited to LPS and IFNγ since IL-1 has been shown to induce iNOS in rat vascular smooth muscle cells and mesangial cells.

TGFß is a deactivating factor being able to suppress iNOS protein expression by both transcriptional and post-transcriptional ways: transcription inhibition, iNOS mRNA destabilization, iNOS mRNA translation inhibition. Suppression of iNOS by TGF-ß has been described in many cells; macrophages, vascular smooth muscle cells, chondrocytes, mesangial cells and epithelial cells. Suppression of iNOS by glucocorticoids may also be exerted at both transcriptional and post-transcriptional steps the latter being described at high concentrations of dexamethasone. Other cytokines have been shown to inhibit iNOS expression, including IL-4, IL-10, IL-8, PDGF, IGF-1 and thrombin. Some ways of post-translational control of iNOS are also reported, through tetrahydrobiopterin availibility. GTP-cyclohydrolase I responsible for de novo tetrahydrobiopterin synthesis can be induced by proinflammatory cytokines in VSMC. iNOS may be subject to negative feedback by NO but the mechanisms of autoregulation are not clear, either binding of NO to the heme moiety of NOS or control by NO of the expression of NOS gene.

Bovine and rat aortic smooth cells express iNOS after 6 hour-exposure to LPS, IL-1 and INFγ. While NO synthase inhibitors prevented IL-1-induced hypocontractility in rat aortic rings, IL-1 was found to increase the production of NO and one of its metabolites (nitrite) by rat aortic smooth muscle cells in culture [67]. Busse et al. detected the formation of NO in the cytosol of cytokine (IL-1, TNF, IFNγ) - treated smooth muscle cells by activation of purified guanylate cyclase [74]. The molecular target of NO in smooth muscle cells is primarily the soluble guanylate cyclase which allows formation of cGMP from GTP. Increase in cGMP has been shown in both in vivo (endotoxemia) and in vitro (vascular exposure to bacterial fragments and cytokines) experiments [72, 74-76]. Increase in cGMP leads to vasodilation through the inhibition of constrictive pathways at different molecular levels : phosphoinositide turnover, Ca^{2+} disponibility, and myosin light chain kinase, resulting from the activition of cGMP-dependent protein kinase.

Very few data concerning inducible NOS activity in human vessels are available. Recently, it has been demonstrated that vascular smooth muscle cells of the human internal mammary artery possess an L-arginine/NO pathway inducible by LPS, but not by IL-1 [77, 78]. Morever, LPS and IL-1β appear to depress the contractility of the human saphenous vein by a mechanism which is not dependent on NOS or guanylate cyclase activities. These results support previous findings by Beasley and McGuiggin [79] showing that IL-1β activates the soluble guanylate cyclase in cultured human smooth muscle cells from saphenous or umbilical veins by a mechanism which is NO-independent.

Vascular cytostasis and cell energetic depletion

A considerable pool of NO is trapped within cytokine-treated vascular SMC, resulting in the nitrosylation of heme and non heme iron proteins, as identified by electron paramagnetic resonance in TNF/IFNγ-activated rat vascular smooth muscle cells [80]. Nitrosylation of heme proteins such as guanylate cyclase is obtained at low doses of cytokine treatment. At high doses or prolonged cytokine treatment, nitrosylation of iron-sulfur proteins such as enzyme complexes I and II of the respiratory chain occurs. NO binding to enzymes complexes of the respiratory chain blocks in VSMC mitochondrial respiration and induces a switch toward anaerobic metabolism, with lactate production and ATP depletion [81]. Vasoplegia is indeed a characteristic feature of septic shock and has now some molecular explanations.

Activation of a nuclear enzyme: poly-ADP-ribose-Synthetase

The nuclear enzyme poly(ADP-ribose) synthetase (PARS), also called poly(ADP-ribose) polymerase (PARP) or poly(ADP-ribose) transferase, is a protein-modifying and nucleotide-polymerizing enzyme present abundantly in the nucleus of human cells, as well as in all mammalian cells [82]. The activation of PARS in response to DNA single strand breakage results in the utilization of NAD+ as a substrate to transfer ADP ribose polymers to a variety of nuclear proteins, including histones, topoisomerasesas well as PARS itself. Nicks and breaks in the DNA strand can be induced by a variety of stimuli, especially oxidant attacks. Massive ADP-ribosylation of nuclear proteins by PARS results in rapid NAD+ depletion followed by energy depletion, cellular dysfunction and cellular death. Indeed, NAD+ serves as a cofactor for glycolysis and the tricarboxylic acid cycle, thus providing ATP for most cellular processes.

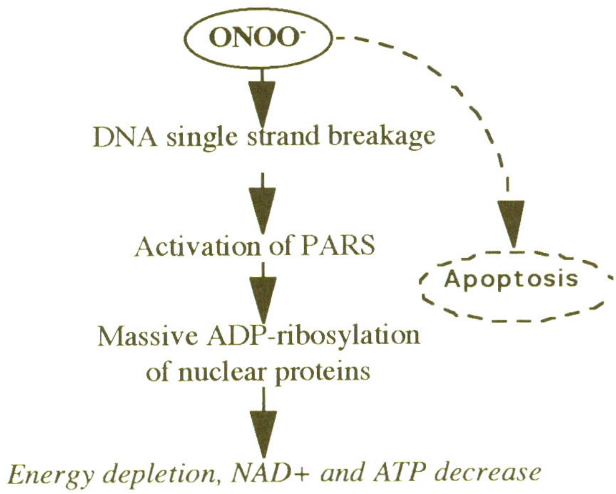

Figure 3. *Peroxynitrite-induced cellular injury via PARS activation*

The oxidative injury associated with simultaneous production of nitric oxide (NO) and oxyradicals is mediated by peroxynitrite. Peroxynitrite is capable of causing DNA single strand breakage and consequent activation of PARS in a number of cell types, especially endothelial and smooth muscle cells. Exposure of rat aortic SMC or HUVEC to peroxynitrite results in the activation of PARS with consequent reduction of intracellular NAD+, ATP, and mitochondrial respiration. Stimulation of rat SMC with LPS and IFNγ results in a rapid (few hours) and sustained (up to 24h) production of superoxide anion, while the production of NO begins to be detected after 24h, and increases over a 48-h period, as is observed for peroxynitrite production [83]. Rat aortic smooth mucle cells exposed to peroxynitrite exhibit a rapid loss in mitochondrial respiration, paralleled to DNA breakage and followed by PARS activation [83]. This results in a decrease in intracellular NAD+ and ATP content [83]. Moreover, vascular contractility is depressed when rat aortic rings are exposed to peroxynitrite, but this is only partially reversed by PARS inhibition. PARS activation is also implicated in oxidant-induced injury to EC [84] and PARS inhibition improves the endothelium-dependent relaxations of vascular rings obtained from LPS-treated rats.

In addition, PARS activation may play a role in peroxynitrite-induced apoptosis in vascular cells although conflicting results are reported. PARS activation has been recently implicated in the process of apoptosis, since PARS may serve as a deah substrate for caspase 3 (CPP32) leading to its cleavage and inactivation. The inactivation of PARS would be critical to enhance apoptosis since PARS is assumed to be a DNA repair enzyme. It is believed now that the roles of PARS in promoting apoptosis might depend on cell type and on the stimulus and substrates, rather than PARS mediating caspase 3-induced apoptosis, as observed for thymocytes from PARS-/- mice becoming apoptotic when exposed to peroxynitrite. By contrast, the energy depletion as a consequence of PARS activation may induce rapid necrotic cell death, rather than the delayed apoptotic cell death. Necrosis is characterized by disruption of cell membrane integrity, release of lactate deshydrogenase and mitochondrial injury. In fact, in some cell types subjected to peroxynitrite injury, the inhibition of PARS may devert the mode of cell death from the necrotic to the apoptotic one, consistent with the requiring of energy for apoptosis. Therefore, vascular cell metabolic suppression may occur during inflammatory activation and involves both PARS-mediated pathways and PARS-independent pathways.

Vascular cell death

Two main forms of cell death have been described [85, 86]:
* an accidental death, referred as "oncosis", or more imprecisely necrosis. Necrotic cells swell and lyse with disruption of cell membrane, feature depletion of intracellular ATP stores, and release their cytoplasmic and nuclear content including toxic components into the extracellular space.
* an active death also called "cellular suicide" and referred as "apoptosis" (see for reviews Haunstetter and Izumo [87]). Apoptosis is regulated by a cascade of proteins encoded by different genes. Cells undergoing apoptosis rapidly shrink, lose their intercellular contacts, and subsequently exhibit chromatin condensation, nuclear fragmentation, cytoplasmic blebbing to finally constitute cellular fragmentation into small apoptotic bodies enclosed onto plasma membranes. Necrosis results from severe extracellular insult in conditions of ATP depletion. Apoptosis results from the intracellular accomplishment of a programmed cell death requiring energy (see chapter on *Apoptosis in normal and pathological vessels*)

Phagocytosis by macrophages or adjacent cells is the final event of the lives of cells undergoing apoptosis. VSMC have been shown to bind and phagocytize adjacent apoptotic VSMC [88].VSMC are also able to recognize exposed phosphatidylserine on other cell types, as judged by their ability to bind erythrocytes having a high degree of exposed phosphatidylserine. Apoptotic cells express cell surface changes which allow recognition and removal before lysis. A number of receptors have been identified in vitro. Vascular cells undergoing apoptosis express phosphatidyl serine in cell surface, labeled by annexin-V.

The removal of apoptotic cells or bodies appears to be critical in the resolution of inflammation. Indeed, the apoptotic bodies are limited by cytoplasmic membranes and this prevents the release of potentially toxic or immunogenic intracellular contents into the surrounding tissue, in contrast to necrosis. Moreover, it is clearly shown now that the ingestion of apoptotic cells by macrophages actively inhibits their production of proinflammatory mediators such as TNF, IL-1, IL-8 as well as leucotriene C4 and thromboxane B2 and this inhibition is mediated by the increasing production of TGFβ, PAF and PGE2.

However, not all features of inflammation are turned off when apoptosis occurs. Indeed, apoptotic ECs and VSMC [89] become procoagulant by the increased expression of phosphatidylserine and the loss of anticoagulant membrane components such as thrombomodulin, heparan sulfates and tissue factor pathway inhibitor. Likewise, apoptosis of vascular cells leads to a dramatic increase of thrombin generation. Thrombin activation was also seen with apoptotic human VSMCs and was inhibited by annexin V. This phenomena is very near that of platelet activation where exposure of PS on the outer leaflet is determinant for efficient activation of coagulation factors. Indeed, platelet derived microparticles influence endothelial functions. Platelet MP derived arachidonic acid can induce endothelial prostacyclin formation via upregulation of COX-2 and increase monocyte adhesiveness to endothelium via upregulation of ICAM-1 on endothelial cells and that of CD-11a and CD-11b on monocytes [90]. Apoptosis of endothelial cells is associated with paracrine induction of adhesion molecules : apoptosis of HUVECs induced hyperadhesiveness of HUVECs for the THP-1 monocytic cell line [91]. Hyperadhesiveness developed in association with induction of ICAM-1 and VCAM-1 on HUVECs, and in association with increased ICE activity and IL-1β release.

Protective state related to the upregulation of protective genes

Response to injury must be considered as an active process. If inflammation is uncontrolled, endothelial cell death occurs. However, endogenous protective genes can be expressed by vascular cells to limit the inflammatory process and the injury. This regulatory response has been called accommodation by Bach et al., and may lead to a "superprotected state" [92]. Indirect arguments suggest that both EC and VSMC may develop an autoprotective phenotype during inflammation. Indeed, the endothelial dysfunction induced by a brief exposure of human superficial hand vein to LPS is transitory and recovers after 7 days [93]. Previously, McKenna [70] reported the spontaneous recovery of vascular tissue contractile function during sustained LPS exposure.

Deactivation of vascular cells

Many autocrine and paracrine factors may attenuate the proinflammatory response of vascular cells, including cytokines, growth factors, NO, heat shock proteins and mechanical forces. Although TGFβ may exert constrasting effects during inflammation, TGFβ may be one of the most potent deactivating factors of vascular cells. Indeed, TGFβ is able to restore endothelial dependent vasodilation impaired by TNFα [94]. In addition,TGFβ has been shown to inhibit VCAM-1 expression by VSMC [95] and to suppress iNOS induction in the vascular wall, leading to the prevention of LPS shock in the rat [96]. TGFβ expressed by vascular cells may also operate as a paracrine antiinflammatory factor: glomerular mesangial cells can express TGFβ in an active form which inhibits production of proinflammatory cytokines by emigrated macrophages and this cross-communication between glomerular mesangial cells and infiltrating macrophages may play an important role in the recovery from glomerulonephritis [97].

Other growth factors exhibit deactivating properties: FGFs suppress transcriptional activation of TF in human EC [22]. The concomitant endothelial expression of iNOS during inflammatory stimulation may be protective for a part, since both dexamethasone and NOS inhibitors greatly enhance the adhesive properties of activated endothelium [98]. This last observation may help to understand the deleterious effects of NOS inhibitors when used in vivo during sepsis: the global pressive state may be restored but some regional circulations may be impaired by agreggations of platelets and leucocytes in the microcirculation. Morever, NO has been shown in vitro to attenuate vascular inflammatory activation: NO inhibits MCS-F synthesis in EC [99] and diminished VCAM-1 expression induced by IFNγ in human VSMC [100] and these antiinflammatory actions of NO are mediated by the inhibition of NF-κB activity.

Physiological levels of shear stress are essentially anti-inflammatory and antiadhesive since prolonged exposure to flow results in down-regulation of ICAM-1 and VCAM-1 [101], whereas low and high shear stresses may initiate an inflammatory response. In addition, shear stress inhibits apoptosis in human EC [102].

Recently, Sata and Walsh [103] reported that Fas ligand (FasL) is expressed constitutively on ECs both in vitro and in vivo, and both on human macro- and microvascular ECs. Local administration of TNFa to arteries down-regulates FasL expression and induces mononuclear cell infiltration, while "forced" FasL expression (by adenovirus-FasL transfection) markedly attenuates TNF-induced cell infiltration. Moreover, adherent mononuclear cells undergo apoptosis rather than diapedesis under these conditions, as a result of Fas-FasL ligation. These results suggest that downregulation of FasL by the endothelium is essential for TNF-induced leukocyte extravasation, and that Fas ligand expression by the endothelium play an active role in inhibiting leukocyte extravasation.

Survival of vascular cells

Bcl-2 was first identified as a frequent translocation occuring in B cell follicular lymphoma and was found to function unlike other oncogenes, by promoting cell survival instead of cell proliferation. The Bcl-2 family now comprises more than 12 proteins and both inhibitors and stimulators exist now in this family, suggesting that apoptosis depends on a ratio between pro-and anti-apoptotic proteins. Indeed, Bcl-2 inhibits EC apoptosis while Bax promotes apoptosis.

Despite recurrent exposure to cellular toxins from the circulation and tissue, endothelial cells are remarkably resistant to cell death. Members of the growing Bcl-2 family are present in endothelial cells to provide protection against apoptosis. In endothelial cells, human Al is rapidly induced by phorbol ester, and by TNFα and IL-1β, but not by FGF-2 or VEGF [104]. Al is the only known Bcl-2 family member that is inducible by inflammatory cytokines, suggesting that it may play a protective role during inflammation. Additionally, vascular smooth muscle cells and various nonhematopoietic tissues express human Al, indicating that human Al is a widely expressed Bcl-2 homologue [104]. As the Bcl-2 family of proteins plays a prominent role in regulating cell survival, some authors attempted to identify Bcl-2 homologues expressed in endothelial cells. In addition to the previously identified A1, four other members of the Bcl-2 family, Bcl-2, Mcl-1, Bcl-X(L), and Bax, are expressed in endothelial cells.

The A20 gene product is a novel zinc finger protein originally described as a TNF-inducible early response gene in HUVECs. A20 is the category of "protective" genes that are induced in response to inflammatory stimuli to protect EC from unfettered activation and from undergoing apoptosis even when NF-κB is blocked. A1 and A20 both products of anti-apoptotic genes reveal a dual function. They inhibit NF-κB activation and protect from apoptosis.

Bcl-2 is induced by FGF-2 and VEGF inhibits endothelial cell apoptosis induced by tumor necrosis factor-alpha. Indeed, a complex balance exists between growth and death signals, and some factors previously named growth factors, must be better considered as survival signals (anti-apoptotic factors) than as true proliferative factors.

Other genes are believed to be protective. IAP (inhibitor of apoptosis protein) is one of the NF-κB regulated genes that operates to prevent programmed cell death of EC in inflammation and can be induced in EC. An inducible form of heme oxygenase can be upregulated in human EC by TNF and IL-1 and may serve a protective function against oxidant injury [105]. TNF and IL-1 are thus potent regulators of vascular gene expression and are paradoxically able to give to the same target cell a cell survival signal in addition to a death signal.

Bach et al. [92] showed that organ xenografts under certain circumstances may survive in the presence of anti-graft antibodies and complement, and found that the endothelial cells (ECs) in hamsterhearts, that accommodate themselves in rats, express certain genes, such as A20 and bcl-2, while hearts that are rejected do not express these genes. A20 and Bcl-2 in vitro protected ECs from apoptosis and prevented upregulation in those cells of proinflammatory genes such as cytokines, procoagulant and adhesion molecules. In addition, vessels of rejected hearts showed florid transplant arteriosclerosis whereas those of accommodated hearts did not. Accommodated xenografts had an ongoing T helper cell type 2 (Th2) cytokine immune response, whereas the rejected grafts have a Th1 response.

Vascular growth

Endothelial regeneration

Under normal circumstances, the endothelium is a stable population of cells, with cell replication rates falling rapidly after birth to a very low overall rate around 0.6% at adult ages [106]. Replication may however occur, focused on small clusters and associated with areas of cell death, and is followed by migration of adjacent endothelial cells and platelet adherence [107, 108].

In contrast to the localized cell death and regeneration, endothelial injury may result in large areas of dead cells, with or without denuding and exposure of the underlying subendothelial matrix. The most studied model of endothelial injury without denuding and associated with increased turnover is the response to endotoxin [109]. The death of endothelial macrovascular cells results in early bulging (within few hours) of endothelial cells into the lumen, followed by desquamation of large areas after 24 hours with little if any denuding. Indeed, endothelial turnover is increased in endotoxin-treated animals. Increased turnover is also seen in areas of increased haemodynamic stress.

Some other endothelial injuries involve exposure of the underlying matrix and result mostly from physical aggression (angioplasty). In most of these cases, endothelial denuded areas are rapidly covered with a sparse monolayer of platelets with occasional leucocyte recruitment, especially monocytes [110].

The process of endothelial regeneration involves initial spreading of adjacent cells, followed by cell migration and replication. When the endothelial monolayer is reestablished, the proliferative endothelial cells come back to quiescence to recover a functional phenotype. DNA synthesis and mitosis are evident at the edges of the wound by 12-24 hours.

The signals responsible for endothelial regeneration are complexes and include mechanical factors such as loss of cell-cell contact as well as local growth factors. Endothelial cell growth is known to be inhibited by cell-cell contact and the loss of cell contacts between dying and adjacent cells is likely to allow the cells to undergo proliferation. Growth factors that are releaed by wounded cells in the vascular wall and present in subendothelial matrix play a major role in endothelial regeneration. In focal endothelial injury, cells are liberated from contact inhibition and under the influence of FGF-2 present in the subendothelial matrix and are released by endothelial dead cells and wounded cells. Adjacent endothelial cells thus migrate, displace and phagocyte (in the case of apoptosis) the dying cells, and proliferate: beginning DNA synthesis and cell division. Frequently, a few platelets and leukocytes are recruited and participate in endothelial reconstruction in at least two ways: first physically, a process of fusion between cells (platelets and andothelial cells) may occur; secondly, growth factors released by these cells sustain the endothelial regeneration (especially, thrombin released by platelets and EGF/IGF-1 released by monocytes/ macrophages). In this situation, the role of FGF-2 is predominant.

In denuding injury, the situation seems to involve a complex set of intercellular cooperation of various growth factors. FGF-2 and VEGF, derived from the vessel wall, act in concert with less potent mitogenic factors such as EGF, IGF, G/GM-CSF. The final stage of re-endothelialization involves inhibition of cell growth and recovery of a quiescent phenotype. A regulatory loop between NO and VEGF has been recently described, where NO released by renerated endothelial cells inhibits VEGF synthesis by local cells. Confluence of endothelial cells is associated by growth inhibition resuling both from cell-cell contacts and growth inhibitory factors actively released by endotelial cells. These inhibitory factors involve cytokines: TNFα and IL-1, growth factors such as TGFβ, chemokines such as PF4 and IP-10, and integrins.

Angiogenesis

The formation of new capillary blood vessels is an important event during inflammation and wound repair. Angiogenesis is driven by proangiogenic cytokines, growth factors, and is tempered by a variety of inhibitors of neovascularisation (see chapter on *Angionesis*).

Macrophages have been suggested to be essential cell mediators during repair phase of inflammation, being sources of cytokines and growth factors, of metalloproteases, their inhibitors and matrix components, but also being "house keepers" able to phagocyte apoptotic bodies of the cellular environnment.

A number of endothelial integrins, primarily β_1 and β_3 integrins have been implicated in angiogenesis. The integrin $\alpha_V\beta_3$ has been reported to be necessary to neovascularization of wounds and promotes endothelial cords in vitro. Cytokine-dependent pathways of angiogenesis may depend on aV subunit: angiogenesis induced by FGF-2 and TNFα depend on $\alpha_V\beta_3$, while that initiated by VEGF and TGFa depend on avb5. Interactions of endothelial cells with other adhesion molecules may also play a role during angiogenesis. For example, the L-selectin ligand CD34 show high EC expression in developing tissues as well as in tumors and during wound healing. Moreover, soluble endothelial adhesion molecules may act as angiogenic factors. Indeed, soluble E-selection and soluble VCAM-1 are shed from the cell surface upon cytokine activation and have been shown to trigger directly neovascularization by binding to their respective EC ligands and recruit adjacent ECs.

Recent studies on the proliferative capacity of human surgical wound fluids revealed that the initial angiogenic stimulus was supplied by FGF-2, followed by a more prolonged angiogenic stimulus due to VEGF. VEGF derived mainly from fibroblasts and macrophages present at wound sites. New vessel formation in healing wounds follows a special time course with newly formed vessels first evident 2 to 3 days after injury with maximal presence at 1 to 2 weeks. This time course is that followed by the upregulation of VEGF.

FGF-2 and VEGF may represent thus 2 components of an angiogenic cascade. FGF-2 may initiate angiogenesis.It is preformed and stored in cells and can be rapidly released after stimulation by factors such as thrombin; its target cells are numerous (many cell types carry FGF receptors) and its actions include endothelial cell proliferation and migration, fibroblast migration, collagen deposition. VEGF induction may follow FGF action and support capillary formation, VEGF receptors being almost exclusive to endothelium. One other interesting possible mediator of VEGF activation would be local hypoxia of wounds, since VEGF is known to be induced by hypoxia and since angiogenic activity in wounds varies inversely with wound oxygen levels.

The mechanisms down-regulating angiogenic activity when neovascularization has restored wound perfusion, are less clear. However, a coordination between eNOS activity and VEGF activity may play a role, as suggested by Tsurumi et al. [111] who showed that NO secreted by a restored endothelium down-regulated VEGF expression following arterial injury. VEGF has been shown to upregulate expression of both ecNOS and COX-1 in EC.

ACUTE AND CHRONIC INFLAMMATORY VASCULAR DISEASES

Septic shock

Septic shock is a complication of sepsis in almost half of these patients with a mortality rate ranging between 40% and 60% despite advances in cardiovascular support and antibiotic therapy. The sepsis syndrome identifies a group of patients with systemic response to infection: fever or hypothermia, tachycardia, tachypnea, leukocytosis or neutropenia, thrombocytopenia, disseminated intravascular coagulation, hyperglycemia. Shock is identified by the development of hypotension, defined as a decrease in systolic arterial pressure below 90 mm Hg in normotensive patients or a decrease by more than 40 mm Hg in hypertensive patients.

The systemic response to infection including hemodynamic alterations is expressed by activated cells of the host, especially leukocytes and vascular cells, which play a major role in amplifying the release of soluble mediators, namely IL-1 and TNF. The major role of TNF in the etiology of septic shock has been supported by studies showing that TNF-specific antibodies can prevent death after subsequent administration of lethal doses of endotoxin. Elevated serum TNF concentrations have been found in patients with Gram-negative bacteremia or in healthy individuals after infusion with endotoxin. However, TNF merely interacts with a multitude of other cytokines. In septic patients, shock is correlated with elevated LPS, TNF and IL-1 concentrations, and monocytes from these septic patients are able to spontaneously produce cytokines. In experimental animal models and in humans, LPS-administration induces successive secretions of cytokines with a TNF peak at 1.5 h followed by an IL-1 peak at 3 h, and IL-6 beginning to be secreted after 6h in parallel to

IL-8. During lethal bacteremia anti-TNF antibodies both prevent from lethality and markedly decrease IL-1 and IL-6 appearance.

Binding activity for NF-κB consensus probes was recently studied in nuclear extracts from peripheral blood mononuclear cells of septic patients, and nonsurvivors could be distinguished from survivors by an increase in NF-κB binding activity; this increase in NF-κB binding activity was comparable to APACHE II gravity score as a predictor of outcome [112]. The same authors selected to prevent NF-κB activation in the mouse LPS model, with intravenous somatic gene transfer with a plasmid overexpressing I kappa B and observed that endothelial cells and monocytes/macrophages were the major target cells for this transfer. Somatic gene transfer with I-κB given before LPS reduced NF-κB binding activity, reduced Tissue factor expression and activation of the plasmatic coagulation system and markedly increased survival. Together these data suggest that the expression by endothelial cells of the balance between NF-κB and I-κB might be an critical event in clinical sepsis.

Tissue factor (TF) is the principal initiator of coagulation in vivo and is closely tied to the host response in sepsis, being a cause of disseminated intravascular coagulation. Several experimental studies report a protective effect of neutralizing anti-tissue factor antibodies in models of sepsis and there is a relationship between TNFα and TF expression in vivo since neutralizing TNF reduces activation of coagulation in vivo. TNF has been recently shown to induce fibrin formation in capillaries via TF induction [113].

Assessment of endothelial cell adhesion molecule expression in vivo is rather difficult. However, the use of radiolabeled monoclonal antibodies against adhesion molecules has allowed us to describe this expression in different vascular beds of rats and mice subjected to endotoxin shock [63]. In response to LPS, P-selectin expression is biphasic, with a low and brief peak at 30 mns as a result of Webel-Palade bodie mobilization and then an increase up to 4h with a decline to a steady-state at 24h resulting from de novo synthesis. E-selectin expression increases within 2h with a peak at 3h and a return to baseline values by 24h. ICAM-1 reaches peak values by 5 h and remained elevated up to 24h while VCAM-1 reaches a plateau a little later and remains elevated over a period of 48h. Presence of soluble adhesion molecules has been reported in plasma of septic patients. In support of the implication of adhesive molecules in shock, antibodies anti-integrin CD18 protect animals from hemorragic shock injury.

Mice and rats can acutely produce NO in response to systemic administration of LPS as well as during severe infections with both Gram negative and Gram positive bacteria. Nitrite/nitrate levels are found elevated in plasma and urine of septic animals. Administration of LPS to rodents produces widespread tissue induction of high levels of iNOS activity: in lungs, heart, kidney, myocardium, aorta and spleen. iNOS expression in the brain was not detectable. Studies on in vivo iNOS immunochemical localization indicate that macrophages are the major site of this expression while in vivo iNOS protein signal is present but weaker in hepatocytes, myocytes, vascular cells, mesangial and epithelial cells. Numerous studies of acute sepsis in rodents now indicate a first minor phase of early activation of the endothelial NOS followed by the dominant and delayed induction of vascular iNOS in response to LPS [114]. Observations with iNOS knocked-out mice established the crucial role of NO as a key agent as LPS-induced death. iNOS mutant mice were more resistant to LPS-induced death than wild-type mice. IL-2 therapy of advanced cancer was the first observation that evidenced the existence of an inducible L-arginine-NO pathway expressed in vivo in humans [115]. A 9-fold increase in plasma levels of nitrate has been reported in patients undergoing immunotherapy with IL-2 and LAK, which was correlated toxic hemodynamic side effects [116]. NO production assessed by plasma nitrate measurement in septic patients were significantly higher (about 6-fold) than in trauma patients [117]. Low systemic vascular resistance and high endotoxin levels were associated with high NO2/NO3 values [117].

Recent studies obtained evidence of the involvement of PARS in the vascular failure of septic shock [82]. PARS inhibition in vivo preserved ex vivo endothelial dysfunction (reduced endothelium-dependent vasodilation) [84], and contractility of aortas from endotoxic rats and improved survival in lethal murine endotoxic shock [118].

Studies on sepsis describe apoptosis in three main cell types: hepatocytes, T lymphocytes in the thymus, and endothelial cells. Indeed, the generalized Shwartzman reaction induced in mice by two consecutive injections of lipopolysaccharide is associated with positive staining by the in situ specific labeling of fragmented DNA in endothelial cells of various organs. A recent important study by Haimovitz-Friedman et al. [119] provides evidence that endotoxic shock in mice results from disseminated endothelial apoptosis. Endothelial apoptosis occurs in different organs such as intestine, lung and thymus, and preceedes nonendothelial tissue damage. This is induced by TNF, as suggested by the protection afforded by TNF-binding protein, and mediated the sphingomyelin pathway. Indeed, LPS or TNFα administration is followed within 1h by tissue generation of the pro-apoptotic lipid ceramide, and acid sphingomyelinase knockout mice are protected against endothelial apoptosis and animal death. It is noteworthy that endothelial apoptosis preceeds the tissue inflammatory response characterized by leukocyte infiltration. Moreover, the same authors demonstrate that exogenous administration of FGF-2 simultaneous to LPS protects mice from both EC death and animal death.

Ischemia-reperfusion

The mechanisms of acute organ damage following ichaemia-reperfusion are thought to involve immediate cellular damage caused by reactive oxygen species (ROS), induced cellular responses such as TNF, and subacute leucocyte recruitment. Many studies indicate that oxygen free radical formation initiates the cascade of events involved in ischemia/reperfusion injury, including cellular necrosis and delayed apoptosis, DNA damage, leucocyte activation and infiltration, cytokine and chemokine release [120]. DNA binding activities of NF-κB and AP-1 are increased during the first hour of the reperfusion phase of injury, and mitochondrial superoxide dismutase gene transfers to the liver before ischemia-reperfusion reduces associated-redox activation of AP-1 and NF-κB and subsequent liver damage [121].

In the model of cerebral ischaemia, TNF, IL-1 and IL-6 expressions have been demonstrated both in the peripheral blood as in the brain parenchyma [122], and IL-6 plasma levels seem to correlate with the extent of tissue damage. TGFβ is sometimes described surrounding the injured tissue, with prolonged expression. TGFβ seems rather to play a protective role in several regional models of ischaemia-reperfusion, in the myocardium, intestine and brain [123]. In a model of myocardial ischemia, TNF is released and impairs endothelium-dependent vasodilation [94]. Inhibition of selectins by the treatment with antibodies against P-selectin or L-selectin or with soluble ligands for these selectins protects against necrosis in a model of myocardial ischaemia. In mice lacking ICAM-1 the brain damage is less severe than in normal mice after occlusion of the middle cerebral artery. Soluble isoforms of adhesion molecules, shed from the surfaces of activated cells, have been detected in serums of patients with infarcts or acute stroke, especially soluble VCAM-1.

Neutrophils have a well-etablished role during ischaemia-reperfusion injury and their role has been best demonstrated by studies showing the beneficial effects of neutrophil depletion using CD11b/CD18 monoclonal antibodies [124]. Injury to vascular endothelium is a critical event, and is mediated by oxidants and serine proteinases released by activated neutrophils [125]. However, recent findings indicate that activated CD4+ T lymphocytes may preceede neutrophil recruitment [126]. Indeed, CD4+ T cell influx to the liver after ischaemia occurs within the first hour or reperfusion and CD4+ T depletion is capable of reducing both the extent of injury and neutrophil infiltration during the subacute of ischaemia/reperfusion liver damage [126].

In a model of cerebral ischemia, the expression of MMP9 increased first in endothelial cells and neutrophils (after 24h of focal ischemia), within and in the periphery of the infarct, and later on (after 5 days) in macrophages [127]. Furthermore, the selective inhibition of MMP9 activity significantly reduced the infarct size.

The role of PARS in ischaemia-reperfusion injury is now well documented. Neural damage following stroke is related to a complex chain of events, including oxyradicals, NO, peroxynitrite, and excitatory neurotransmitters. In PARS (-/-) KO mice, a markedly reduced infarct volume is observed after a transient middle cerebral artery occlusion, associated with the absence of ADP-ribose formation and the preservation of NAD+ levels, in contrast to what occur in wild type mice [128]. Interestingly, cerebral cortical cultures from PARS(-/-) mice are completely resistant to neurotoxicity induced either by NMDA or NO donors or by oxygen and glucose deprivation. In addition, te 80% reduction in infarct volume of PARP-/- mice exceeds protection reported with any known treatment. PARS inhibition also prevents neural damage following vascular strokes in rats. Eliasson et al. [128] propose the following sequence of events leading to cell death during cerebral ischaemia:

"ischaemia following cerebral vessel occlusion reduces the resting membrane potential of glia and neurons in the brain. Potassium leaks out of cells and depolarizes neurons leading to a massive release of glutamate. Acting via NMDA receptors, glutamate triggers a release of NO, which combines with superoxide to form peroxynitrite. Peroxynitrite damages DNA and its fragments activate PARS. Massive activation of PARS depletes NAD+ via ADP-ribose polymer formation. ATP is depleted in an effort to provide new NAD+, leading to energy depletion and necrotic cell death."

The activation of PARS also play a role in the development of reperfusion injury in the heart. PARS inhibitors given immediately before the reperfusion of the ischaemic myocardium dramatically reduce the infarct size [129]. The cardioprotective effects of PARS inhibitors include the histologic profile, the metabolic status, the creatine phosphokinase activity and the myeloperoxidase activity associated with a reduced infiltration of neutrophils into the reperfusion myocardium. Moreover, a marked reduction in peroxynitrite production caused by PARS inhibtion suggests a self-amplifying circle. The role of PARS activation in ischemia/reperfusion is not confined to the brain or the heart since in other animal studies, the inhibition of PARS also provides protection in the reperfused gut, skeletal muscle and retina.

Hypoxic-ischemic injury in many organs results in both necrotic and apoptotic cell death [86]. In vitro hypoxia/reperfusion injury to endothelial cells and neurons induces apoptosis. Olivetti et al. showed in humans that myocardial infarction is associated with activation of programmed myocyte cell death in the surviving portion of the heart [130], and Ikeda et al. observed in the rat that apoptosis is a major mode of cell death in a model of ischemia-reperfusion injury to the small intestine [131]. In addition, inhibition of apoptosis has been shown to be an effective neuroprotective strategy in different cerebral ischemia models [132].

In reponse to hypoxia/ischaemia, vascular growth factors can be upregulated in cells of the injured tissue: especially, in preexisting vascular cells and macrophages. For example, the expression of VEGF and its two tyrosine kinase receptors (flk-1, flt-1) is increased severalfold in the ischemic heart, with most of the increase of VEGF mRNA levels occuring in myocytes and macrophages of the myocardium. In parallel, the extracellular matrix is modified and the expression of the growth factor specific receptors is increased on endothelial cells. Indeed, reduction in oxygen tension promotes enhanced expression of VEGF, FGFs, PDGF and their receptors, as well as endothelial cell growth.

Kaposi sarcoma

Kaposi's sarcoma (KS) is a proliferative disease of vascular origin frequent and aggressive in HIV-patients. KS is characterized by spindle cell proliferation, neoangiogenesis with endothelial cell activation, inflammatory cell infiltration with macrophages and lymphocytes, and edema. A new herpesvirus, termed KS-associated virus or human herpesvirus-8 (HHV-8) has been identified in KS lesions and may have a determinant role in KS pathogenesis. HHV-8 may act as an inductive factor of KS. Indeed, AIDS-KS tissue samples are positive for human herpesvirus 8 sequences [133], found especially in EC and KS spindle cells [134]. HHV-8 encodes for several genes which products are homologs of human cytokines: IL-6, chemokines (MIP), cytokine receptors (IL-8R) and antiapoptotic genes (bcl-2). Most of these viral gene products have been shown to be functional and may participate in KS transformation. Extracellular HIV-1-Tat protein released by infected cells may act as a progression factor. Besides the crucial role of viruses, KS seems to be a cytokine-mediated vascular disease (for review see [135]). Since Gallo et al. findings in 1988 showing that conditioned media from HTLV-II infected CD4+ exerted important growth stimulation for KS cells, several mitogens have been identified including FGF-2, VEGF, OSM, IL-6, scatter factor [135]. Both FGF-2 and VEGF play a key role in KS proliferation. Spindle cells of primary KS lesions and KS-derived spindle cell cultures coexpress high levels of FGF-2 and VEGF. TNF, IL-1 and IFNγ induce production and release of FGF-2, which stimulates angiogenesis, endothelial cell growth and spindle cell growth in an autocrine manner. KS cells also show synthesis of VEGF isoforms that are mitogenic to endothelial cells but not to KS cells, suggesting a prevailing paracrine effect of this cytokine partially due to the level of expression of flt-1 that is down-regulated in KS cells as compared with endothelial cells. In addition, VEGF and FGF-2 synergize to induce vascular permeability and edema.

KS spindle cell growth and spread have been reported to be modulated also by HIV gene product Tat, which acts in synergy with cytokines. Moreover, HIV-1 Tat has been shown recently to act like a cytokine and binds to the Flk-1/KDR receptor for VEGF-A, which is expressed by KS cells [136]. KS cells also overexpress several integrins that facilitate cellular interactions with extracellular matrix components and subsequent tumor angiogenesis, by influencing the process of apoptosis. Indeed, there is evidence that several cell survival gene products are expressed by both EC and KS cells and are implicated in the proliferative features of this vascular tumor. Both cultured ECs and Kaposi's sarcoma tumor cells express high levels of Bcl-xL and Bcl-2. In addition, Pammer et al. report the expression of CD40 on the spindle tumor cells as well as on the adjacent EC. This latter expression may lead to an anti-apoptotic response, as CD40 signaling has been shown to induce various cell survival gene products, such as Bcl-xL.

All these impressive findings suggest that tumorigenesis and angiogenesis in KS are regulated by a complex balance between pro- and anti-apoptotic signals under the influence of various cytokines and growth factors expressed within the tumor.

REFERENCES

1. Libby P, Ordovas JM, Birinyi LK, Auger KR, Dinarello CA. Inducible interleukin-1 gene expression in human vascular smooth muscle cells. J Clin Invest 1986;78:1432-8.

2. Libby P, Ordovas JM, Auger KR, Robbins AH, Birinyi LK, Dinarello CA. Endotoxin and tumor necrosis factor induce interleukin-1 gene expression in adult human vascular endothelial cells. Am J Pathol 1986;124:179-85.

3. Loppnow H, Libby P. Comparative analysis of cytokine induction in human vascular endothelial and smooth muscle cells. Lymphokine Res 1989;8:293-9.

4. Warner SJ, Auger KR, Libby P. Interleukin 1 induces interleukin 1. II. Recombinant human interleukin 1 induces interleukin 1 production by adult human vascular endothelial cells. J Immunol 1987;139:1911-7.

5. Warner SJ, Auger KR, Libby P. Human interleukin 1 induces interleukin 1 gene expression in human vascular smooth muscle cells. J Exp Med 1987;165:1316-31.

6. Warner SJ, Libby P. Human vascular smooth muscle cells. Target and source of Tumor Necrosis Factor. J Immunol 1989;142:100-109.

7. Schonbeck U, Herzberg M, Petersen A, et al. Human vascular smooth muscle cells express interleukin-1 beta-converting enzyme (ICE), but inhibit processing of the interleukin-1 beta precursor by ICE. J. Exp. Med. 1997;185:1287-1294.

8. Mantovani A, Bussolino F, Introna M. Cytokine regulation of endothelial cell function: From molecular level to the bedside. Immunol. Today 1997;18:231-240.

9. Oppenheimer-Marks N, Brezinschek RI, Mohamadzadeh M, Vita R, Lipsky PE. Interleukin 15 is produced by endothelial cells and increases the transendothelial migration of T cells In vitro and in the SCID mouse- human rheumatoid arthritis model In vivo. J Clin Invest 1998;101:1261-72.

10. Tilg H, Dinarello CA, Mier JW. IL-6 and APPs: anti-inflammatory and immunosuppressive mediators. Immunol Today 1997;18:428-32.

11. Vora M, Yssel H, Devries JE, Karasek MA. Antigen presentation by human dermal microvascular endothelial cells - immunoregulatory effect of IFN-gamma and IL-10. J. Immunol. 1994;152:5734-5741.

12. Introna M, Mantovani A. Early activation signals in endothelial cells. Stimulation by cytokines. Arterioscler Thromb Vasc Biol 1997;17:423-128.

13. Bradley JR, Thiru S, Pober JS. Hydrogen peroxide-induced endothelial retraction is accompanied by a loss of the normal spatial organization of endothelial cell adhesion molecules. Am. J. Pathol. 1995;147:627-641.

14. Modur V, Li Y, Zimmerman GA, Prescott SM, McIntyre TM. Retrograde inflammatory signaling from neutrophils to endothelial cells by soluble interleukin-6 receptor alpha. J Clin Invest 1997;100:2752-6.

15. Nathan C, Sporn M. Cytokines in context. J. Cell Biol. 1991;113:981-986.

16. Pugin J, Ulevitch RJ, Tobias PS. A critical role for monocytes and CD14 in endotoxin induced endothelial cell activation. J. Exp. Med. 1993;178:2193-2200.

17. Pugin J, Ulevitch RJ, Tobias PS. Tumor necrosis factor-alpha and interleukin-1 beta mediate human endothelial cell activation in blood at low endotoxin concentrations. Journal of Inflammation 1995;45:49-55.

18. Sterpetti AV, Cucina A, Dangelo LS, Cardillo B, Cavallaro A. Shear stress modulates the proliferation rate, protein synthesis, and mitogenic activity of arterial smooth muscle cells. Surgery 1993;113:691-699.

19. Lee S, Felts KA, Parry GCN, Armacost LM, Cobb RR. Inhibition of 5-lipoxygenase blocks IL-1 beta-induced vascular adhesion molecule-1 gene expression in human endothelial cells. J. Immunol. 1997;158:3401-3407.

20. De Caterina R, Libby P, Peng HB, et al. Nitric oxide decreases cytokine-induced endothelial activation. Nitric oxide selectively reduces endothelial expression of adhesion molecules and proinflammatory cytokines. J Clin Invest 1995;96:60-8.

21. Spiecker M, Peng HB, Liao JK. Inhibition of endothelial vascular cell adhesion molecule-1 expression by nitric oxide involves the induction and nuclear translocation of IkappaBalpha. J Biol Chem 1997;272:30969-74.

22. Pendurthi UR, Williams JT, Rao LV. Acidic and basic fibroblast growth factors suppress transcriptional activation of tissue factor and other inflammatory genes in endothelial cells. Arterioscler Thromb Vasc Biol 1997;17:940-6.

23. Mach F, Schonbeck U, Sukhova GK, et al. Functional CD40 ligand is expressed on human vascular endothelial cells, smooth muscle cells, and macrophages: implications for CD40-CD40 ligand signaling in atherosclerosis. Proc Natl Acad Sci U S A 1997;94:1931-6.

24. Karmann K, Min W, Fanslow WC, Pober JS. Activation and homologous desensitization of human endothelial cells by CD40 ligand, tumor necrosis factor, and interleukin 1. J. Exp. Med. 1996;184:173-182.

25. Schonbeck U, Mach F, Bonnefoy JY, Loppnow H, Flad HD, Libby P. Ligation of CD40 activates interleukin 1 beta-converting enzyme (caspase-1) activity in vascular smooth muscle and endothelial cells and promotes elaboration of active interleukin 1 beta. J. Biol. Chem. 1997;272:19569-19574.

26. Henn V, Slupsky JR, Gräfe M, et al. CD40 ligand on activated platelets triggers an inflammatory reaction of endothelial cells. Nature 1998;391:591-594.

27. Laman JD, de Smet BJGL, Schoeneveld A, van Meurs M. CD40-CD40L interactions in atheroclerosis. Immunol Today 1997;18:272-277.

28. Palmer-Crocker RL, Hughes CWC, Pober JS. IL-4 and IL-13 activate the JAK2 tyrosine kinase and STAT 6 in cultured human vascular endothelial cells through a common pathway that does not involve the gc chain. J Clin Invest 1996;98:604-609.

29. Bernard C, Merval R, Lebret M, et al. Activation of human vascular smooth muscle cells by oncostatin M. Synergy with IL-1. Circ Res 1998:in press.

30. Hirota H, Yoshida K, Taga T, Kishimoto T. gp130 signaling pathways: recent advances and implications for cardiovascular disease. Tr Cardiovasc Med 1996;6:109-115.

31. Parry GCN, Mackman N. NF-kB mediated transcription in human monocytic cells and endothelial cells. Trends Cardiovasc Med 1998;8:138-142.

32. Collins T, Read MA, Neish AS, Whitley MZ, Thanos D, Maniatis T. Transcriptional regulation of endothelial cell adhesion molecules: NF-kappa B and cytokine-inducible enhancers. Faseb J. 1995;9:899-909.

33. Jiang C, Ting AT, Seed B. PPAR-gamma agonists inhibit production of monocyte inflammatory cytokines. Nature 1998;391:82-6.

34. Ricote M, Li AC, Willson TM, Kelly CJ, Glass CK. The peroxisome proliferator-activated receptor-gamma is a negative regulator of macrophage activation. Nature 1998;391:79-82.

35. Marx N, Sukhova G, Murphy C, Libby P, Plutzky J. Macrophages in human atheroma contain PPARgamma: differentiation- dependent peroxisomal proliferator-activated receptor gamma(PPARgamma) expression and reduction of MMP-9 activity through PPARgamma activation in mononuclear phagocytes in vitro. Am J Pathol 1998;153:17-23.

36. Staels B, Koenig W, Habib A, et al. Activation of human aortic smooth-muscle cells is inhibited by PPARa but not by PPARg activators. Nature 1998;393:790-793.

37. Yao LB, Pan JL, Setiadi H, Patel KD, Mcever RP. Interleukin 4 or oncostatin M induces a prolonged increase in P-selectin mRNA and protein in human endothelial cells. J. Exp. Med. 1996;184:81-92.

38. Mayadas TN, Johnson RC, Rayburn H, Hynes RO, Wagner DD. Leukocyte rolling and extravasation are severely compromised in P selectin-deficient mice. Cell 1993;74:541-54.

39. Labow MA, Norton CR, Rumberger JM, et al. Characterization of E-selectin-deficient mice: demonstration of overlapping function of the endothelial selectins. Immunity 1994;1:709-20.

40. Furie MB, Randolph GJ. Chemokines and tissue injury. Am J Pathol 1995;146:1287-301.

41. Taub DD. Chemokine-leukocyte interactions. The voodoo that they do so well. Cytokine Growth Factor Rev 1996;7:355-76.

42. Wang JM, Shen W, Su S. Chemokines and their role in cardiovascular diseases. Trends Cardiovasc Med 1998;8:169-174.

43. Luster AD. Chemokines-chemotactic cytokines that mediate inflammation. N Engl J Med 1998;338:436-445.

44. Gupta SK, Lysko PG, Pillarisetti K, Ohlstein E, Stadel JM. Chemokine receptors in human endothelial cells. Functional expression of CXCR4 and its transcriptional regulation by inflammatory cytokines. J Biol Chem 1998;273:4282-7.

45. Hayes IM, Jordan NJ, Towers S, et al. Human vascular smooth muscle cells express receptors for CC chemokines. Arterioscler Thromb Vasc Biol 1998;18:397-403.

46. Goebeler M, Yoshimura T, Toksoy A, Ritter U, Brocker EB, Gillitzer R. The chemokine repertoire of human dermal microvascular endothelial cells and its regulation by inflammatory cytokines. J Invest Dermatol 1997;108:445-51.

47. Tenaglia AN, A.J. B, Wilkins RG, et al. Levels of expression of P-selectin, E-selectin, and intercellular adhesion molecule-1 in coronary atherectomy specimens from patients with stable and unstable angina pectoris. Am J Cardiol 1997;79:742-747.

48. Campbell JJ, Hedrick J, Zlotnik A, Siani MA, Thompson DA, Butcher EC. Chemokines and the arrest of lymphocytes rolling under flow conditions. Science 1998;279:381-4.

49. Sgadari C, Angiolillo AL, Tosato G. Inhibition of angiogenesis by interleukin-12 is mediated by the interferon-inducible protein 10. Blood 1996;87:3877-3882.

50. Boring L, Gosling J, Cleary M, Charo IF. Decreased lesion formation in CCR2-/- mice reveals a role for chemokines in the initiation of atherosclerosis. Nature 1998;394:894-7.

51. Bais C, Santomasso B, Coso O, et al. G-protein-coupled receptor of Kaposi's sarcoma-associated herpesvirus is a viral oncogene and angiogenesis activator. Nature 1998;391:86-9.

52. Massova I, Kotra LP, Fridman R, Mobashery S. Matrix metalloproteinases: structures, evolution, and diversification. Faseb J 1998;12:1075-95.

53. Schonbeck U, Mach F, Sukhova GK, et al. Regulation of matrix metalloproteinase expression in human vascular smooth muscle cells by T lymphocytes - A role for CD40 signaling in plaque rupture? Circ. Res. 1997;81:448-454.

54. Yang JH, Briggs WH, Libby P, Lee RT. Small mechanical strains selectively suppress matrix metalloproteinase 1 expression by human vascular smooth muscle cells. J Biol Chem 1998;273:6550-5.

55. Fabunmi RP, Sukhova GK, Sugiyama S, Libby P. Expression of tissue inhibitor of metalloproteinases-3 in human atheroma and regulation in lesion-associated cells: a potential protective mechanism in plaque stability. Circ Res 1998;83:270-8.

56. Pober JS, Cotran RS. Cytokines and endothelial cell biology. Physiol Rev 1990;70:427-451.

57. Cines DB, Pollak ES, Buck CA, et al. Endothelial cells in physiology and in the pathophysiology of vascular disorders. Blood 1998;91:3527-3561.

58. Harlan JM, Harker LA, Reidy MA, Gajdusek CM, Schwartz SM, Striker GE. Lipopolysaccharide-mediated bovine endothelial cell injury in vitro. Lab Invest 1983;48:269-74.

59. Young JS, Headrick JP, Berne RM. Endothelial-dependent and -independent responses in the thoracic aorta during endotoxic shock. Circ Shock 1991;35:25-30.

60. Robert R, Chapelain B, Jean T, Néliat G. Interleukin-1 impairs both vascular contraction and relaxation in rabbit isolated aorta. Biochem Biophys Res Commun 1992;182:733-739.

61. Kessler P, Bauersachs J, Busse R, Schini-Kerth VB. Inhibition of inducible nitric oxide synthase restores endothelium- dependent relaxations in proinflammatory mediator-induced blood vessels. Arterioscler Thromb Vasc Biol 1997;17:1746-55.

62. Gamble JR, Khewgoodall Y, Vadas MA. Transforming growth factor-beta inhibits E-selectin expression on human endothelial cells. J. Immunol. 1993;150:4494-4503.

63. Eppihimer MJ, Granger DN. Endothelial cell adhesion molecule expression in acutely inflamed tissues. In: Vinent JL, ed. Update in intensive care and emergency medicine. Berlin Heidelberg: Springer Verlag, 1997:53-62.

64. Gimbrone MA, Nagel T, Topper JN. Biomechanical activation: An emerging paradigm in endothelial adhesion biology. J. Clin. Invest. 1997;99:1809-1813.

65. Nagel T, Resnick N, Atkinson WJ, Dewey CF, Gimbrone MA. Shear stress selectively upregulates intercellular adhesion molecule-1 expression in cultured human vascular endothelial cells. J. Clin. Invest. 1994;94:885-891.

66. Lucas R, Lou JN, Morel DR, Ricou B, Suter PM, Grau GE. TNF receptors in the microvascular pathology of acute respiratory distress syndrome and cerebral malaria. J. Leukocyte Biol. 1997;61:551-558.

67. Beasley D, Cohen R, Levinsky N. Endotoxin inhibits contraction of vascular smooth muscle in vitro. Am J Physiol 1990;258:H1187-H1192.

68. Julou-Schaeffer G, Gray G, Fleming I, Schott C, Parrat J, Stoclet J. Loss of vascular responsiveness induced by endotoxin involves L-arginine pathway. Am J Physiol 1990; 259:H1038-H1043.

69. Auguet M, Lonchampt MO, Delaflotte S, Goulinschulz J, Chabrier PE, Braquet P. Induction of nitric oxide synthase by lipoteichoic acid from staphylococcus-aureus in vascular smooth muscle cells. Febs Lett. 1992;297:183-185.

70. McKenna TM. Recovery of vascular tissue contractile function during sustained endotoxin exposure. Am J Physiol 1992;263:H1628-H1631.

71. Beasley D, Cohen RA, Levinski NG. Interleukin 1 inhibits contraction of vascular smooth muscle. J Clin Invest 1989;83:331-335.

72. Beasley D, Schwartz JH, Brenner BM. Interleukin 1 induces prolonged L-arginine-dependent cyclic guanosine monophosphate and nitrite production in rat vascular smooth muscle cells. J Clin Invest 1991;87(2):602-8.

73. Bernard C, Szekely B, Philip I, Wollman E, Payen D, Tedgui A. Activated macrophages depress the contractility of rabbit carotids via an L-arginine:nitric oxide dependent effector mechanism. Connection with amplified cytokine release. J. Clin. Invest. 1992;89:851-860.

74. Busse R, Mülsch A. Induction of nitric oxide synthase by cytokines in vascular smooth muscle cells. FEBS Lett 1990;275:87-90.

75. Fleming I, Julouschaeffer G, Gray GA, Parratt JR, Stoclet JC. Evidence that an l-arginine/nitric oxide dependent elevation of tissue cyclic gmp content is involved in depression of vascular reactivity by endotoxin. Br. J. Pharmacol. 1991;103:1047-1052.

76. French JF, Lambert LE, Dage RC. Nitric oxide synthase inhibitors inhibit interleukin-1-beta-induced depression of vascular smooth muscle. J. Pharmacol. Exp. Therap. 1991;259:260-264.

77. Thorintrescases N, Hamilton CA, Reid JL, et al. Inducible L-arginine nitric oxide pathway in human internal mammary artery and saphenous vein. Am. J. Physiol. Heart Circ. Physiol. 1995;37:H1122-H1132.

78. Berrazueta JR, Salas E, Amado JA, Devega MJS, Poveda JJ. Induction of nitric oxide synthase in human mammary arteries in vitro. Eur. J. Pharmacol. 1994;251:303-305.

79. Beasley D, McGuiggin M. Interleukin-1 activates soluble guanylate cyclase in human vascular smooth muscle cells through a novel nitric oxide-independent pathway. J Exp Med 1994;179:71-80.

80. Geng YJ, Petersson AS, Wennmalm A, Hansson GK. Cytokine-induced expression of nitric oxide synthase results in nitrosylation of heme and nonheme iron proteins in vascular smooth muscle cells. Exp. Cell Res. 1994;214:418-428.

81. Geng Y, Hansson GK, Holme E. Interferon-gamma and tumor necrosis factor synergize to induce nitric oxide production and inhibit mitochondrial respiration in vascular smooth muscle cells. Circ. Res. 1992;71:1268-1276.

82. Szabo C, Dawson VL. Role of poly(ADP-ribose) synthetase in inflammation and ischaemia-reperfusion. Trends in Pharmacol Sci 1998;19:287-298.

83. Szabo C, Zingarelli B, Oconnor M, Salzman AL. DNA strand breakage, activation of poly(ADP-ribose) synthetase, and cellular energy depletion are involved in the cytotoxicity in macrophages and smooth muscle cells exposed to peroxynitrite. Proc. Natl. Acad. Sci. USA 1996;93:1753-1758.

84. Szabo C, Ohshima H. DNA damage induced by peroxynitrite: subsequent biological effects. Nitric Oxide 1997;1:373-85.

85. Schwartz SM, Bennett MR. Death by any other name. Am J Pathol 1995;147:229-234.

86. Majno G, Joris I. Apoptosis, oncosis, and necrosis. Am J Pathol 1995;146:3-15.

87. Haunstetter A, Izumo S. Apoptosis: basic mechanisms and implications for cardiovascular disease. Circ Res 1998;82:1111-1129.

88. Bennett MR, Evan GI, Schwartz SM. Apoptosis of human vascular smooth muscle cells derived from normal vessels and coronary atherosclerotic plaques. J Clin Invest 1995;95:2266-2274.

89. Bennett MR, Littlewood TD, Schwartz SM, Weissberg PL. Increased sensitivity of human vascular smooth muscle cells from atherosclerotic plaques to p53-mediated apoptosis. Circ Res 1997;81:591-599.

90. Barry OP, Pratico D, Savani RC, FitzGerald GA. Modulation of monocyte-endothelial cell interactions by platelet microparticles. J Clin Invest 1998;102:136-44.

91. Hebert MJ, Gullans SR, Mackenzie HS, Brady HR. Apoptosis of endothelial cells is associated with paracrine induction of adhesion molecules: evidence for an interleukin-1beta-dependent paracrine loop. Am J Pathol 1998;152:523-32.

92. Bach FH, Ferran C, Hechenleitner P, et al. Accommodation of vascularized xenografts: Expression of "protective genes" by donor endothelial cells in a host Th2 cytokine environment. Nature Med. 1997;3:196-204.

93. Bhagat K, Collier J, Vallance P. Local venous responses tea endotoxin in humans. Circulation 1996;94:490-497.

94. Lefer AM, Tsao P, Aoki N, Palladino MA, Jr. Mediation of cardioprotection by transforming growth factor-beta. Science 1990;249:61-4.

95. Gamble JR, Bradley S, Noack L, Vadas MA. TGF-beta and endothelial cells inhibit VCAM-1 expression on human vascular smooth muscle cells. Arteriosclerosis Thrombosis and Vascular Biology 1995;15:949-955.

96. Perrella MA, Patterson C, Tan L, et al. Suppression of interleukin-1 beta-induced nitric-oxide synthase promoter/enhancer activity by transforming growth factor-beta 1 in vascular smooth muscle cells - Evidence for mechanisms other than NF-kappa B. J. Biol. Chem. 1996;271(23):13776-13780.

97. Kitamura M, Suto T, Yokoo T, Shimizu F, Fine LG. Transforming growth factor-beta 1 is the predominant paracrine inhibitor of macrophage cytokine synthesis produced by glomerular mesangial cells. J. Immunol. 1996;156(8):2964-2971.

98. Radomski MW, Vallance P, Whitley G, Foxwell N, Moncada S. Platelet adhesion to human vascular endothelium is modulated by constitutive and cytokine induced nitric oxide. Cardiovasc. Res. 1993;27:1380-1382.

99. Peng HB, Rajavashisth TB, Libby P, Liao JK. Nitric oxide inhibits macrophage-colony stimulating factor gene transcription in vascular endothelial cells. J. Biol. Chem. 1995;270:17050-17055.

100. Shin WS, Hong YH, Peng HB, Decaterina R, Libby P, Liao JK. Nitric oxide attenuates vascular smooth muscle cell activation by interferon-gamma The role of constitutive NF-kappa B activity. J. Biol. Chem. 1996;271:11317-11324.

101. Sampath R, Kukielka GL, Smith CW, Eskin SG, Mcintire LV. Shear stress-mediated changes in the expression of leukocyte adhesion receptors on human umbilical vein endothelial cells in vitro. Ann. Biomed. Eng. 1995;23:247-256.

102. Dimmeler S, Haendeler J, Nehls M, Zeiher AM. Suppression of apoptosis by nitric oxide via inhibition of interleukin-1 beta-converting enzyme (ICE)-like and cysteine protease protein (CPP)-32-like proteases. J. Exp. Med. 1997;185:601-607.

103. Sata M, Perlman H, Muruve DA, et al. Fas ligand gene transfer to the vessel wall inhibits neointima formation and overrides the adenovirus-mediated T cell response. Proc Natl Acad Sci USA 1998;95:1213-1217.

104. Karsan A, Yee E, Kaushansky K, Harlan JM. Cloning of a human Bcl-2 homologue: Inflammatory cytokines induce human A1 in cultured endothelial cells. Blood 1996;87:3089-3096.

105. Terry CM, Clikeman JA, Hoidal JR, Callahan KS. Effect of tumor necrosis factor-a and interleukin-1a on heme oxygenase-1 expression in human endothelial cells. Am J Physiol 1998;274:H883-H891.

106. Brindle NPJ. Growth factors in endothelial regeneration. Cardiovasc. Res. 1993;27:1162-1172.

107. Hansson GK, Schwartz SM. Evidence for cell death in the vascular endothelium in vivo and in vitro. Am J Pathol 1983;112:278-86.

108. Hansson GK, Chao S, Schwartz SM, Reidy MA. Aortic endothelial cell death and replication in normal and lipopolysaccharide-treated rats. Am J Pathol 1985;121:123-7.

109. Reidy MA, Schwartz SM. Endothelial injury and regeneration. IV. Endotoxin: a nondenuding injury to aortic endothelium. Lab Invest 1983;48:25-34.

110. Stemerman MB, Spaet TH, Pitlick F, Cintron J, Lejnieks I, Tiell ML. Intimal healing. The pattern of reendothelialization and intimal thickening. Am J Pathol 1977;87:125-42.

111. Tsurumi Y, Murohara T, Krasinski K, et al. Reciprocal relation between VEGF and NO in the regulation of endothelial integrity. Nature Med 1997;3:879-886.

112. Böhrer H, Qiu F, Zimmermann T, et al. Role of NFkB in the mortality of sepsis. J Clin Invest 1997;100:972-985.

113. Zhang Y, Deng Y, Wendt T, et al. Intravenous somatic gene transfer with antisense tissue factor restores blood flow by reducing tumor necrosis factor-induced tissue factor expression and fibrin deposition in mouse meth-A sarcoma. J Clin Invest 1996;97:2213-24.

114. Szabo C, Salzman AL, Ischiropoulos H. Endotoxin triggers the expression of an inducible isoform of nitric oxide synthase and the formation of peroxynitrite in the rat aorta in vivo. Febs Lett. 1995;363:235-238.

115. Hibbs JB, Westenfelder C, Taintor R, et al. Evidence for cytokine-inducible nitric oxide synthesis from l-arginine in patients receiving interleukin-2 therapy. J. Clin. Invest. 1992;89:867-877.

116. Ochoa JB, Curti B, Peitzman AB, et al. Increased circulating nitrogen oxides after human tumor immunotherapy correlation with toxic hemodynamic changes. J. Nat. Cancer Inst. 1992;84:864-867.

117. Ochoa JB, Udekwu AO, Billiar TR, et al. Nitrogen oxide levels in patients after trauma and during sepsis. Ann. Surg. 1991;214:621-626.

118. Szabo C. The pathophysiological role of peroxynitrite in shock, inflammation, and ischemia reperfusion injury. Shock 1996;6:79-88.

119. Haimovitz-Friedman A, Cordon-Cardo C, Bayoumy S, et al. Lipopolysaccharide induces disseminated endothelial apoptosis requiring ceramide generation. J Exp Med 1997;186:1831-1841.

120. Abe JI, Berk BC. Reactive oxygen species as mediators of signal transduction in cardiovascular disease. Trands Cardiovasc Med 1998;8:59-64.

121. Zwacka RM, Zhou W, Zhang Y, et al. Redox gene therapy for ischaemia/reperfusion injury of the liver reduces AP1 and NFkB activation. Nature Med 1998;4:698-704.

122. Pantoni L, Sarti C, Inzitari D. Cytokines and cell adhesion molecules in cerebral ischemia: experimental bases and therapeutic perspectives. Arterioscler Thromb Vasc Biol 1998;18(4):503-13.

123. Gross V, Leser HG, Heinisch A, Scholmerich J. Inflammatory mediators and cytokines new aspects of the pathophysiology and assessment of severity of acute pancreatitis. Hepatogastroenterology 1993;40:522-530.

124. Bautista AP, Spolarics Z, Jaeschke H, Smith CW, Spitzer JJ. Monoclonal antibody against the CD18 adhesion molecule stimulates glucose uptake by the liver and hepatic nonparenchymal cells. Hepatology 1993;18:1026.

125. Varani J, Ward PA. Mechanisms of endothelial cell injury in acute inflammation. Shock 1994;2:311-319.

126. Zwacka RM, Zhang Y, Halldorson J, Schlossberg H, Dudus L, Engelhardt JE. CD4+ T-Lymphocytes mediates ischemia-reperfusion-induced inflammatory responses in mouse liver. J Clin Invest 1997;100:279-289.

127. Romanic AM, White RF, Arleth AJ, Ohlstein EH, Barone FC. Matrix metalloproteinase expression increases after cerebral focal ischaemia in rats: inhibition of matrix metalloproteinase-9 reduces infarct size. Stroke 1998;29:1020-1030.

128. Eliasson MJL, Sampei K, Mndir AS, et al. Poly(ADP-ribose) polymerase gene disruption renders mice resistant to cerebral ischaemia. Nature Med 1997;3:1089-1095.

129. Zingarelli B, Salzman AL, Szabo C. Gene disruption of poly(ADP-ribose) synthetase inhibits the expression of P-selectin and intercellular adhesion molecule-1 in myocardial ichaemia/reperfusion injury. Circ Res 1998;83:85-94.

130. Olivetti G, Quaini F, Sala R, et al. Acute myocardial infarction in humans is associated with activation of programmed myocyte cell death in the surviving portion of the heart. J Mol Cell Cardiol 1996;28:2005-2016.

131. Ikeda H, Suzuki Y, Suzuki M, et al. Apoptosis is a major mode of cell death caused by ischaemia and ischaemia/reperfusion injury to the rat intestinal epithelium. Gut 1998;42:530-537.

132. Silverstein FS. Can inhibition of apoptosis rescue ischemic brain. J Clin Invest 1998;101:1809-1810.

133. Chang Y, Cesarman E, Pessin MS, et al. Identification of herpesvirus-like DNA sequences in AIDS-associated Kaposi's sarcoma. Science 1994;266:1865-1869.

134. Boshoff C, Schulz TF, Kennedy MM, et al. Kaposi's sarcoma-associated herpesvirus infects endothelial and spindle cells. Nature Med. 1995;1:1274-1278.

135. Nickoloff BJ, Foreman KE. Charting a new course through the chaos of KS (Kaposi's sarcoma). Am J Pathol 1996;148:1323-9.

136. Albini, Soldi A, Giunciuglio R, et al. The angiogenesis induced by HIV-1 Tat protein is mediated by the Flk-1/KDR receptor on vascular endothelial cells. Nature Med. 1996;2:1371-1375.

10. APOPTOSIS IN NORMAL AND PATHOLOGICAL VESSELS

Ziad Mallat and Alain Tedgui
INSERM U141, Lariboisière Hospital, Paris, France

Vascular pathologists have long been intrigued by the observation of a form of active vascular cell death within the arterial wall in atherosclerosis [1, 2] or during normal aging [3] in the absence of overt necrosis. Rational explanations for these observations may now be at hand after the description by Kerr and colleagues in the early 1970s of a novel form of cell death distinct from necrosis, which they designated "apoptosis" (from the Greek word for falling) [4]. It has since been increasingly recognized that apoptosis may be involved in a number of critical events occurring during normal development and may play a key role in a wide variety of diseases [5].

In contrast to the cellular and organelle swelling and to the blebbing and membrane rupture associated with the classic form of induced cell death or "oncosis", apoptosis was initially described in cells that are programmed to die on the basis of characteristic morphology [4]. Cells undergoing apoptosis show cell shrinkage, chromatin margination and condensation with subsequent internucleosomal fragmentation of DNA, membrane redistribution of phospholipids [6] and budding (emission of pseudopodia), with maintenance of membrane integrity. The process tends to break off into apoptotic bodies that are recognized and rapidly engulfed by professional macrophages or adjacent neighboring cells without inducing an inflammatory response. However, cell corpses that escape phagocytosis and remain free undergo secondary necrosis [7].

INDUCERS OF APOPTOSIS IN THE VASCULAR WALL

Apoptosis can be initiated by *intrinsic signals* during normal development, for example. This occurs when cells activate an internal program of self-destruction (cell suicide, programmed cell death) in response to an internal clock, withdrawal of survival factors, changes in hemodynamic parameters or loss of contact. Every cell already contains all the components of the suicide machinery and is ready to engage in self-destruction unless it is actively signaled not to do so.

On the other hand, apoptosis can be initiated by a wide variety of *extrinsic signals* such as cytokines, hormones, oxidized lipids, chemotherapeutic, ionizing or viral agents [8]. In contrast to the more chronic and progressive loss of cells that occurs during normal development, apoptosis triggered by extrinsic signals is generally more acute and massive. Therefore, the capacity for removal of apoptotic cells may be overcome and secondary necrosis of unremoved apoptotic cells is frequent. This may lead to chronic accumulation of cellular debris with the potential for inducing inflammatory and/or auto-immune responses [9-11].

MOLECULAR MECHANISMS OF APOPTOSIS

Apoptosis is controlled by an ordered, stereotyped cascade of cellular events. Some death-commitment pathways are ubiquitous (for example, those involved in radiation-induced DNA damage) while others are present only in specific cells therefore ensuring their effective removal without the elimination of other cell types adjacent to them. It is the relative abundance of pro-apoptotic and anti-apoptotic molecules within the cell at a given time that will determine the cell's fate.

The Different Stages of the Apoptotic Process

The programmed cell death cascade can be divided into at least three main, functionally distinct, stages [12-14] (Figure 1):

1) The initiation or signalization phase during which cells receive the death-inducing signal. This may be accomplished by attachment of death-promoting molecules (i.e., Tumor Necrosis Factor (TNF)-α, Fas ligand) to death receptors (TNFR1, Fas) on the cell surface, with subsequent recruitment of death domain proteins (FADD, TRADD, RIP) required for caspase-8 activation (which initiates the lethal proteolytic cascade) [15]. The initiation phase may also be triggered by shortage of metabolic supply or removal of obligatory survival-promoting factors allowing an endogenous default death program to be turned on, or by subnecrotic damage due to ishemia or irradiation for example, with subsequent activation of the p53 gene and, eventually, stimulation of downstream cyclin-dependent kinase inhibitors (CDKI) [16, 17].

2) The control and effector phase during which activation of caspases occurs with loss of mitochondrial membrane potential [18]. Caspases, the mamalian CED-3 homologs, are a family of cysteine proteases with aspartate specificity that have been implicated in the transduction and execution of the apoptotic program [19]. Activation of caspase-8 (and subsequent activation of the caspase cascade) occurs following ligation of death-signal-transmitting receptors as described above. Additionally, other pathways may be activated sequentially or in parallel. Binding of cytochrome c (released from the mitochondrion) to apoptotic protease activating factor-1 (Apaf-1) allows Apaf-1 to bind and to activate caspase-3 (and also caspase-9 and -8) therefore propagating the caspase cascade (reviewed in [8] and [8bis]). Several second messengers may be involved in the intracellular transduction of the apoptotic signals and these may include ceramide generation from acid sphyngomyelinases, nitric oxide production and formation of peroxynitrite. Phosphorylation (or dephosphorylation) of different proteins and protein kinases are important events in the variety of transduction pathways and modulate the apoptotic process.

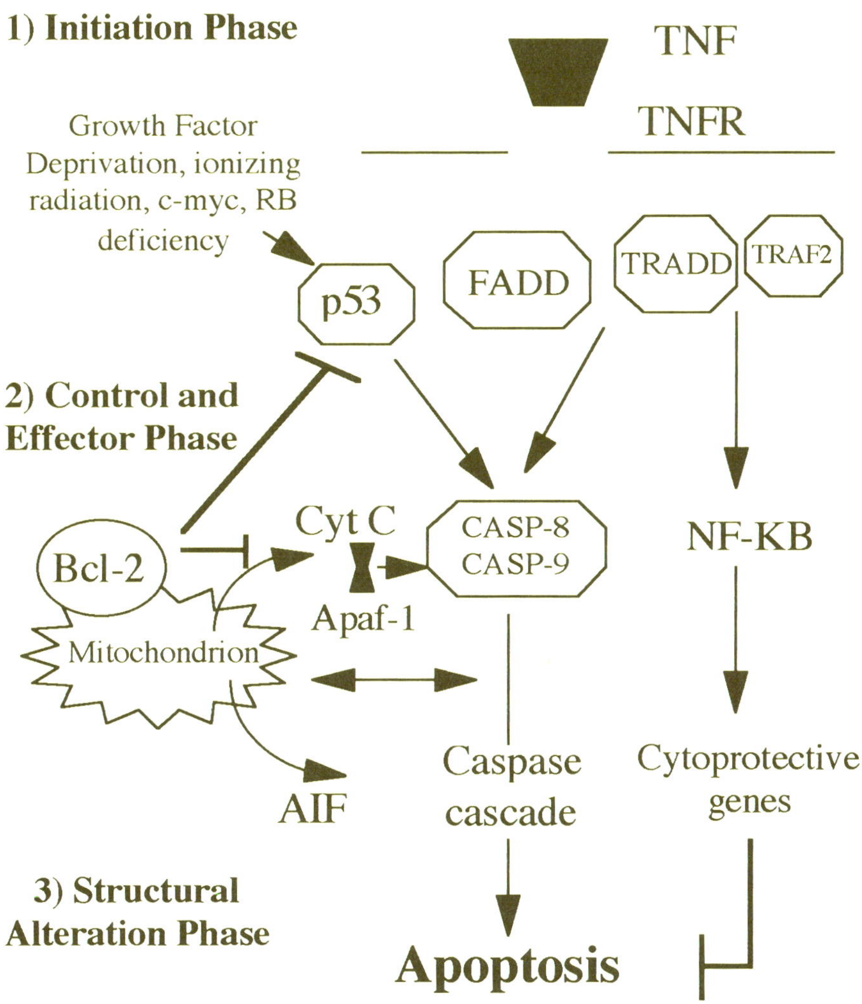

***Figure 1**. A simplified representation of the programmed cell death cascade.*

The execution phase is controlled by the Bcl-2 family members [20] acting upstream from the caspases, through inhibition of cytochrome c and/or apoptosis inducing factor (AIF) release from the mitochondrion or through binding and sequestration of Apaf-1 away from caspase-3. The Bcl-2 family of proteins contains both inhibitors (i.e., Bcl-2, Bcl-x_L) and inducers (i.e., Bax, Bcl-x_S) of apoptosis acting as homodimers and heterodimers.

3) The structural alterations and DNA degradation phase, where caspase activation leads to the cleaving of laminin, the intermediate filament of the nuclear envelope, of poly(ADP)ribose polymerase [21] and of the inhibitor of caspase-activated deoxyribonucluease (ICAD) (cleaved by caspase-3), allowing CAD to enter the nucleus and degrade chromosomal DNA [22, 23]. After the completion of the apoptotic process, dead cells are removed from the tissue as a result of specific recognition and phagocytosis by adjacent professional and/or non-professional cells through a variety of mechanisms that may implicate phosphatidylserine, thrombospondin, the vitronectin receptor or CD14, for example [24-26].

APOPTOSIS OF VASCULAR CELLS

Apoptosis of Vascular Endothelial Cells

Vascular endothelial cells form the inner lining of all blood vessels. During both normal development and pathology, the formation of new vessels and the regression of preexisting ones are dependent on the balance between endothelial cell proliferation and endothelial cell apoptosis. In mature vessels, endothelial cell turnover is also under the control of these tightly regulated phenomena. Since the vascular endothelium is involved in various physiological processes, endothelial cell apoptosis (and dysfunction) may constitute an initial step in a variety of pathological situations such as atherosclerosis and hypertension. As for other cell types, it has been hypothesized that interactions of endothelial cells with their microenvironment may be critical for their survival.

Hemodynamic factors and endothelial cell apoptosis

Vascular endothelial cells are continuously exposed to a range of hemodynamic forces which have a great impact on their cellular structure and function. Variations in blood flow play an important role in vessel growth or regression, and in the focal development of atherosclerosis. A link between mechanical stimulation and cell survival or death has been therefore suggested and demonstrated by several groups. Human umbilical vein endothelial cells (HUVEC) cultured under static conditions undergo a basal low level of apoptosis [27]. The empty space left between cells induces proliferation of adjacent cells so that an equilibrium state is reached with low levels of apoptosis and proliferation. Exposure to flow in a perfusion chamber or in an ex vivo organ culture directly inhibits the apoptotic process and indirectly supresses proliferation, leading to a "truly" quiescent monolayer [27]. NO donors have been shown to inhibit caspase-3 activation, which is essential to the completion of the endothelial cell apoptotic process, suggesting that shear stress-dependent release of NO may be involved in shear stress-dependent anti-apoptotic activity [28]. This anti-apoptotic pathway involves the phosphorylation of Akt/PKB by shear stress [29]. The suppression of both endothelial cell apoptosis and caspase-3 activation by physiological levels of shear stress in other models (growth factor withdrawal or cytokine-induced apoptosis) is prevented by pharmacological inhibtion of glutathione biosynthesis or nitric oxide synthase [30, 31]. This suggests that the reduction in oxidative flux by shear stress may be essential to prevent the activation of caspases and to inhibit endothelial cell apoptosis.

Oxidized LDLs, oxysterols and endothelial cell apoptosis

The evidence available from several groups suggests that oxidized LDLs (oxLDLs) play a central role in atherogenesis beginning in the earliest phases of this threatening process. In vitro studies had already shown that oxLDLs exhibit cytotoxic effects on cultured endothelial cells, but the type of cell death elicited remained unknown [32]. However, several authors recently reported increased apoptotic cell death of bovine aortic endothelial cells exposed to cholesterol oxides in culture [33] and increased apoptosis (and possibly secondary necrosis) of HUVEC after exposure to oxLDLs [34, 35]. OxLDL-induced apoptosis is subsequent to a sustained and delayed peak in cytosolic calcium and is prevented by chelating extracellular calcium or by inhibiting calcium influx [35]. Furthermore, ceramide generation by acid sphingomyelinase and activation of caspase-3 are indispensable for oxLDL-induced apoptosis [36]. OxLDLs are known to increase the generation of reactive oxygen species which are implicated in the apoptotic process in various cell types. Addition of antioxidants (vitamins C and E) completely prevents the activation of caspase-3 by oxLDLs and inhibits the apoptotic process [34].

Inflammation and endothelial cell apoptosis

Endothelial dysfunction is central to many inflammatory diseases affecting the vessel wall (atherosclerosis, for example) and several other organs (septic shock). Also, inflammatory cytokines such as TNF-α are increased in the circulating blood of patients with terminal heart failure and patients with septic shock and may participate to the vascular dysfunction observed in these diseases. Incubation of HUVEC with TNF-α markedly increases endothelial cell apoptosis via activation of caspase-3, a process that can be completely abrogated by inhibitors of caspases or by shear stress [30].

Peripheral blood mononuclear monocytes (PBMCs) induce apoptotic death in cultured HUVECs when preactivated by bacterial lipopolysaccharide (LPS) or by ionizing radiation [37]. This process is dependent at least in part on TNF-α since addition of an anti-TNF-α monoclonal antibody blocks the cell-to-cell contact-mediated apoptosis. Interestingly, the anti-inflammatory cytokine interleukin (IL)-10 has anti-apoptotic effects in this model, suggesting that the balance between endothelial cell survival and death may depend on the balance between pro- and anti-inflammatory cytokines. Incubation of HUVEC with LPS in culture induces apoptosis and secondary necrosis in a time-dependent manner. In this setting, apoptosis is independent of TNF-α release but is associated with the induction of Bax and is prevented by incubation with antioxidants (that enhance Bcl-2 and decrease Bax) [38]. It should be noted that TNF-α, like other inflammatory cytokines, is capable of inducing human A1, a human Bcl-2 homologue [39] and may also activate the NF-κB pathway [40]. TNF-α may therefore initiates both pro-apoptotic and anti-apoptotic pathways.

Systemic administration of LPS induces the endotoxic shock syndrome with associated systemic inflammation, multiple organ damage, circulatory collapse and death. This process is mediated by the release of TNF-α and other cytokines. Since TNF-α and LPS can cause apoptotic death of endothelial cells, it has been hypothesized that the primary tissue target in septic shock may be the endothelium. Indeed, it was demonstrated that injection of LPS and TNF-α in mice induces massive apoptosis in the endothelium of several organs before any other form of tissue damage [41]. Unlike studies performed in vitro, LPS-induced endothelial apoptosis in vivo is dependent on TNF-α since it is blocked by the TNF-binding protein. This latter molecule also prevents LPS-induced ceramide generation, suggesting that systemic TNF is required for both responses [41]. The indispensable role of ceramide generation in disseminated endothelial apoptosis is demonstrated by the observation that acid sphingomyelinase knockout mice are protected from endothelial cell apoptosis and from death despite the normal increase in serum TNF-α in response to LPS [41]. It would be interesting to examine whether the beneficial effects of IL-10 in reducing mortality from septic shock in animals [42] are related, at least in part, to its potential anti-apoptotic properties.

(Patho)physiologically relevant anti-apoptotic factors

Loss of cell-to-cell contact leads to the activation of a default death program in a variety of cell types. Cell-matrix interactions are important for endothelial cell survival [43-45]. Indeed interaction and adhesion via the integrin $\alpha v\beta 3$ inhibits p53 activity and decreases the expression of p21 and of the pro-apoptotic protein Bax, therefore promoting cell survival [46]. This is particularly important during migration of endothelial cells in angiogenesis as well as in the maintenance of tumor vasculature. TNF and interferon (IFN)-γ interact with $\alpha v\beta 3$ to cause selective disruption of the tumor vasculature in patients with melanoma. Indeed, exposure of human endothelial cells to TNF and IFN-γ results in supression of endothelial cell $\alpha v\beta 3$ activity leading to a decreased $\alpha v\beta 3$-dependent endothelial cell adhesion and survival [47].

Hyperoxia-induced regression of newly formed retinal vessels in premature infants is due to withdrawal of a vascular survival factor identified as vascular endothelial growth factor (VEGF) [48]. Similarly, VEGF withdrawal induces endothelial cell apoptosis and shedding leading to regression of hemangioblastoma-like vessels [49]. In vitro, VEGF inhibits endothelial cell apoptosis in response to a variety of stimuli such as hyperoxia [48], growth factor withdrawal, disruption of extracellular matrix [50], ionizing radiation [51], peroxynitrite and stimulation with TNF-α [52]. This is accomplished in part through upregulation of the anti-apoptotic proteins Bcl-2 and A1 [53], and stimulation of protein kinase C or PI3-kinase [52].

Basic fibroblast growth factor (FGF-2) is another intravascular survival factor for endothelial cells in vivo. Intravenous administration of FGF-2 in a murine model of endotoxic shock blocks LPS-induced ceramide generation and endothelial cell apoptosis. Consequently, animal mortality is reduced despite persistent elevation in serum TNF-α [41]. In vitro studies have shown that FGF-2 specifically induces Bcl-2, but not other members of the Bcl-2 family, and inhibits serum-deprivation-induced endothelial cell death [54]. However, other Bcl-2-independent mechanisms, such as tyrosine phosphorylation, may also account for the inhibition of endothelial apoptosis by FGF-2 [54].

Estrogen has been shown to be anti-atherogenic and to exert beneficial effects in the regulation of vascular tone in atherosclerotic arteries. Estradiol treatment of HUVEC in culture results in a dose-dependent, receptor-mediated inhibition of TNF-α-induced apoptosis and activation of the caspase-1 pathway [55], suggesting that the beneficial atheroprotective effects of estradiol may result, at least in part, from the preservation of endothelial integrity.

Apoptosis of Vascular Smooth Muscle Cells

After decades of intensive research on the mechanisms of vascular smooth muscle cell (VSMC) proliferation, recent studies have been undertaken to understand the mechanisms and roles of smooth muscle cell survival/apoptosis in normal vessel development and pathology. Thus, VSMC growth is now viewed as the result of the opposing effects of cell proliferation and apoptosis.

Many known growth factors for VSMCs (insulin-like growth factor-1 or IGF-1, platelet-derived growth factor-BB or PDGF-BB and FGF-2) have now been shown to act as survival factors rather than true mitogens. These growth factors partially prevent apoptosis of human or rat VSMCs cultured under low serum conditions [56, 57]. Induction of proliferation and prevention of apoptosis by the same factors are likely to be mediated by different downstream signalling pathways.

Besides growth factors, vasoactive substances such as angiotensin II (Ang II) and nitric monoxyde (NO) have been implictaed in chronic vascular remodeling. Forced local generation of Ang II in the vessel wall (after overexpression of the angiotensin-converting enzyme, ACE) promotes VSMC growth and wall thickening [58], while local production of high levels of NO inhibits cell growth and neointimal formation following balloon injury [59]. Recent experiments suggest that these effects on vascular remodeling may be the result of a modulation of the apoptotic process in VSMCs. NO induces apoptosis in VSMCs and this process is prevented by the inhibition of the cGMP-dependent protein kinase Iα, as well as by the addition of Ang II (acting via the angiotensin receptor type 1), emphasizing the importance of the balance between NO and Ang II in the modulation of VSMC growth [60]. It should be noted that the pro-apoptotic effects of NO occur at supraphysiological (pathological or inflammatory) levels while its anti-apoptotic effects (prevention of caspase activation) are essentially observed at physiological (endothelial) concentrations.

Production of high levels of NO via the inducible nitric oxide synthase (iNOS) also account for the pro-apoptotic effects of pro-inflammatory cytokines (TNF-α, IFN-γ and IL-1) in rat but not in human VSMCs [61]. Stimulation of VSMCs in culture by pro-inflammatory cytokines also induces the expression of Fas (and increases their sensitivity to apoptosis) [62], suggesting that the production of such cytokines by immune cells in atherosclerosis may promote VSMC death through this pathway.

VSMC apoptosis may be induced in vitro by many other factors including inhibitors of protein kinase C, calcium channel blockers, angiotensin-converting enzyme inhibitors, stimulators of cAMP-dependent protein kinase [63], tissue inhibitor of metalloproteinase type 3 [64], oxidized LDLs [65], oxysterols [66] and even antioxidants [67]. The clinical relevance of the pro-apoptotic effects of these factors remains to be established.

Genes involved in the regulation of VSMC apoptosis

Deregulated expression of the *c-myc* oncogene and stable infection with the adenovirus E1A induce apoptosis in rat VSMCs under low serum conditions [68, 69]. On the other hand, overexpression of p53 induces apoptosis in rat VSMCs infected with *c-myc* and E1A but not in normal cells [69]. Prevention of *c-myc* and E1A-mediated apoptosis by overexpression of *bcl-2* is independent of p53 [69].

These observations are important for several reasons: the *c-myc* oncogene is overexpressed in VSMCs of human atherosclerotic plaques and these cells display a lower growth rate in culture in comparison with VSMCs obtained from normal arteries, especially under low serum conditions [70]. As expected, this intrinsic defect in growth has recently been shown to be, at least in part, the result of increased spontaneous cell death by apoptosis [56]. The characteristics of plaque VSMCs (overexpression of *c-myc*) may also explain in part their hypersensitivity to p53-mediated apoptosis [71]. However, basal p53 activity is similar in plaque and normal VSMCs and suppression of basal p53 activity blocks growth arrest in both cell types but does not prevent apoptosis suggesting that the mechanisms of p53-mediated apoptosis of plaque VSMCs may be distinct from those inducing growth arrest [71]. The slower rates of proliferation and earlier senescence of plaque VSMCs may be explained by the predominance of the hypophosphorylated form of the retinoblastoma gene product (RB) and the low E2F transcriptional activity in these cells [72]. Both inactivation of RB and inhibition of p53 activity are required for plaque VSMCs to proliferate without apoptosis [72].

ROLE OF APOPTOSIS IN NORMAL VESSEL DEVELOPMENT

Physiological remodeling of blood vessels before and after birth has been shown to be the result of a balance between cell apoptosis and proliferation. Developmentally programmed capillary regression is initiated by macrophage-mediated endothelial cell apoptosis [73]. The dying cells are projected into the capillary lumen causing temporary or permanent block to blood flow. Following cessation of flow, secondary synchronous apoptosis of vascular endothelial cells occurs, leading to capillary regression [74].

Apoptosis of smooth muscle cells occurs during the development of the human ductus arteriosus [75]. The role of apoptosis has also been investigated in vessel remodeling that occurs as arteries adapt to changes in cardiovascular function after birth. The abdominal aorta of neonatal lambs undergoes excessive remodeling in association with a profound decrease in blood flow after birth. DNA accumulation in this vessel is far below that predicted from proliferation rates if all cells were assumed to survive. Cho et al. have demonstrated that apoptosis significantly contributes to postpartum arterial remodeling and that changes in cell death rates alone may be sufficient to induce profound changes in the vessel wall mass [76].

ROLE OF APOPTOSIS IN VASCULAR PATHOLOGY

Apoptosis in Atherosclerosis

Observations that cell death occurs in atherosclerosis have already been made by Virchow [1], and experimental studies in cholesterol-fed swines undertaken by Thomas et al. [2] more than 20 years ago showed that cell death is a major event occurring during plaque development and may be associated with cell proliferation. More recently, Parkes et al. observed that plaque-derived smooth muscle cells (p-VSMC) are characterized by *c-myc* overexpression and have a growth disadvantage in culture in comparison with VSMCs from normal arteries, especially under low serum conditions [70]. At that time, cancer researchers have shown that overexpression of *c-myc* induces cell death by apoptosis in the absence of serum and that cells can be rescued by overexpression of *bcl-2* [77]. On the basis of these results, it has been suggested that apoptosis may occur in human atherosclerosis. Bennett et al. have subsequently demonstrated that deregulated expression of *c-myc* in VSMCs induces apoptosis [68] and that human p-SMCs have a markedly elevated rate of apoptosis in vitro, a process that can be reversed by expression of *bcl-2* or by other survival factors such as IGF-1 and PDGF [56]. This was the first indication that apoptosis may occur in human atherosclerosis.

In situ distribution of apoptosis

Several studies have shown evidence for in situ apoptotic cell death in animal and human atherosclerotic plaques [78-85]. Apoptosis, which is almost absent in normal arteries, becomes barely detectable in fatty streaks and is more abundant in advanced plaques. All cell types are involved, including SMCs, macrophages and T-lymphocytes. Removal of apoptotic cells in human plaques may be inefficient, as a significant percentage of secondary necrosis is observed. Foam cell apoptosis (and secondary necrosis) is thought to contribute to the formation of the acellular lipid core [85], whereas VSMCs die within their cages of thickened basal lamina in the hypocellular fibrous cap and are split into a myriad of membrane vesicles [86]. Apoptosis is rarely observed in the endothelial cell layer covering atherosclerotic plaques, perhaps due to the low level of endothelial cell death or to the rapid shedding of dead cells into the bloodstream.

The distribution of apoptosis is heterogeneous within the plaque, being more frequent in regions with a high density in macrophages, suggesting that these cells may be participating in the induction of apoptosis. The rate of apoptotic cell death in the plaque (2-10%) (81, 87, 88) may, in many instances, exceed the reported rate of cell proliferation (<1%) [89, 90]. In the light of observations that human atherosclerotic plaques are generally more prone to progression rather than regression, this finding suggests that apoptosis may actually be implicated in natural plaque progression (rather than regression), through development of the acellular lipid core for example. Indeed, macrophage apoptosis has frequently been identified at the edges of the lipid core, suggesting that it may actively contribute to its formation. Moreover, the heterogeneous distribution of apoptosis implies that some regions of the plaque may show substantially higher levels of cell death, possibly predisposing to plaque rupture (and vessel occlusion or plaque progression). Finally, apoptosis greatly enhances tissue factor activity in the atherosclerotic plaque (see below) and may therefore predispose to thrombotic complications and plaque progression.

Pro-and anti-apoptotic factors in atherosclerotic plaques

Many factors with in vitro pro- and anti-apoptotic activities in macrophages and smooth muscle cells have been shown to be expressed in human atherosclerosis. Pro-inflammatory cytokines with pro-apoptotic potential such as IL-1, TNF-α, and IFN-γ are present in the human plaque and may be involved in the local induction of the apoptotic process, although no studies have been performed showing direct correlation between their expression and the occurrence of apoptosis. However, one of their mediators, inflammatory NO, produced as the result of upregulation of inducible nitric oxide synthase (iNOS), may exert potent cell cytotoxicity. We have recently found a positive association between iNOS expression and TUNEL labeling for apoptotic cells in advanced human atherosclerosis [88].

Cytokine-induced apoptosis may also be mediated in part by the Fas receptor. In vitro studies have shown that pro-inflammatory cytokines induce upregulation of Fas in human VSMCs and this fits well with the more prominent expression of Fas in the human plaque intima than in the media [62]. Moreover, cytokine-primed VSMCs, but not unprimed cells, are sensitive to anti-Fas stimulation in vitro and undergo apoptosis. This may explain the association between Fas-positive cells and apoptotic VSMCs in regions of plaques that are rich in activated T-lympocytes and macrophages [62]. Fas-mediated apoptotic cytotoxicity was also reported in human transplant coronary artery disease [91]. Interestingly, decreased Fas expression in a subset of plaque-SMCs has been reported, which could lead to resistance to apoptosis via the Fas system [92].

Smooth muscle cells die within the human plaque despite the presence of survival factors such as IGF-1 and PDGF [56]. Deregulated expression of *c-myc* in the absence of significant Bcl-2 expression [68], and increased sensitivity to p53 [71] might offer some explanations. Recent data have also shown increased expression of the pro-apoptotic protein Bax in human fatty streaks and advanced plaques. Bax was undetectable in normal arteries where expression of the anti-apoptotic protein Bcl-x_L predominated [86]. Therefore, it seems that the balance between pro-and anti-apoptotic proteins in atherosclerosis is in favor of the former, suggesting that plaque-SMCs are programmed to die and effectively undergo apoptosis when additional pro-apoptotic stimuli are present (inflammatory cells and cytokines).

Unlike the pro-inflammatory cytokines, little is known about the expression of anti-inflammatory mediators in human atherosclerosis and their potential role in the modulation of apoptosis. We have recently shown that high levels of expression of the anti-inflammatory IL-10 protein are associated with significantly reduced levels of iNOS expression (Table 1) [88] and apoptotic cell death (Table 2) [88], suggesting that apoptosis may result from an excessive inflammatory reaction and may be modulated by interfering with the inflammatory response.

Table 1*. Distribution of inducible nitric oxide synthase (iNOS) expression according to the level of IL-10 expression.*

	Level of IL-10 expression in regions of atherosclerotic plaques			
	0%	< 10%	10-50%	> 50%
%iNOS positive fields	82%	40%	27%	30%
%iNOS negative fields	18%	60%	73%	70%

Table 2*. Distribution of TUNEL positivity according to the level of IL-10 expression.*

	Level of IL-10 expression in regions of atherosclerotic plaques			
	0%	< 10%	10-50%	> 50%
%TUNEL positive fields	31%	15%	10%	0%
%TUNEL negative fields	69%	85%	90%	100%

The execution phase of apoptosis is caspase-dependent. In accordance with this concept, we and others [79, 84] have shown that caspases, especially caspase-3, colocalized with apoptotic cells, particularly macrophages, in advanced human atherosclerotic plaques (Table 3). However, caspase-3 positive but TUNEL-negative T-lymphocytes may be observed, and this is consistent with the observation that caspase-3 cleavage in this cell type may occur without the induction of apoptosis [93].

***Table** 3. Percentages of specific cell types staining for TUNEL and/or caspase-3.*

	TUNEL+*	TUNEL-	
Cell type	Caspase-3+	Caspase-3+	Caspase-3-
Macrophage	51 ± 6	2 ± 1	47 ± 5
T cell	27 ± 8	17 ± 2	56 ± 7
SMC	4 ± 2	0	96 ± 2

* TUNEL+/Caspase-3- cells were debris

Potential roles of apoptosis in human atherosclerosis

Much interest has focused on the regulation of apoptosis but only speculations have been raised about the potential roles, beneficial or harmful, of this process of cell death in atherosclerosis. Death of VSMCs by apoptosis may be viewed as harmful because it may weaken the fibrous cap by decreasing the synthesis of extracellular matrix and therefore may lead to plaque rupture. However, others may argue that the plaque stabilization observed with anti-β3 integrin antibodies [94] may be the result of selective apoptosis of intimal smooth muscle cells that express $\alpha_V\beta_3$ integrin and whose attachment to the extracellular matrix are disrupted by the anti- β3 integrin antibody treatment.

Death of T-lymphocytes and macrophages by apoptosis may be viewed as beneficial if apoptosis is not accompanied by an inflammatory reaction. Indeed, removal of these cells from the plaque could attenuate the inflammatory response, decrease the synthesis of matrix metalloproteinases (MMPs) and the consequent breakdown of the extracellular matrix, therefore favoring plaque stabilization. It may also be argued that apoptotic death of any cell type in the plaque may favor plaque regression. However, removal of apoptotic cells from the atherosclerotic tissue may not be very efficient, leaving unremoved apoptotic bodies with potentially high immunogenic properties. Moreover, unremoved apoptotic cells are prone to undergo secondary necrosis and this may lead to accumulation of extracellular lipids and to perpetuation of the inflammatory response. Whatever the potential role proposed for apoptosis, there is still no direct evidence in humans to support any of these speculations.

Pollman et al. have recently reported regression of atherosclerotic plaques of cholesterol-fed rabbits after inhibition of neointimal cell Bcl-x expression and induction of neointimal cell apoptosis [95]. It still unknown, however, whether human atherosclerotic plaques will behave in the same manner. Rabbit atherosclerotic plaques do not rupture whereas human plaques are very prone to rupture. On the other hand, it could be hazardous to induce apoptotic cell death in human atherosclerotic plaques without ensuring high anti-coagulation levels given the increased propensity of apoptotic cells to induce thrombin generation [96, 97]. Indeed, phosphatidylserine (PS) exposure on the outer cell surface, a hallmark of apoptosis, greatly enhances tissue factor activity and is therefore highly thrombogenic [98]. This procoagulant potential of apoptotic cells may be deleterious in atherosclerosis since it may be involved in determining plaque thrombogenicity following plaque rupture and may therefore favor the occurrence of acute ischemic events and infarction.

Indeed, Flynn et al. have shown that thrombin generation is increased at the contact of plaque-derived smooth cells undergoing apoptosis in culture [97]. Moreover, we have recently observed significant extracellular tissue factor (TF) expression around apoptotic cells in some regions of human atherosclerotic plaques suggesting that TF may be shed from apoptotic cells via apoptotic microparticles [99]. To examine whether apoptosis may be involved in TF activity in vivo, we examined advanced human atherosclerotic plaques for the presence of shed membrane apoptotic microparticles (captured by biotinylated annexin V insolubilized onto streptavidin-coated microtitration plaques) and determined the procoagulant potential of these microparticles [99]. We found that high levels of shed membrane microparticles of monocytic and lymphocytic origins were produced in the atherosclerotic plaques and were associated with increased TF activity. Interestingly, removing the microparticles from the supernatants resulted in a 97% reduction in TF activity, indicating that almost all of the TF activity of plaque extracts was associated with the shed membrane microparticles [99]. These experiments provide direct evidence that apoptosis is involved in the increased thrombogenic potential of human atherosclerotic plaques. Moreover, preliminary experiments in our laboratory show that plasma levels of procoagulant microparticles are increased in patients with acute coronary syndromes and may be associated with coronary reocclusion [Mallat et al, unpublished observations].

Apoptosis in Restenosis

Animal models of neointima formation following balloon injury have shown that apoptotic cell death occurs at different points in time and at variable levels during the repair process. Apoptotic death of smooth muscle cells is observed as early as 30 minutes after arterial injury and is associated with decreased expression of the anti-apoptotic protein Bcl-x [100]. After then, a second window of apoptosis is observed and is associated with increased smooth muscle cell proliferation [80]. Persistent apoptotic cell death at later stages, in the absence of cell proliferation, has been implicated in the regulation of intimal thickening [101].

Studies available in human restenotic plaques are somewhat contradictory. Isner et al. observed an increased apoptotic rate in restenotic versus primary coronary plaques [78]. This finding was recently challenged, however, by Bauriedel et al. [87] who observed decreased levels of apoptosis in restenotic versus primary plaques with no differences in necrosis levels. High cellularity is a key finding in late human restenosis and the findings of Bauriedel et al. are more consistent with the paradigm that lower rates of apoptosis result in hyperplasia. This also fits well with the finding of decreased p53 activity (and potentially p53-mediated apoptosis) in restenotic material from human coronary lesions [102].

The observation that apoptosis may particpite in cellularity regulation in animal models of neointima formation and in human restenotic lesions prompted researchers to examine the effects of pro-apoptotic strategies on this process. Steg et al. [103] used suicide gene therapy by transfer of a replication-defective adenoviral vector expressing the thymidine kinase gene combined with ganciclovir treatment. In vitro transfection and treatment of SMCs induced their death by apoptosis, and this mechanism was presumably responsible for the reduction of restenosis after angioplasty in tranfected and treated atheromatous rabbit arteries. Pollman et al. [95] recently studied another approach interfering with the anti-apoptotic genes expressed in the neointima. This consisted in the local application of antisense oligonucleotides encoding a sequence complementary to the coding region for the anti-apoptotic *bcl-x* gene. Using this strategy, they were able to decrease neointimal expression of Bcl-x, inducing selective apoptosis of neointimal cells and regression of neointimal hyperplasia. Other investigators have used a direct pro-apoptotic strategy delivering Fas ligand into the vessel wall and inducing apoptosis of Fas-bearing VSMCs [104]. This also led to significant inhibition of neointima formation adding to the evidence that apoptosis is a key determinant of neointimal thickening evolution after arterial injury. It remains to be determined, however, whether such impressive results may be achieved in humans without serious side effects (such as increase in the acellular lipid core and vessel thrombosis).

Apoptosis in Other Vascular Diseases

Hypertension

Recent studies have shown that remodeling of target organs of hypertension is associated with reduced apoptotic cell death early after birth accounting for an important organ hyperplasia. However, in the course of hypertension, significant death by apoptosis can be detected contributing to continuous remodeling of these organs [105]. Endothelial cell death by apoptosis may also be involved in microvascular rarefaction that is associated with hypertension [106]. An imbalance has been found between the levels of pro- and anti-apoptotic proteins of the Bcl-2 family which may account, at least in part, for these findings [107]. These studies prompted a reevaluation of the effects of anti-hypertensive therapeutic interventions at the level of apoptosis. As a result, it was shown that ACE inhibitors, angiotensin II receptor (AT1) antagonists and calcium channel blockers stimulate VSMC apoptosis before reducing blood pressure [108]. A better understanding of the mechanisms and role of apoptosis in the setting of hypertension may help in the development of novel therapeutic startegies to control hypertensive vessel and organ remodeling.

Diabetes

Diabetic retinopathy is characterized by the progressive obliteration of retinal microvessels leading to ischemia and unregulated angiogenesis with sight-threatening consequences. Diabetes leads to an accelerated death by apoptosis of both pericytes and endothelial cells before any histologic evidence of retinopathy appears. Upon exhaustion of the replicative life span of these cells, capillary obliteration occurs [109]. It is likely that diabetes compromises the survival of vascular cells in other vessel types and contributes, through vascular cell apoptosis, to atherosclerosis and vascular thrombosis.

CONCLUSION

The observation that cell death and cell proliferation may occur independently and the growing understanding of the molecular basis for these phenomena will stimulate the development of interventions selectively aimed at cell death to prevent the progression of a variety of vessel diseases.

REFERENCES

1. Virchow R. Cellular Pathology as Based Upon Physiological and Pathological Histology, ed 2. Birmingham, AL, Classics of Medicine Library, 1858: 361.

2. Thomas WA, Reiner JM, Florentin FA, Lee KT, Lee WM. Population dynamics of arterial smooth muscle cells. V. Cell proliferation and cell death during initial three months in atherosclerotic lesions induced in swine by hypercholesterolemic diet and intimal trauma. *Exp Mol Pathol.* 1976;24:360-374.

3. Cliff WJ. The aortic tunica media in aging rats. *Exp Mol Pathol.* 1970;13:172-189.

4. Kerr JFR, Wyllie AH, Currie AR. Apoptosis: a basic biological phenomenon with wide-ranging implications in tissue kinetics. *Br J cancer.* 1972;26:239-257.

5. Hetts SW. To die or not to die. An overview of apoptosis and its role in disease. *JAMA.* 1998;279:300-307.

6. Martin SJ, Reutelingsperger CPM, McGahon AJ, Rader JA, van Schie RCA, LaFace DM, Green DR. Early redistribution of plasma membrane phosphatidylserine is a general feature of

apoptosis regardless of the initiating stimulus: inhibition by overexpression of Bcl-2 and Abl. *J Exp Med.* 1995;182:1545-1556.

7. Majno G, Joris I. Apoptosis, oncosis, and necrosis. *Am J Pathol.* 1995;146:3-15.

8. Haunstetter A, Izumo S. Apoptosis: basic mechanisms and implications for cardiovascular disease. *Circ Res.* 1998;82:1111-1129.

8bis. Green DR. Apoptotic pathways: the roads to ruin. *Cell.* 1998;94:695-698.

9. Casciola-Rosen LA, Anhalt G, Rosen A. Autoantigens targeted in systemic lupus erythematosus are clustered in two populations of surface structures on apoptotic keratinocytes. *J Exp Med.* 1994;179:1317-1330.

10. Tan EM. Autoimmunity and apoptosis. *J Exp Med.* 1994;179:1083-1086.

11. Rosen A, Casciola-Rosen L, Ahearn J. Novel packages of viral and self-antigens are generated during apoptosis. *J Exp Med.* 1995;181:1557-1561.

12. Kroemer G, Zamzami N, Susin SA. Mitochondrial control of apoptosis. *Immunology Today.* 1997;18:44-51.

13. Golstein P. Controlling cell death. *Science.* 1997;275:1081-1082.

14. Green DR, Reed JC. Mitochondria and apoptosis. *Science.* 1998;281:1309-1311.

15. Ashkenazi A, Dixit VM. Death receptors: signaling and modulation. *Science.* 1998;281:1305-1308.

16. Hartwell LH, Kastan MB. Cell cycle control and cancer. *Science.* 1994;266:1821-1828.

17. Polyak K, Xia Y, Zweier JL, Kinzler KW, Vogelstein B. A model for p53-induced apoptosis. *Nature.* 1997;389:300-305.

18. Green D, Kroemer G. The central executioners of apoptosis: caspases or mitochondria? *Trends Cell Biol.* 1998;8:267-271.

19. Thornberry NA, Lazebnik Y. Caspases: enemies within. *Science.* 1998;281:1312-1316.

20. Kroemer G. The proto-oncogene Bcl-2 and its role in regulating apoptosis. *Nat Med.* 1997;3:614-620.

21. Nicholson DW, Ali A, Thornberry NA, Vaillancourt JP, Ding CK, Gallant M, Gareau Y, Griffin PR, Labelle M, Lazebnik YA, Munday NA, Raju SM, Smulson ME, Yamin T-T, Yu VL, Miller DK. Identification and inhibition of the ICE/CED-3 protease necessary for mammalian apoptosis. *Nature.* 1995;376:37-43.

22. Sakahira H, Enari M, Nagata S. Cleavage of CAD inhibitor in CAD activation and DNA degradation during apoptosis. *Nature.* 1998;391:96-99.

23. Enari M, Sakahira H, Yokoyama H, Okawa K, Iwamatsu A, Nagata S. A caspase-activated DNase that degrades DNA during apoptosis, and its inhibitor ICAD. *Nature.* 1998;391:43-50.

24. Fadok VA, Savill JS, Haslett C, Bratton DL, Doherty DE, Campbell PA, Henson PM. Different populations of macrophages use either the vitronectin receptor or the phosphatidylserine receptor to recognize and remove apoptotic cells. *J Immunol.* 1992;149:4029-4035.

25. Savill J, Fadok V, Henson P, Haslett C. Phagocyte recognition of cells undergoing apoptosis. *Immunol Today.* 1993;14:131-136.

26. Devitt A, Moffatt OD, Raykundalia C, Capra JD, Simmons DL, Gregory CD. Human CD14 mediates recognition and phagocytosis of apoptotic cells. *Nature.* 1998;392:505-509.

27. Kaiser D, Freyberg MA, Friedl P. Lack of hemodynamic forces triggers apoptosis in vascular endothelial cells. *Biochem Biophys Res Commun.* 1997;231:586-590.

28. Dimmeler S, Haendeler J, Nehls M, Zeiher AM. Suppression of apoptosis by nitric oxide via inhibition of interleukin-1 beta-converting enzyme (ICE)-like and cysteine protease protein (CPP)-32-like proteases. *J Exp Med.* 1997;185:601-607.

29. Dimmeler S, Assmus B, Hermann C, Haendeler J, Zeiher AM. Fluid shear stress stimulates phosphorylation of Akt in human endothelial cells: involvement in suppression of apoptosis. *Circ Res.* 1998;83:334-341.

30. Dimmeler S, Haendeler J, Rippmann V, Nehls M, Zeiher AM. Shear stress inhibits apoptosis of human endothelial cells. *Febs Lett.* 1996;399:71-74.

31. Hermann C, Zeiher AM, Dimmeler S. Shear stress inhibits H2O2-induced apoptosis of human endothelial cells by modulation of the glutathione redox cycle and nitric oxide synthase. *Arterioscler Thromb Vasc Biol.* 1997;17:3588-3592.

32. Hessler JL, Robertson AL, Chisolm GM. LDL-induced cytotoxicity and its inhibition by HDL in human vascular smooth muscle and endothelial cells in culture. *Atherosclerosis.* 1979;32:213-229.

33. Lizard G, Deckert V, Dubrez L, Moisant M, Gambert P, Lagrost L. Induction of apoptosis in endothelial cells treated with cholesterol oxides. *Am J Pathol.* 1996;148:1625-1638.

34. Dimmeler S, Haendeler J, Galle J, Zeiher AM. Oxidized low-density lipoprotein induces apoptosis of human endothelial cells by activation of CPP32-like proteases: A mechanistic clue to the 'response to injury' hypothesis. *Circulation.* 1997;95:1760-1763.

35. Escargueil-Blanc I, Meilhac O, Pieraggi M-T, Arnal J-F, Salvayre R, Nègre-Salvayre A. Oxidized LDLs induce massive apoptosis of cultured human endothelial cells through a calcium-dependent pathway: prevention by aurintricarboxylic acid. *Arterioscler Thromb Vasc Biol.* 1997;17:331-339.

36. Harada-Shiba M, Kinoshita M, Kamido H, Shimokado K. Oxidized low density lipoprotein induces apoptosis in cultured human umbilical vein endothelial cells by common and unique mechanisms. *J Biol Chem.* 1998;273:9681-9687.

37. Lindner H, Holler E, Ertl B, Multhoff G, Schreglmann M, Klauke I, Schultz-Hector S, Eissner G. Peripheral blood mononuclear cells induce programmed cell death in human endothelial cells and may prevent repair : role of cytokines. *Blood.* 1997;89:1931-1938.

38. Haendeler J, Zeiher AM, Dimmeler S. Vitamin C and E prevent lipopolysaccharide-induced apoptosis in human endothelial cells by modulation of Bcl-2 and Bax. *Eur J Pharmacol.* 1996;317:407-411.

39. Karsan A, Yee E, Kaushansky K, Harlan JM. Cloning of a human Bcl-2 homologue: Inflammatory cytokines induce human A1 in cultured endothelial cells. *Blood.* 1996;87:3089-3096.

40. Van Antwerp DJ, Martin SJ, Kafri T, Green DR, Verma IM. Suppression of TNF-alpha-induced apoptosis by NF-KappaB. *Science.* 1996;274:787-789.

41. Haimovitz-Friedman A, Cordon-Cardo C, Bayoumy S, Garzotto M, McLoughlin M, Gallily R, Edwards III CK, Schuchman EH, Fuks Z, Kolesnick R. Lipopolysaccharide induces disseminated endothelial apoptosis requiring ceramide generation. *J Exp Med.* 1997;186:1831-1841.

42. Howard M, Muchamel T, Andrade S, Menon S. Interleukin 10 protects mice from lethal endotoxemia. *J Exp Med.* 1993;177:1205-1208.

43. Meredith J, Fazeli B, Schwartz M. The extracellular matrix as a survival factor. *Mol Biol Cell.* 1993;4:953-961.

44. Re F, Zanetti A, Sironi M, Polentarutti N, Lanfrancone L, Dejana E, Colotta F. Inhibition of anchorage-dependent cell spreading triggers apoptosis in cultured human endothelial cells. *J Cell Biol.* 1994;127:537-546.

45. Brooks PC, Montgomery AMP, Rosenfeld M, Reisfeld RA, Hu TH, Klier G, Cheresh DA. Integrin alpha(v)beta(3) antagonists promote tumor regression by inducing apoptosis of angiogenic blood vessels. *Cell.* 1994;79:1157-1164.

46. Stromblad S, Becker JC, Yebra M, Brooks PC, Cheresh DA. Suppression of p53 and p21 expression by vascular cell integrin avβ3 during angiogenesis in vivo. *J Clin Invest.* 1996;98:426-433.

47. Rüegg C, Yilmaz A, Bieler G, Bamat J, Chaubert P, Lejeune FJ. Evidence for the involvement of endothelial cell integrin avβ3 in the disruption of the tumor vasculature induced by TNF and IFN-g. *NatMed.* 1998;4:408-413.

48. Alon T, Hemo I, Itin A, Pe'er J, Stone J, Keshet E. Vascular endothelial growth factor acts as a survival factor for newly formed retinal vessels and has implications for retinopathy of prematurity. *NatMed.* 1995;1:1024-1028.

49. Benjamin LE, Keshet E. Conditional switching of vascular endothelial growth factor (VEGF) expression in tumors: induction of endothelial cell shedding and regression of hemangioblastoma-like vessels by VEGF withdrawal. *Proc NatlAcad Sci USA.* 1997;94:8761-8766.

50. Watanabe Y, Dvorak HF. Vascular permeability factor/vascular endothelial growth factor inhibits anchorage-disruption-induced apoptosis in microvessel endothelial cells by inducing scaffold formation. *Exp CellRes.* 1997;233:340-349.

51. Katoh O, Tauchi H, Kawaishi K, Kimura A, Satow Y. Expression of the vascular endothelial growth factor (VEGF) receptor gene, KDR, in hematopoietic cells and inhibitory effect of VEGF on apoptotic cell death caused by ionizing radiation. *Cancer Res.* 1995;55:5687-5692.

52. Karsan A. Tumor necrosis factor and endothelial cell death. *Trends Cardiovasc Med.* 1998;8:19-24.

53. Gerber HP, Dixit V, Ferrara N. Vascular endothelial growth factor induces expression of the antiapoptotic proteins Bcl-2 and A1 in vascular endothelial cells. *J Biol Chem.* 1998;273:13313-13316.

54. Karsan A, Yee E, Poirier GG, Zhou P, Craig R, Harlan JM. Fibroblast growth factor-2 inhibits endothelial cell apoptosis by Bcl-2 dependent and independent mechanisms. *Am J Pathol.* 1997;151:1775-1784.

55. Spyridopoulos I, Sullivan AB, Kearney M, Isner JM, Losordo DW. Estrogen-receptor-mediated inhibition of human endothelial cell apoptosis. Estradiol as a survival factor. *Circulation.* 1997;95:1505-1514.

56. Bennett MR, Evan GI, Schwartz SM. Apoptosis of human vascular smooth muscle cells derived from normal vessels and coronary atherosclerotic plaques. *J Clin Invest.* 1995;95:2266-2274.

57. Fox JC, Shanley JR. Antisense inhibition of basic fibroblast growth factor induces apoptosis in vascular smooth muscle cells. *J Biol Chem.* 1996;271:12578-12584.

58. Morishita R, Gibbons GH, Ellison KE, Lee W, Zhang LN, Yu H, Kaneda Y, Ogihara T, Dzau VJ. Evidence for direct local effect of angiotensin in vascular hypertrophy: in vivo gene transfer of angiotensin converting enzyme. *J Clin Invest.* 1994;94:978-984.

59. von der Leyen H, Gibbons GH, Morishita R, Lewis NP, Zhang L, Kaneda Y, Cooke JP, Dzau VJ. Gene therapy inhibiting neointimal vascular lesion: in vivo transfer of endothelial cell nitric oxide synthase gene. *Proc Natl Acad Sci USA*. 1995;92:1137-1141.

60. Pollman MJ, Yamada T, Horiuchi M, Gibbons GH. Vasoactive substances regulate vascular smooth muscle cell apoptosis. Countervailing influences of nitric oxide and angiotensin II. *Circ Res*. 1996;79:748-756.

61. Geng YJ, Wu Q, Muszynski M, Hansson GK, Libby P. Apoptosis of vascular smooth muscle cells induced by in vitro stimulation with interferon-gamma, tumor necrosis factor-alpha, and interleukin-1 beta. *Arterioscler Thromb Vasc Biol*. 1996;16:19-27.

62. Geng YJ, Henderson LE, Levesque EB, Muszynski M, Libby P. Fas is expressed in human atherosclerotic intima and promotes apoptosis of cytokine-primed human vascular smooth muscle cells. *Arterioscler Thromb Vasc Biol*. 1997;17:2200-2208.

63. Leszczynski D, Zhao Y, Luokkamäki M, Foegh ML. Apoptosis of vascular smooth muscle cells. Protein kinase C and oncoprotein Bcl-2 are involved in regulation of apoptosis in non-transformed rat vascular smooth muscle cells. *Am J Pathol*. 1994;145:1265-1270.

64. Baker AH, Zaltsman AB, George SJ, Newby AC. Divergent effects of tissue inhibitor of metalloproteinase-1, -2 or -3 overexpression on rat vascular smooth muscle cell invasion, proliferation, and death in vitro. *J Clin Invest*. 1998;101:1478-1487.

65. Joringe S, Crisby M, Thyberg J, Nilsson J. DNA fragmentation and ultrustructural changes of degenerating cells in atherosclerotic lesions and smooth muscle cells exposed to oxidized LDL in vitro. *Arterioscler Thromb Vasc Biol*. 1997;17:2225-2231.

66. Nishio E, Watanabe Y. Oxysterols induced apoptosis in cultured smooth muscle cells through CPP32 protease activation and BCL-2 protein downregulation. *Biochem Biophys Res Commun*. 1996;226:928-934.

67. Tsai JC, Jain M, Hsieh CM, Lee WS, Yoshizumi M, Patterson C, Perrella MA, Cooke C, Wang H, Haber E, Schlegel R, Lee ME. Induction of apoptosis by pyrrolidinedithiocarbamate and N-acetylcysteine in vascular smooth muscle cells. *J Biol Chem*. 1996;271:3667-3670.

68. Bennett MR, Evan GI, Newby AC. Deregulated expression of the *c myc* oncogene abolishes inhibition of proliferation of rat vascular smooth muscle cells by serum reduction, interferon-g, heparin, and cyclic nucleotide analogues and induces apoptosis. *Circ Res*. 1994;74:525-536.

69. Bennett MR, Evan GI, Schwartz SM. Apoptosis of rat vascular smooth muscle cells is regulated by p53-dependent and -independent pathways. *Circ Res*. 1995;77:266-273.

70. Parkes JL, Cardell RR, Hubbard FCJ, Hubbard D, Meltzer A, Penn A. Cultured human atherosclerotic plaque smooth muscle cells retain transforming potential and display enhanced expression of the myc protooncogene. *Am J Pathol*. 1991;138:765-775.

71. Bennett MR, Littlewood TD, Schwartz SM, Weissberg PL. Increased sensitivity of human vascular smooth muscle cells from atherosclerotic plaques to p53-mediated apoptosis. *Circ Res*. 1997;81:591-599.

72. Bennett MR, Macdonald K, Chan SW, Boyle JJ, Weissberg PL. Cooperative interactions between RB and p53 regulate cell proliferation, cell senescence, and apoptosis in human vascular smooth muscle cells from atherosclerotic plaques. *Circ Res*. 1998;82:704-712.

73. Lang RA, Bishop JM. Macrophages are required for cell death and tissue remodeling in the developing mouse eye. *Cell*. 1993;74:453-462.

74. Meeson A, Palmer M, Calfon M, Lang R. A relationship between apoptosis and flow during programmed capillary regression is revealed by vital analysis. *Development*. 1996;122:3929-3938.

75. Slomp J, Gittenbergerdegroot AC, Glukhova MA, Vanmunsteren JC, Kockx MM, Schwartz SM, Koteliansky VE. Differentiation, dedifferentiation, and apoptosis of smooth muscle cells during the development of the human ductus arteriosus. *Arterioscler Thromb Vascr Biol.* 1997;17:1003-1009.

76. Cho A, Courtman DW, Langille BL. Apoptosis (programmed cell death) in arteries of the neonatal lamb. *Circ Res.* 1995;76:168-175.

77. Evan GI, Wyllie AH, Gilbert CS, Littlewood TD, Land H, Brooks M, Waters CM, Penn LZ, Hancock DC. Induction of apoptosis in fibroblasts by c-myc protein. *Cell.* 1992;69:119-128.

78. Isner JM, Kearney M, Bortman S, Passeri J. Apoptosis in human atherosclerosis and restenosis. *Circulation.* 1995;91:2703-2711.

79. Geng Y-J, Libby P. Evidence for apoptosis in advanced human atheroma. Colocalization with interleukin-1β-converting enzyme. *Am J Pathol.* 1995;147:251-266.

80. Han DKM, Haudenschild CC, Hong MK, Tinkle BT, Leon MB, Liau G. Evidence for apoptosis in human atherogenesis and in a rat vascular injury model. *Am J Pathol.* 1995;147:267-277.

81. Kockx MM, Demeyer GRY, Muhring J, Bult H, Bultinck J, Herman AG. Distribution of cell replication and apoptosis in atherosclerotic plaques of cholesterol-fed rabbits. *Atherosclerosis.* 1996;120:115-124.

82. Björkerud S, Björkerud B. Apoptosis is abundant in human atherosclerotic lesions, especially in inflammatory cells (macrophages and T cells), and may contribute to the accumulation of gruel and plaque instability. *Am J Pathol.* 1996;149:367-380.

83. Hegyi L, Skepper JN, Cary NR, Mitchinson MJ. Foam cell apoptosis and the development of the lipid core of human atherosclerosis. *J Pathol.* 1996;180:423-442.

84. Mallat Z, Ohan J, Lesèche G, Tedgui A. Colocalization of CPP-32 with apoptotic cells in human atherosclerotic plaques. *Circulation.* 1997;96:424-428.

85. Cai W, Devaux B, Schaper W, Schaper J. The role of Fas/APO 1 and apoptosis in the development of human atherosclerotic lesions. *Atherosclerosis.* 1997;131:177-186.

86. Kockx MM, De Meyer GRY, Muhring J, Jacob W, Bult H, Herman AG. Apoptosis and related proteins in different stages of human atherosclerotic plaques. *Circulation.* 1998;97:2307-2315.

87. Bauriedel G, Schluckebier S, Hutter R, Welsch U, Kandolf R, Lüderitz B, Prescott MF. Apoptosis in restenosis versus stable-angina atherosclerosis. Implications for the pathogenesis of restenosis. *Arterioscler Thromb Vasc Biol.* 1998;18:1132-1139.

88. Mallat Z, Heymes C, Ohan J, Faggin E, Lesèche G, Tedgui A. Expression of interleukin-10 in human atherosclerotic plaques. Relation to inducible nitric oxide synthase expression and cell death. *Arterioscler Thromb Vasc Biol.* 1999;(in press).

89. Gordon D, Reidy MA, Benditt EP, Schwartz SM. Cell proliferation in human coronary arteries. *Proc Natl Acad Sci USA.* 1990;87:4600-4604.

90. Brandl R, Richter T, Haug K, Wilhelm MG, Maurer PC, Nathrath W. Topographic analysis of proliferative activity in carotid endarterectomy specimens by immunocytochemical detection of the cell cycle-related antigen Ki-67. *Circulation.* 1997;96:3360-3368.

91. Dong C, Wilson JE, Winters GL, B.M. M. Human transplant coronary artery disease: pathological evidence for Fas-mediated apoptotic cytotoxicity in allograft arteriopathy. *Lab Invest.* 1996;74:921-931.

92. Han DKM, Wright ME, Dixit VM, Pruit R, Son Soe M, Lynch DH, Schwartz SM. Evidence for escape of apoptosis by a loss of Fas in atherosclerotic plaque smooth muscle cells. *Circulation.* 1996;94 (suppl I):I-397.

93. Miossec C, Dutilleul V, Fassy F, Diuhercend A. Evidence for CPP32 activation in the absence of apoptosis during T lymphocyte stimulation. *J Biol Chem*. 1997;272:13459-13462.

94. Topol EJ, Califf RM, Weisman HF, Ellis SG, Tcheng JE, Worley S, Ivanhoe R, George BS, Fintel D, Weston M, et al. Randomized trial of coronary intervention with antibody against platelet IIb/IIIa integrin for reduction of clinical restenosis: results at six months. The EPIC Investigators. *Lancet*. 1994;343:881-886.

95. Pollman MJ, Hall JL, Mann MJ, Zhang L, Gibbons GH. Inhibition of neointimal cell bcl-x expression induces apoptosis and regression of vascular disease. *Nat Med*. 1998;4:222-227.

96. Bombeli T, Karsan A, Tait JF, Harlan JM. Apoptotic vascular endothelial cells become procoagulant. *Blood*. 1997;89:2429-2442.

97. Flynn PD, Byrne CD, Baglin TP, Weissberg PL, Bennett MR. Thrombin generation by apoptotic vascular smooth muscle cells. *Blood*. 1997;89:4378-4384.

98. Bach R, Gentry R, Nemerson Y. Factor VII binding to tissue factor in reconstituted phospholipid vesicles: induction of cooperativity by phosphatidylserine. *Biochemistry*. 1986;25:4007-4020.

99. Mallat Z, Hugel B, Ohan J, Lesèche G, Freyssinet JM, Tedgui A. Shed membrane microparticles with procoagulant potential in human atherosclerotic plaques. A role for apoptosis in plaque thrombogenicity. *Circulation*. 1999;99:(in press).

100. Perlman H, Maillard L, Krasinski K, Walsh K. Evidence for the rapid onset of apoptosis in medial smooth muscle cells after balloon injury. *Circulation*. 1997;95:981-987.

101. Bochaton-Piallat ML, Gabbiani F, Redard M, Desmoulière A, Gabbiani G. Apoptosis participates in cellularity regulation during rat aortic intimal thickening. *Am J Pathol*. 1995;146:1059-1064.

102. Speir E, Modali R, Huang ES, Leon MB, Shawl F, Finkel T, Epstein SE. Potential role of human cytomegalovirus and p53 interaction in coronary restenosis. *Science*. 1994;265:391-394.

103. Steg PG, Tahlil O, Aubailly N, Caillaud JM, Dedieu JF, Berthelot K, Le Roux A, Feldman L, Perricaudet M, Denefle P, Branellec D. Reduction of restenosis after angioplasty in an atheromatous rabbit model by suicide gene therapy. *Circulation*. 1997;96:408-411.

104. Sata M, Perlman H, Muruve DA, Silver M, Ikebe M, Libermann TA, Oettgen P, Walsh K. Fas ligand gene transfer to the vessel wall inhibits neointima formation and overrides the adenovirus-mediated T cell response. *Proc Natl Acad Sci USA*. 1998;95:1213-1217.

105. Hamet P, Moreau P, Dam TV, Orlov SN, Tea BS, Deblois D, Tremblay J. The time window of apoptosis: A new component in the therapeutic strategy for cardiovascular remodeling. *J Hypertens*. 1996;14:S65-S70.

106. Gobe G, Browning J, Howard T, Hogg N, Winterford C, Cross R. Apoptosis occurs in endothelial cells during hypertension-induced microvascular rarefaction. *J Struct Biol*. 1997;118:63-72.

107. Diez J, Panizo A, Hernandez M, Pardo J. Is the regulation of apoptosis altered in smooth muscle cells of adult spontaneously hypertensive rats? *Hypertension*. 1997;29:776-780.

108. Deblois D, Tea BS, Dam TV, Tremblay J, Hamet P. Smooth muscle apoptosis during vascular regression in spontaneously hypertensive rats. *Hypertension*. 1997;29:340-349.

109. Mizutani M, Kern TS, Lorenzi M. Accelerated death of retinal microvascular cells in human and experimental diabetic retinopathy. *J Clin Invest*. 1996;97:2883-2890.

11. STIFFNESS OF WALL MATERIAL IN HUMAN HYPERTENSION

Michel E. Safar
Department of Internal medecine and INSERM U337,
Broussais Hospital, Paris, France

Damage of the large arteries is strongly associated with cardiovascular accidents, mainly in the cerebral, the coronary and the renal circulation [1,2]. Because such accidents are observed both in normotensive and hypertensive populations, it has been suggested that a common trigger factor, atherosclerosis, plays a major contributive role in the mechanism of the vascular alterations. However, independently of atherosclerosis, hypertensive vessels are thicker and stiffer than those of corresponding normotensive controls [2-4]. Thus the question is raised to know whether the specific alterations of hypertensive vessels play a role in cardiovascular morbidity and mortality.

Hypertrophy of hypertensive conduit arteries has been largely documented in the literature [4-6]. Carotid hypertrophy may be considered as a cardiovascular risk factor in subjects with hypertension [4]. However, not only structure but arterial function should be taken in consideration. Over 50 years of age, a disproportionate increase in systolic over diastolic blood pressure, and therefore of pulse pressure, has been noticed in patients with hypertension and is usually attributed to an increased rigidity of the arterial wall [7]. However, till recent years, there was no systematic and quantitative determination of the stiffness of wall material of large vessels.

Nowadays, using non invasive techniques in humans, it is possible to measure with a high degree of reproducibility the diameter and the wall thickness of superficial large arteries, such as the common carotid artery and the radial artery [3,4]. Subsequently, the incremental elastic modulus, a quantitative evaluation of the stiffness of the wall material, may be determined with precision.

The purpose of the present report was : (i) to describe the characteristics of arterial geometry and structure in various models of human and animal hypertension, (ii) to indicate the modalities of determination of arterial incremental modulus in situations involving either in vivo or in vitro measurements, and (iii) to apply this basic knowledge to the study of large peripheral arteries in subjects with hypertension, particularly to the mechanism of the disproportionate increase in systolic blood pressure observed in older subjects.

ARTERIAL HYPERTROPHY AND COMPOSITION OF THE VASCULAR WALL IN HYPERTENSION

Based on histomorphometry, arterial hypertrophy has been recognized for many years in animal hypertension. Nowadays the same finding may be observed in populations of human hypertensive subjects in which wall thickness may be measured transcutaneously using echo-Doppler techniques. Thus, it is important to describe the structural characteristics of the hypertensive arterial wall in animals and men [8].

The medial layer normally constitutes most of the thickness of an artery. During growth and development, there is, in parallel with the rise in blood pressure (and particularly in animal hypertension), a rapid increase in the number of medial smooth muscle cells and their organization into lamellar units. In genetic hypertension in rats, medial hypertrophy occurs very early and may be observed even at birth. When comparing adults of different species, the number of lamellar units and the overall wall thickness reflect the size of the animal and the diameter of the vessel [9]. The calculated tension per lamellar unit is relatively constant throughout a wide range of species and vessel sizes. Hypertension causes an increase in thickness in the arterial media that serves to counteract the rise in wall tension. During hypertension, the number of lamellar units remains relatively constant and the increase in wall thickness is achieved by changes both in cellular mass and in connective tissue content [8, 9], which have of course different particularities.

The increase in blood vessel wall mass in chronic hypertension (i.e. arterial hypertrophy) is mainly due to vascular smooth muscle cells and can be seen in vessels of all sizes, ranging from large conduit arteries to resistance arterioles. The enhanced muscle mass might be caused by an increase in cell size (hypertrophy) or by an increase in cell number (hyperplasia) [10]. In animal studies, smooth muscle hypertrophy appears to predominate in the large conduit vessels, whereas hyperplasia is observed in the small arteries and arterioles. Many investigators have also reported an increase in cellular DNA content associated with the increase in cell mass (polyploidy) [10].

The smooth muscle cells of the arterial media are encircled by a dense network of connective tissue which, in fully relaxed vessels, constitutes the major component of the mechanical properties of the arteries. Experimental hypertension causes an increase in arterial collagen, elastin and in a number of other proteins [8-12] presumably as a result of stimulation of their production by smooth muscle cells. The rise in collagen and elastin levels is approximately proportionate to the increase in arterial weight. Recent data suggest that the organization of connective tissue might be influenced by a complex family of receptor proteins called integrins, which bind specifically to individual connective tissue components [11]. Another set of connective tissue proteins receiving considerable recent attention are the fibronectins. These large glycoproteins with a relative molecular mass of approximately 500 000 are composed of two subunits with repeating sequences that bind fibrinogen, heparin and collagen [12]. Finally proteoglycans [10] are polyanionic glycosaminoglycans capable of binding a great quantity of sodium and calcium ions, particularly in response to a change in the osmotic composition of the extra-cellular space, thus modifying the arterial mechanical properties and the local pressure.

In some cases, an invasion of inflammatory cells is observed and might modulate the remodeling process by the secretion of growth factors, proteases and cytokines. However foam cells are not observed in the absence of hyperlipemia and atherosclerosis. There are arguments for endothelial dysfunction in hypertensive vessels, although this abnormality may be somewhat controversial in humans [13, 14]. Finally, there is intimal and subintimal thickening and in some cases, vascular smooth muscle cell-like cells have been reported to migrate into the subintimal space.

With aging, arteries progressively stiffen, due to thickening of media and intima with accumulation of collagenous fibers, and deposition of calcium with degeneration of the elastic laminae [2]. The loss of distensibility is associated with a progressive dilatation of arteries predominantly due to the fragmentation and rupture of elastin fibers, a specific aspect of the aging process independent of high blood pressure [2]. As arteries stiffen, the pulse wave velocity, the most classical marker of arterial stiffness, progressively increases. These changes, which are most pronounced in the aorta, are often attributed to the fatiguing effect of cyclic stress acting over many decades [2]. Experimental [15, 16] and clinical [3] studies have indicated that cyclic stress might contribute to the development of extracellular matrix and subsequently to arterial damage. A major manifestation of these changes is the rise in systolic and pulse pressure [2,3], a finding largely shown by longitudinal epidemiological studies in humans [7]. Whereas the deleterious effect of mechanical stress on the arterial wall seems to be well established, the possible role of vascular endothelium should also be considered nowadays. In contrast to hypertension-induced endothelium changes, the deleterious effect of age on endothelium function is universally accepted and might play a major role to vascular alterations [14].

Because it is often postulated that the vascular remodeling induced by high blood pressure might predispose subjects to the end-organ damage observed in hypertension, one may speculate that pharmacological or nutritional interventions would have not only to modify blood pressure, but also to produce changes in arterial structure. In the past, most of the pharmacological interventions published in the literature were performed using prevention protocols in animals, the principal goal of which being the pathophysiological approach of hypertension. Recently, therapeutic protocols have been performed, allowing us to demonstrate that antihypertensive drugs might reverse the structural changes of the large arteries in animal models of hypertension. Thus, the effects of sodium diet and diuretics [17], converting enzyme inhibitors [18,19], calcium entry blockers [20,21] and β-blocking agents [22] have been investigated in detail.

Regarding sodium, it has been reported that, in genetic hypertension in rats, high sodium intake is associated with increased thickness of the arterial wall and accumulation of connective tissue [17,23]. At the exception of Dahl sensitive rats, this pathological process is not accompanied by significant changes in blood pressure. This lack of change in blood pressure is particularly observed in stroke-prone spontaneously hypertensive rats [17]. Furthermore, in this strain, decreased sodium intake or administration of diuretics may reverse the structural arterial alterations independently of blood pressure level and induce a disappearance of cerebrovascular accidents.

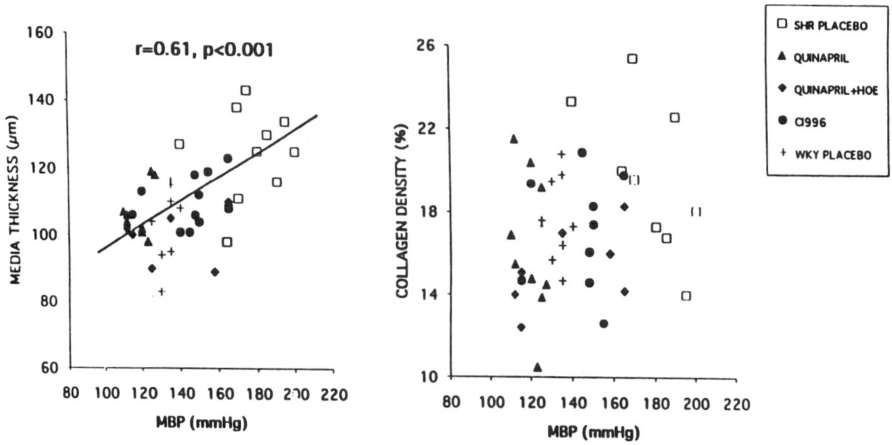

Figure 1. *Relationship of mean blood pressure (MBP) with media thickness (on the left) and collagen density (on the right), in a population of normotensive (WKY) and hypertensive rats (SHR), either treated by placebo or by the converting enzyme inhibitor Quinapril in the presence or absence of the bradykinin B_2 inhibitor Hoe 140, or by the AT_1 Angiotensin II receptor antagonist (CI 996) [18]. Note that only media-thickness, and not collagen density was correlated with blood pressure.*

Regarding antihypertensive agents, many studies have shown that in SHRs carotid hypertrophy was reversed in conjunction with the decrease in blood pressure. Whereas the decrease in arterial smooth muscle mass seems to be principally mediated by the reduction in blood pressure, there is less information concerning the behavior of extracellular matrix (Figure 1). A decrease in the collagen content of the arterial wall has been obtained with long-term treatments involving calcium entry blockers or converting enzyme inhibitors [18-21]. In this latter case, the decrease in collagen was not directly related to the blood pressure reduction because it was observed even using non-antihypertensive doses (Figure 1) [19]. In hypertensive rats, blockade of AT_1 receptors of Angiotensin II, and not bradykinin, has been shown to be responsible for the reduction in aortic collagen in vivo [18].

Taken together, these findings indicate that arterial hypertrophy, a characteristic feature of various models of human and animal hypertension, may be reduced by administration of antihypertensive agents. There is less information on the reversibility of the changes in extra-cellular matrix and in the mechanical properties of the arterial wall. Such structural alterations are related either to the effect of mechanical stimuli such as the blood pressure itself [either in its steady (mean arterial pressure) or pulsatile (pulse pressure) component]; or to the contribution of humoral and/or autocrine-paracrine growth factors (such as those related to the renin-angiotensin system); or to a combination of both. The respective contribution of each of these different mechanisms acting on the arterial wall is yet difficult to establish.

ARTERIAL HYPERTROPHY AND STIFFNESS OF WALL MATERIAL BASIC DEFINITIONS

Compliance and distensibility are the most widely used parameters to evaluate arterial function. Both of them relate to the storage capacities of the vessels, i.e. the ability to change the cyclic flow coming from the heart into a continuous flow at the capillary level [2]. These storage capacities are influenced not only by the level of blood pressure but also by the mechanical properties of the vessels, which relate to several mechanisms. Not only the degree of vascular hypertrophy, the tone of vascular smooth muscle and the histological composition of the arterial wall are involved, but also the stiffness of wall material. Nowadays, a quantitative evaluation of wall stiffness is usually obtained from the determination of the arterial incremental elastic modulus, also called Young modulus.

Definition of the incremental elastic modulus (Einc)

The law of Laplace states that the tension (T, force per unit length) in the wall of a cylindrical vessel is related to transmural pressure (PT) and radius (r): T = PT x r [24]. The circumferential stress on the wall depends on the wall thickness, so that the circumferential stress exerted by the tissue is:

$$\sigma\,\theta = PT \times ri/h \qquad (1)$$

where ri is internal radius and h thickness of the artery.

Circumferential stress is used to determine the incremental elastic modulus provided arterial wall motion is detected and therefore the strain-stress relation of the artery may be established. The circumferential strain ($\in_\theta$) of an artery may be :

$$\in_\theta = (d - do)/do \qquad (2)$$

where d is the observed internal diameter and do is the baseline internal diameter. Usually, do is defined either as the diameter of the retracted, totally unloaded vessel, or as diameter at low or 0 mmHg pressure or as unstressed diameter [2,24,25]. The latter determination is theoretically the most appropriate for materials having linear strain-stress relations. The ratio of stress to strain may be used to compute the Young's modulus of elasticity, i.e. the stretching force per unit cross-sectional area required to elongate a strip of a vessel wall by 100 %. However, for materials with non-linear strain-stress relations as arterial vessels, the slope of the strain-stress curve should be used to determine an incremental elastic modulus. Thus, for a cylindrical vessel with an equal wall stiffness in all directions, Einc is:

$$E_{inc} = \Delta \in_{\theta} \text{ x } 0.75/\Delta \in_{\theta} \qquad (3)$$

Note that, for a given vessel, the assumption of equal stiffness in all directions may be highly questionable, as indicated below.

In vitro measurement of Einc

In vitro, the static stress-strain relationship is determined using specific preparations in which the distending pressure is measured by increasing steps, and, at each level of distending pressure, the internal radius of the vessel and the wall thickness are evaluated. Stress is calculated according to the Laplace law (equation 1) and strain, the fractional increase in circumferential length, using equation 2.

As an example [26], an arterial vessel is immediately placed after surgical excision, in a cold (4°C) Krebs solution aerated with a gas mixture of 95 % O_2 / 5% CO_2 at pH 7.4. After removing fat and loose tissue from the adventitia, each terminal portion of the segment is cannulated and then perfused-fixed in a perfusion system consisting of a reservoir connected to the vessel and to a pressure transducer. Endothelium function should be preserved, a key point which is often difficult to assess [26]. Pressurization of the vessel is achieved by inflating a cuff around the reservoir and closing the vessel by turning a stopcock positioned distally to the artery. Since all arteries displays longitudinal retraction when they are harvested, the length of each arterial segments is measured carefully before excision. Because the longitudinal retraction in general approximates 10 ± 8%, the vessel is stretched after cannulation to 110 % of its retracted length and maintained at this fixed length throughout the study. Before measuring the arterial diameter and wall thickness, the artery is inflated several times to a pressure of up 250 mmHg in order to exclude the excessive hysteresis found on first inflation.

Pressure-diameter curves are generally obtained by increasing intraluminal pressure in 25 mmHg steps to 200 mmHg. The transmural pressure is maintained at each level until the vessel segment exhibits a steady diameter for at least 2 min. At each level of transmural pressure, internal diameter and wall thickness are measured in baseline conditions and, thereafter, in constricted and then fully relaxed vessels (Figure 2). Note that, in such preparations, diameter may be measured at zero pressure, enabling a true application of equation 2. Nowadays ultrasound techniques (as in Figure 2) are widely used to measure diameter and thickness with a high degree of precision and reproducibility [3]. Previously, wall thickness was simply calculated from the dry weight of the artery, using a cylindrical model of the vessel [2].

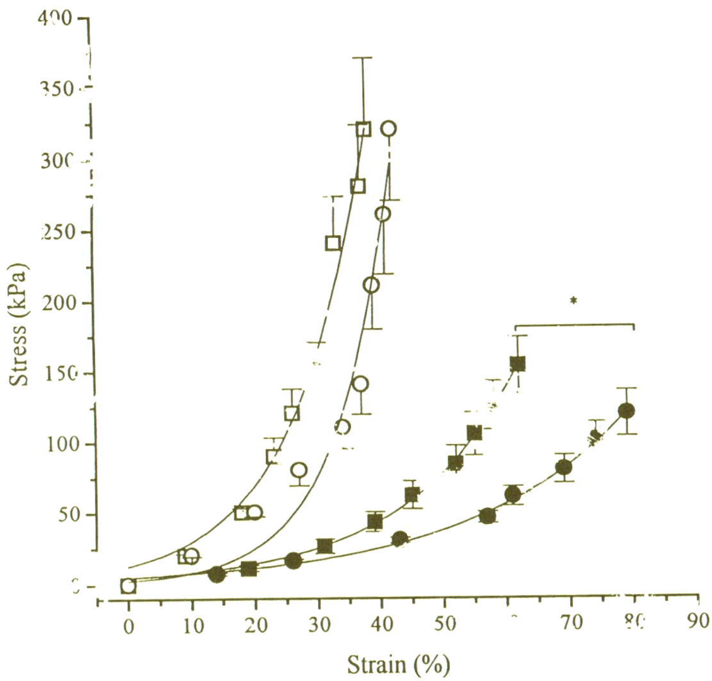

Figure 2. Stress-strain relationship of the radial artery (■ ●) and of the internal mammary artery (❑ ◯) in baseline conditions (❑ ■) and under norepinephrine (◯ ●). E_{inc} is the slope of the strain-stress curve [26]. Note that E_{inc} is steeper for the internal mammary than for the radial artery and that, under norepinephrine, the curve is shifted to the right. The shift is significant only for the radial artery, indicating a decrease in the stiffness of wall material under norepinephrine ($P < 0.05$).*

In vivo measurement of Einc

In animals and humans, the stiffness of vascular wall can be estimated using a non invasive approach of the pressure-diameter curve within the systolic-diastolic range of the operational blood pressure. The determination, which is performed in living conditions, respects blood flow, smooth muscle tone and the integrity of endothelium. Diastolic diameter is usually accepted as representing baseline diameter. Nowadays, systolic-diastolic changes of internal diameter of various large arteries like the common carotid artery or the radial artery can be recorded using high resolution echotracking techniques [2, 3]. Systolic-diastolic changes of blood pressure can be measured during the same experiment using aplanation tonometry or photoplethysmography [2]. As examples, mean wall thickness can be determined with a high degree of reproducibility at the site of the common carotid artery and of the radial artery, and systolic-diastolic changes in wall thickness can be measured at the site of the radial artery [3]. Nevertheless, such clinical approaches have two disadvantages. Firstly, unstressed diameter cannot be evaluated. Secondly, the mechanical properties of the arterial wall under fully relaxed conditions cannot be determined, a major difference with in vitro preparations.

INCREMENTAL ELASTIC MODULUS AND THE COMPONENTS OF THE ARTERIAL WALL BEARING STRESS

For a long time, the mechanical characteristics of fully relaxed vessels have been attributed to the properties of the fibrous connective tissues in the wall, specifically to those of elastin and collagen. Elastin and collagen are fibrous in nature, although elastin tends to occur in more or less continuous sheets [8]. Elastin is capable of extending more than 100 % [24, 25]. The elastic modulus of elastin, a measure of stiffness, gradually rises from very low values to about 4 0 x 10^6 dyne/cm^2 in the fully stretched state [24]. By contrast collagen is much stiffer. When the slack in the constituent fibers has been taken up, collagen can be stretched only 3-4%, exhibiting an elastic modulus of 0.3-2.5 x 10^{10} dyne/cm [24]. Collagen and elastin are complexed with an amorphous substance formed of mucoprotein. This substance is not itself elastic but contributes in an unquantified extent to the elastic properties of the artery. Another constituent of the arterial wall is smooth muscle which is involved in the production of extracellular matrix, including collagen and elastin. Smooth muscle, while contributing markedly to the tension in the wall in vivo, cannot be properly regarded as a true elastic material [24].

Dobrin and collaborators [24,25], from enzymatic-degradation studies in dog carotid arteries, delineated the contributions of elastin and collagen to three-dimensional elastic properties of arteries. They suggested that elastin bears a constant portion of longitudinal stress at all vessel lengths within the physiological range, a gradually increasing portion of circumferential stress with vessel distension, and a gradually increasing portion of radial stress with progressive compression of the wall. Collagen bears little longitudinal or radial stress with physiological levels of deformation but does bear a gradually increasing portion of circumferential stress with vessel distension beginning at about 60 mmHg.

The circumferential stress given by equation 1 is the mean circumferential stress. However this stress is not uniformly distributed across the arterial wall. It has been shown by several investigators that the point stress is high at the lumen and declines curvilinearly across the wall thickness [24,25]. This distribution closely resembles the density of the elastic lamellae at different points across the wall; elastic lamellae are close together near the lumen and are more widely spaced near the adventitia. Dobrin [24] suggested that an adaptive relationship may exist between morphology and mechanical load at different locations across the wall. Wolinsky & Glagov [9] previously proposed a similar argument as they noted a virtually constant relationship between the total number of elastic lamellae in the aortic wall and the mean wall tension for a wide variety of mammalian species. Another observation is that the circumferential stress is greater on the inner radius of the aortic arch than on the outer radius, this corresponding with greater wall thickness on the inner curve.

Recent works [11,27-29] have drawn attention to the possible involvement of the integrin-type fibronectin receptors of the rat arterial smooth muscle cell in the interconversion between contractile and synthetic phenotypes [11] and mostly in the linkage between the extracellular matrix and smooth muscle cells [28,29]. The latter change may contribute to the distribution of the mechanical forces within the arterial wall through the deformation of cell cytoskeleton, mechanical deformation of transmembrane proteins at the apical surface of the cell, and the activation of signal transduction pathways by the transmission of mechanical loads to focal adhesion sites [27,28]. The integrin family of protein has been shown to be involved in the adhesion of fibronectin to smooth muscle cells [11] and could thus participate to the adhesion of smooth muscle cells to collagen and elastin fibers [27,28]. In addition to integrins, numerous other mechanisms, as those related to proteoglycans (see earlier), to the presence of stretch activated ions channels or to specific receptors have been proposed as playing a role in the distribution of mechanical forces within the arterial wall, and hence in the mechanism of the visco-elastic properties of conduit arteries [29]. For instance, it has been suggested that chronic blockade of AT_1 angiotensin II receptors results in normal arterial pressure in angiotensin II induced hypertensive rats, but leads to significant aortic hypertrophy and fibrosis [30, 31]. On the other hand, chronic blockade of AT_2 receptors has no effect on arterial pressure, although it antagonizes the effect of angiotensin II on arterial hypertrophy and fibrosis. Thus in vivo, only the AT_1 receptors of angiotensin II seem to be sensitive to tensile forces. Further studies are needed in order to determine whether these different mechanisms may influence the elastic modulus of the arterial wall through an increase in the mass of its different constituents or through their reorganization inside the arterial wall or through changes in vasomotor tone and endothelial function.

APPLICATIONS TO CLINICAL HYPERTENSION

For the evaluation of arterial hypertrophy in subjects with clinical hypertension, two different straight, superficial and cylindrical arteries have been widely studied = the common carotid artery and the radial artery [4]. The latter is composed almost exclusively of vascular smooth muscle. For this artery, there is no pathology, no atherosclerotic plaque and little role of age. Thus, when, in the presence of hypertension, radial artery hypertrophy is observed, this hypertrophy may be entirely attributed to high blood pressure, exactly as it is observed in spontaneously hypertensive rats. The situation is quite different for the musculo-elastic common carotid artery. Not only hypertension may be responsible for carotid hypertrophy but also a lot of cofactors as atherosclerosis, cholesterol excess, plasma glucose, insulin, tobacco consumption etc. [4]. Finally, when carotid hypertrophy is present, it is more difficult to attribute hypertrophy to high blood pressure itself. Despite these difficulties, both type of arteries should be investigated in subjects with clinical hypertension since the radial artery is a model of peripheral muscular artery and the carotid artery rather a model of central musculo-elastic artery.

Arterial hypertrophy with normal stiffness of wall material in subjects with essential hypertension of the middle age

Recent studies by Laurent *et al* [32] have shown that in subjects with hypertension of the middle age, carotid and radial hypertrophy is an adaptive phenomenon as a consequence of the Laplace law. In the presence of high blood pressure and of an increased (carotid artery) or normal (radial artery) lumen diameter, hypertrophy compensates to maintain a normal wall stress [32, 33]. In such hypertensive subjects, the stiffness of wall material (E_{inc}) was found normal in operational conditions (Figure 3), and even decreased in isobaric conditions [34]. Conversely, different values of compliance and distensibility were found according to the topography of the vessels : for the radial artery, normal (operational conditions) or increased (isobaric conditions) values have been reported while, for the carotid artery, reduced (operational conditions) or normal (isobaric conditions) values are observed [32-34]. In hypertensive subjects of the middle age, the mechanisms for the normal or decreased values of Einc remains difficult to explain in the presence of arterial hypertrophy. As we mentioned earlier, transmission of mechanical forces within the arterial wall require sophisticated adaptive processes involving the presence of specific receptors and/or stretch activated ions channels [29]. The most important mechanisms are certainly those related to forces elicited as matrix attachment sites and involving integrins. One finding of major importance was the demonstration that the cell cytoskeleton may be made to stiffen in response to mechanical stimuli through specific ligands implying integrin receptors [see review in 11, 12,29].

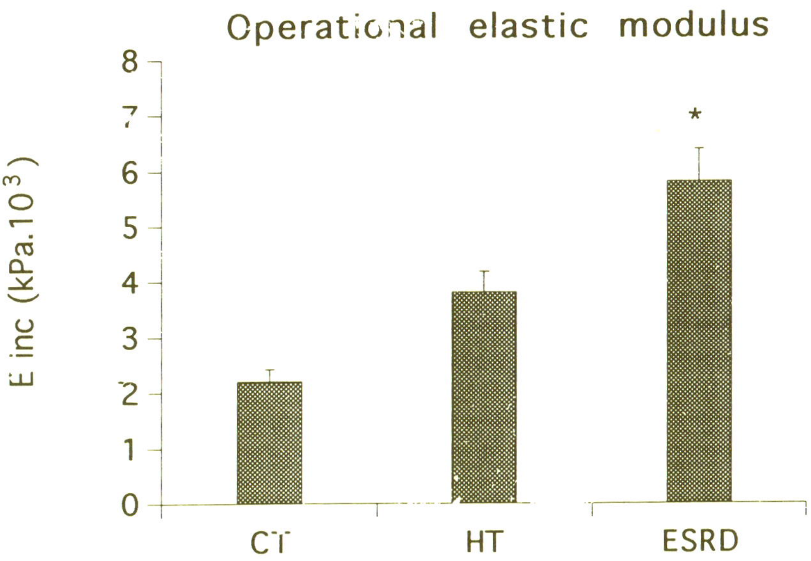

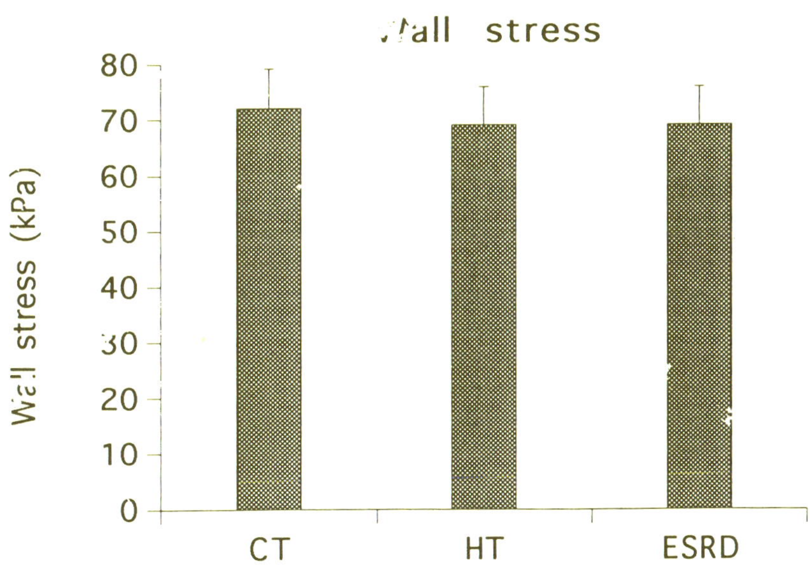

Figure 3. *Wall stress and incremental elastic modulus (E_{inc}) in control subjects (CT), in subjects with essential hypertension in the middle age (HT) and in patients with end-stage renal disease (ESRD) [34]. (*$P < 0.005$). Note that for the same wall stress, E_{inc} is increased only in ESRD patients.*

Arterial hypertrophy and increased stiffness of wall material in various models of human hypertension

Although the stiffness of wall material remains within the normal range in hypertensive subjects of the middle age, there are specific circumstances in which vascular hypertrophy is associated with an increased stiffness of wall material. This alteration is observed in two different populations: patients with end - stage renal disease and hypertensive subjects in the elderly.

In end - stage renal disease, the stiffness of wall material is increased at the site of both the carotid and the radial artery [34, 35]. The increase is independent of blood pressure level and/or wall stress (Figure 3) [34, 35]. Vascular hypertrophy is associated with increased (carotid artery) or normal (radial artery) diameter and reduced isobaric and operational compliance and distensibility. Structural damage of the arterial wall may be responsible for the increased arterial stiffness. In subjects with end stage renal disease, the role of diabetes and/or of calcifications as factors stiffening the arterial wall has been already recognized at the site of central arteries. Diabetes may particularly contribute to the formation of glycation products. On the other hand, the contribution of biochemical compounds related to advanced renal failure has also been suggested to favor increased stiffness. Water logging related to overhydratation, hyperhomocysteinemia and endothelin excess are among the principal factors which have been shown to be strongly associated with increased pulse wave velocity and E_{inc} [36].

In hypertension in the elderly, the situation is more complex. An increased stiffness of the wall material has been recognized, but only at the site of the carotid, and not of the radial artery [37]. The increased stiffness occurs even in the presence of a normal or low circumferential wall stress, implying a major role for non hemodynamic functional and structural factors. As we mentioned earlier, it is well accepted that, at the site of the musculo elastic carotid artery, the increase in lumen diameter with age is due to the alteration of elastic fibers, whereas the decreased distensibility is considered to be due to the accumulation of rigid material as collagen fibers. It has even been suggested that the lumen enlargement contributes to maintain the distensibility and the Windkessel function of the central conduit arteries despite the presence of a significant decrease in compliance. Thus, with aging, the increased rigidity of wall material may contribute to maintain an unchanged wall stress at the price of a reduced strain (see Figure 2 as an example). However, these classical alterations are commonly described in the totality of the population of old subjects, whether they are normotensives or hypertensives. Thus, more specific mechanisms should be involved to explain the increased stiffness of wall material in subjects with systolic hypertension in the elderly. At first approximation, several environmental and/or genetic factors could contribute to stiffen the arterial wall in old hypertensive subjects. First, we have already mentioned that sodium excess may be responsible for structural alterations of arterial vessels, including increased wall thickness and accumulation of connective tissue [17, 23]. Increased sodium sensitivity is a classical feature in old subjects with hypertension. Second, the loss of estrogens after menopause is also associated with increased amounts of rigid connective tissue [38]. Pulse wave velocity is lower in women than in men before menopause and becomes equal after

[2]. Third, genetic factors may be involved in the mechanism of arterial stiffening. A polymorphism of AT_1 receptor genes of Angiotensin II has been found to be significantly associated with high pulse wave velocity in populations of subjects with essential hypertension. This polymorphism is significantly more pronounced in older than in younger subjects [39]. Finally, atherosclerosis itself, independently of aging and hypertension, may favor the development of arterial collagen tissue, particularly at its advanced stage, when diffuse alterations and calcifications are present [2,3]. In contrast, early changes of atherosclerosis, particularly those in which foam cells predominate, do not contribute to increase the stiffness of arterial wall material in subjects with hypertension.

Taken together, these findings suggest that increased rigidity of wall material is as important as vascular hypertrophy to consider in subjects with hypertension. Indeed, vascular hypertrophy is present in all kinds of hypertension and occurs very early in the course of the disease. Nevertheless, only the most severe forms of hypertension, as those related to end-stage renal disease or aging, are susceptible to involve an increased stiffness of wall material in association with vascular hypertrophy.

INTERACTIONS BETWEEN STIFFNESS OF WALL MATERIAL AND THE LEVEL OF BLOOD PRESSURE.

It is well established that, in subjects with hypertension of the middle age, both mean arterial pressure and pulse pressure are augmented. However, in this population, the augmentation of systolic, diastolic and mean arterial pressure is proportional, in association with an unchanged elasticity of vascular wall material. In contrast, in hypertension in the elderly, mean arterial pressure may be either augmented or even normal, but the increase in systolic and diastolic blood pressure are no longer proportional. It results in a specific pattern of disproportionate increase in systolic over diastolic blood pressure or even of isolated systolic hypertension. In this population, the elasticity of wall material is increased [2,3,37,40]. For a long time, it has been suggested that the increased rigidity of the arterial wall may be responsible per se for the high systolic peak and the disproportionate increase in pulse pressure in hypertensive subjects in the elderly. Alternatively, it has been proposed that the increase in pulse pressure is the principal trigger mechanism favoring vascular hypertrophy and increased rigidity of wall material [2]. Thus, it is important in subjects with hypertension to dissociate the two different phenomenons, vascular hypertrophy and increased stiffness of wall material, and to relate each of them to the corresponding patterns of elevated blood pressure. For this, it is necessary to recall several aspects of pulsatile arterial hemodynamics as observed in population of hypertensive subjects.

It is widely accepted that the aortic blood pressure curve is due to the mathematical summation of [2] : (i) an incident pressure wave, which, following ventricular ejection, propagates along the arterial tree from the heart toward peripheral vessels, and (ii) a reflected wave which returns from peripheral (resistant) vessels toward the heart. Whereas the forward wave is simply influenced by the pattern of ventricular ejection and the aortic stiffening, the reflected wave depends on three different parameters: the value of reflection coefficients, the degree of arterial stiffening, and the distance between reflection points and the heart. For a given value of reflection coefficients, increased pulse wave velocity and, additively, reflection sites closer to the heart, cause an earlier return of the backward pressure wave toward the heart, with a resulting higher aortic pulse pressure and systolic peak. This latter pattern, which is the dominant characteristic of hypertension in the elderly, clearly indicates that the major factor contributing to increase selectively pulse pressure and to cause a disproportionate increase in systolic over diastolic pressure is the elevation of pulse wave velocity and the subsequent alteration in the timing of reflected wave. Pulse wave velocity is indeed influenced by blood pressure itself, but also by the rigidity of wall material : at a given blood pressure, the more rigid the arterial wall material, the higher the pulse wave velocity.

According to the Moens Korteweg equation, pulse wave velocity (PWV) depends on wall thickness and Einc, but also on arterial radius [2]. Thus, for a given value of elastic modulus and wall thickness, PWV may be lower if arterial radius is increased. Thus, in the presence of increased arterial radius and diameter, a disappearance of the disproportionate increase of systolic over diastolic blood pressure may be expected through a decrease in PWV, and this might be obtained without any change in Einc and wall thickness. This hemodynamic alteration is readily obtained by the use of nitrates. Animals and human studies have shown that nitrates exert a preferential action in larger as opposed to smaller arteries [2]. Arterial dilation occurs even in the presence of a significant decrease in mean arterial pressure and without any significant change in ventricular ejection, vascular resistance and blood flow velocity. Nitrate - induced smooth muscle relaxation is associated with an increase in isobaric distensibility, without any change in isobaric E_{inc} [41]. In subjects with systolic hypertension in the elderly, acute nitroprusside is able to increase systemic compliance and to decrease pulse wave velocity, thus causing a selective decrease in systolic and pulse pressure (Figure 4) [42]. O'Rourke and collaborators [2] have detailed the pharmacological mechanism by which the selective decrease in systolic and pulse pressure may be obtained. First, nitrate-induced smooth muscle relaxation occurs mainly at the site of peripheral muscular medium-size arteries, causing a decrease in PWV. Second, as a consequence of the PWV reduction, there is a delay in the backward pressure wave, which returns during the diastolic and not the systolic component of the aortic pressure curve. Subsequently, peak systolic blood pressure decreases selectively, with a more pronounced effect in central (Figure 5 upper panel) than in peripheral arteries (Figure 5 lower panel).

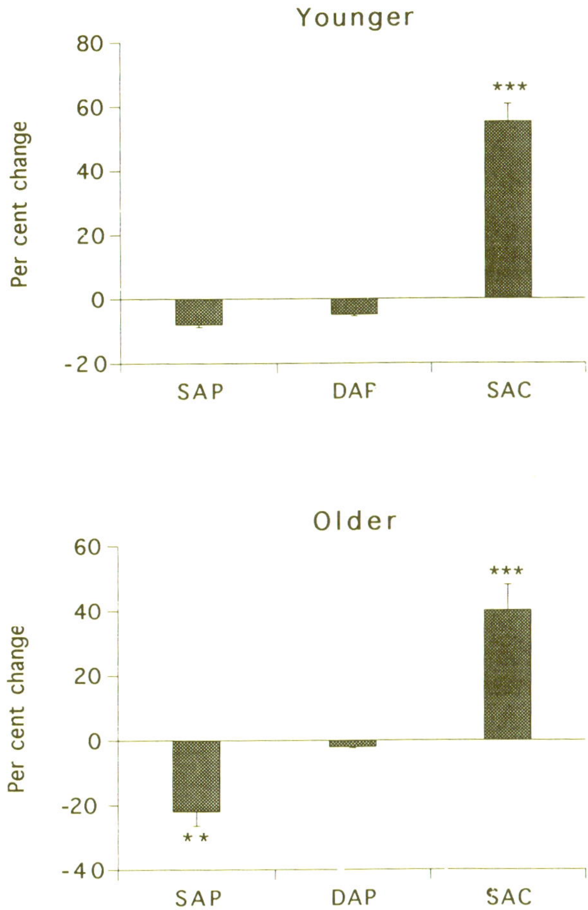

*Figure 4. Systolic hypertension in the elderly : decrease in systolic arterial pressure (SAP), and not in diastolic arterial pressure (DAP) obtained with acute nitroprusside. Note that a comparable result is not obtained in younger subjects despite the same increase in systemic arterial compliance (SAC), probably as a consequence of sympathetic activation [42]. (** $P < 0.01$; *** $P < 0.001$).*

Taken together, such findings show that a pharmacological reduction of increased pulse pressure may be obtained acutely using nitrates, without altering E_{inc} or the composition of the arterial wall. It is so demonstrated that the disproportionate increase in systolic over diastolic blood pressure in hypertension is not the direct consequence of arterial hypertrophy but rather of the increase in pulse wave velocity and change in the timing of wave reflections, two major consequences of increased stiffness of wall material.

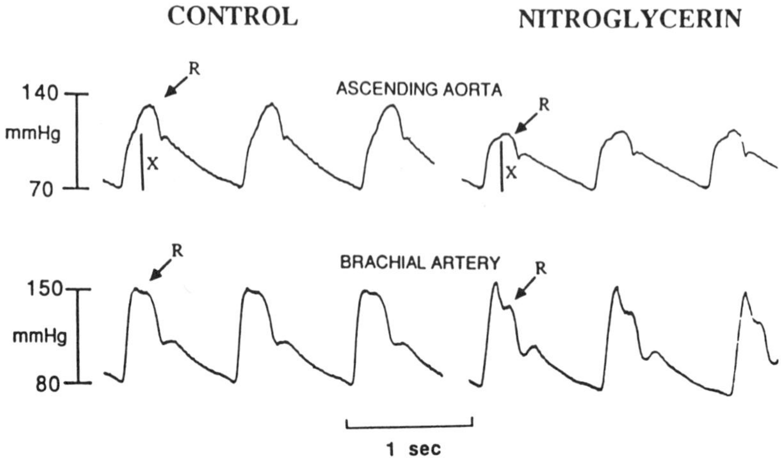

***Figure** 5. Administration of nitroglycerine with selective decrease in systolic blood pressure at the site of the aorta but not the brachial artery [2], as a consequence of the change in the timing of wave reflections.*

CONCLUSIONS

Epidemiological studies have shown that cardiovascular morbidity and mortality increase with age and high blood pressure [1,40]. However, the increase in morbidity and mortality becomes much more pronounced above 50 years of age, a period during which an increase prevalence of disproportionate increase in systolic over diastolic blood pressure is observed [1-3,7]. The disproportionate increase in systolic blood pressure is associated with : (i) an increased stiffness of wall material unrelated to wall stress, and (ii) a selective increase in pulse wave velocity with resulting alterations in wave reflections.

Through modification in the timing of wave reflections, it is possible to reduce pharmacologically the disproportionate increase in systolic blood pressure using acute administration of nitrates. The reduction of systolic blood pressure is also obtained in the long term using antihypertensive agents as diuretics or calcium entry blockers [43,44]. However, these agents not only decrease systolic blood pressure but also diastolic blood pressure, indicating that the major contributive factor to the disproportionate increase in systolic blood pressure, i.e., the increased stiffness of wall material, remains unmodified. Much work remains to be performed in order to alter the structural alterations of arterial wall material which are responsible for systolic hypertension in the elderly.

ACKNOWLEDGEMENTS

This study was performed with the help of INSERM. We thank Mrs. Anne Safar for her skilful technical help.

REFERENCES

1. KANNEL WB and STOKES JLL,1985, Hypertension as a cardiovascular risk factor. Handbook of hypertension. Epidemiology of Hypertension, B.C., Editor. Elsevier Science: Amsterdam, pp. 15-34.

2 NICHOLS WW and O'ROURKE MF, 1998, McDonald's blood flow in arteries. Theoretical, experimental and clinical principles, *in* Fourth Edition, ARNOLD E, Editor: London, Sydney, Auckland. pp. 54-113, 201-222,284-292, 347-401.

3. SAFAR ME, 1989. Pulse pressure in essential hypertension: clinical and therapeutical implications. *J of Hypertens*, 7: 769-776.

4. SAFAR ME, GIRERD X and LAURENT S, 1996. Structural changes of large conduit arteries in hypertension. *J of Hypertens*, . 14: 545-555.

5. GIRERD X, MOURAD J-J, COPIE X, MOULIN C, ACAR C, SAFAR M and LAURENT S, 1994. Non invasive detection of an increased vascular mass in untreated hypertensive patients. *Am J Hypertens*, . 7: 1076-1084.

6. ROMAN MJ, SABA PS, PINI R, SPITZER M, PICKERING TG, ROSEN S, ALDERMAN MH and DEVEREUX RB, 1992. Parallel cardiac and vascular adaptation in hypertension. *Circulation*, . 86: 1909-1918.

7. FRANKLIN SS, GUSTIN W IV, WONG ND, LARSON MG, WEBER MA, KANNEL WB and LEVY D, 1997. Hemodynamic patterns of age-related changes in blood pressure. The Framingham heart study. *Circulation*, . 96: 308-315.

8. GLAGOV S,1972, Hemodynamic risk factors: mechanical stress, mural architecture, medial nutrition and vulnerability of arteries to atherosclerosis, *in* The pathogenesis of atherosclerosis, Wissler RW and Geer JC, Editors. Williams & Wilkins CO: Baltimore. pp. 164-199.

9. WOLINSKI H and GLAGOV S, 1967. A lamellar unit or aortic medial structure and function in mammals. *Circ Res*, 111: 20-99.

10. SCHWARTZ SM, HEIMARK RL and MAJESTY MW, 1990. Developmental mechanisms underlying pathology of arteries. *Physiol Rev*, 70: 1177-1198.

11. HYNES RO, 1987. Integrins : A family of cell surface receptors. *Cell*, . 48: 549-554.

12. KORNERBLIHTT AR, CHAN CT, FRANZBLAU C, FARIS B and CHOBANIAN AV, 1978. Cell specific alternative mRNA splicing generates polypeptide chains differing in the number of internal repeats. *Nucleid Acids Res*, 12: 5853-5868.

13. VAN ZWIETEN PA, SAFAR M, LAURENT S, PFAFFENDORF M, MAARTEN GC, HENDRIKS MGC and BRUNING TA, 1995. New insights into the role of endothelial dysfunction in hypertension. *J of Hypertens*, . 13: 713-716.

14. VANHOUTTE PM and BOULANGER CM, 1995. Endothelium-dependent responses in hypertension. *Hypertens Res*, 18: 87-98.

15. BAUMBACH GL, SIEMS JE and HEISTAD DD, 1991. Effects of local reduction in pressure on distensibility and composition of cerebral arterioles. *Circ Res*, 68: 339-351.

16. CHRISTENSEN KL, 1991. Reducing pulse pressure in hypertension may normalize small artery structure. *Hypertension*, 18: 722-727.

17. LEVY BI, POITEVIN P, DURIEZ M, GUEZ DC, SCHIAVI PD and SAFAR ME, 1997. Sodium, survival and the mechanical properties of the carotid artery in stroke-prone hypertensive rats. *J of Hypertens*, 15: 251-258.

18. BENETOS A, LEVY BI, LACOLLEY P, TAILLARD F, DURIEZ M, and SAFAR ME, Role of angiotensine II and bradykinine on aortic collagen following converting enzyme inhibition in spontaneously hypertensive rats. 1997 , *Arterioscler Thromb Vasc* Biol. 17: 3196-3201.

19. ALBALADEJO P, BOUAZIZ H , DURIEZ M , GOHLKE P, LEVY B , SAFAR M and BENETOS A , 1994. Angiotensin converting enzyme inhibition prevents the increase in aortic collagen in rats. *Hypertension*, . 23: 74-82.

20. LACOLLEY P, GHODSI N, GLAZ□ER E, CHALLANDE P, BRISSAC AM, SAFAR ME and LAURENT S, 1995. Influence of graded changes in vasomotor tone on the carotid arterial mechanics in live spontaneously hypertensive rats. *Br J Pharmacol*, 115: 1235-1244.

21. LEVY BI, DURIEZ M, PHILLIPE M, POITEVIN P and MICHEL JB, 1994. Effect of chronic dihydropyridine (Isradipine) on the large arterial walls of spontaneously hypertensive rats. *Circulation*, . 90: 3024-3033.

22. BENETOS A, POITEVIN P, PROST PL, SAFAR ME, and LEVY BI, Life survival and cardiovascular structures following selective beta-blockade in spontaneously hypertensive rats. 1994 , *Am J Hypertens*. 7: 186-192.

23. LIMAS C, WESTRUM B, LIMAS CJ and COHN JN, 1980. Effect of salt on the vascular lesions of spontaneously rats. *Hypertension*, 2: 477-489.

24. DOBRIN PB, 1978. Mechanical properties of arteries. *Physiol Rev*, 58: 397-460.

25. DOBRIN PB and MRKVICKA R, 1992. Estimating the elastic modulus of non-atherosclerotic elastic arteries. *J. ofHypertens.*, 10 (suppl. 6): S7-S10.

26. CHAMIOT-CLERC P, COPIE X, RENAUD JF, SAFAR M and GIRERD X, 1997. Comparative reactivity and mechanical properties of human isolated internal mammary and radial arteries. *Cardiovasc Res*, In press.

27. LANGILLE BL, 1993. Remodeling of developing and mature arteries: endothelium, smooth muscles, and matrix. *J of CardiovasculPharmacol*, 21 (suppl I): S11-S17.

28. DAVIES PF, 1995. Flow-mediated endothelial mechanotransduction. *Physiol Rev*, 75: 519-560.
29. OSOL G, 1995. Mechanotransduction by vascular smooth muscle. *J Vasc Res*, 32: 275-292.

30. LEVY BI, BENESSIANO J, HENRION D, CAPUTO L, HEYMES C, DURIEZ M, POITEVIN P, and SAMUEL JL, Chronic blockade of AT2-subtype receptors prevents the effect of angiotensin II on the rat vascular structure. 1996 , *J Clin Invest*. 98: 418-425.

31. LEVY BI, MICHEL JL, SALZMANN JL, AZIZI M, POITEVIN F, SAFAR ME, and CAMILLERI JP, Effects of chronic inhibition of converting enzyme on mechanical and structural properties of arteries in rat renovascular hypertension. 1988 , *Cir Res*. 63: 227-229.

32. LAURENT S, CAVIEZEL B, BECK L, GIRERD X, BILLAUD E, BOUTOUYRIE P, HOEKS A and SAFAR M, 1994. Carotid artery distensibility and distending pressure in hypertensive humans. *Hypertension*, 23 [part II]: 878-883.

33. LAURENT S, GIRERD X, MOURAD JJ, LACOLLEY P, BECK L, BOUTOUYRIE P, MIGNOT JP and SAFAR M, 1994. Elastic modulus of arterial artery wall material is not increased in hypertension. *Arteriosclerosis and Thrombosis*, 14: 1231-1233.

34. MOURAD JJ, GIRERD X, BOUTOUYRIE P, LAURENT S, SAFAR M, and LONDON G, Increased stiffness of radial artery wall material in end-stage renal disease. 1997 , *Hypertension*. 30: 1425-1430.

35. LONDON G, GUERIN A, MARCHAIS S, P. B, SAFAR M, DAY M and METIVIER F, 1996. Cardiac and arterial interaction in end-stage renal disease. *Kidney International*, 50: 600-608.
36. BLACHER J, DEMUTH K, GUERIN AP, SAFAR ME, MOATTI N and LONDON GM, 1997. Influence of biochemical alterations on arterial stiffness in patients with end-stage renal disease. *Arteriosclerosis*, . In press.

37. MOURAD JJ, GIRERD X, HANON O, BOUTOUYRIE P, LAURENT S and SAFAR ME, 1997. Influence of age and gender on stiffness of carotid and radial wall material in essential hypertension. *Submitted*.

38. FISCHER GM and SWAIN ML, 1978. In vivo effects of sex hormones on aortic elastin and collagen dynamics in castrated and intact male rats. *Endocrinology*, 10: 92-97.

39. BENETOS A, GAUTHIER S, RICARD S, TOPOUCHIAN J, ASMAR R, POIRIER O, LAROSA E, GUIZE L, SAFAR M, SOUBRIER F, and CAMBIEN F, Influence of angiotensin converting enzyme and angiotensine II Type I receptor gene polymorphisms on aortic stiffness in normotensive and hypertensive patients. 1996 , *Circulation*. 94: 698-703.

40. KANNEL W, WOLF PA, Mc GEE DL, DAWBER TR, MAC NAMRA P and CASTELLI WP, 1981. Systolic blood pressure, arterial rigidity and risks of strokes : the Framingham study. *JAMA*, . 245: 1225-1229.

41. BANK AJ, WILSON RF, KUBO SH, HOLTE JE, DRESING TJ and WANG H, 1995. Direct effects of smooth muscle relaxation and contraction on in vivo human brachial artery elastic properties. *Circ Res*, 77: 1008-1016.

42. SIMON AC, SAFAR ME, LEVENSON JA, KHEDER AM and LEVY BI, 1979. Systolic hypertension: hemodynamic mechanism and choice of antihypertensive treatment. *Am J Cardiol*, . 44: 505-511.

43. SHEP COOPERATIVE RESEARCH GROUP, 1991. Prevention of stroke by antihypertensive drug treatment in older persons with isolated systolic hypertension: final results of the Systolic Hypertension in the Elderly Program (SHEP). *J.A.M.A.*., 265: 3255-3264.

44. STAESSEN JA, FAGARD R, THIJS L, CELIS H, ARABIDZE GG, BIRKENHÄGER WH, BULPITT CJ, DE LEEUW PW, DLLERY CT, FLETCHER AE, FORETTE F, LEONETTI G, NACHEV C, O'BRIEN ET, ROSENFELD J, RODICIO JL, TUOMILEHTO J, ZANCHETTI A and For the Systolic Hypertension in Europe (Syst-Eur) Trial Investigators, 1997. Randomised double-blind comparison of placebo and active treatment for older patients with isolated systolic hypertension. *Lancet*, .350: 757-764.

12. PATHOBIOLOGY OF ATHEROSCLEROSIS

Alain Tedgui, Catherine Bernard, Ziad Mallat
INSERM U 141, Lariboisière hospital, Paris, France

Developments in characterizing the cellular components of the atherosclerotic plaque [1] and in the description of its natural history [2, 3] give a better knowledge of the agents of atherogenesis. Histologically, the foam cells of the fatty streak which characterizes the plaque at an early stage are derived from macrophages (see Stary's classification below). At a later stage the lipidic mass is covered with a fibrous tissue which is mainly made of smooth muscle cells, and thus forms the fibrolipidic plaque. Rather large amounts of T-lymphocytes - about 20% - are found as well, surrounding the plaque and in the fibrous cap [4]. The hypothesis of a primary endothelial lesion appearing before the inflammatory recruitment does not seem to be correct. The endothelium actually remains morphologically intact during the development of atherosclerosis, although it is activated and directly involved in the immuno-inflammatory response. The response develops once plasma LDL have first accumulated in the intima and have been oxidized (by free radicals, endothelial cells, smooth muscle cells or macrophages). Modified LDL are then recognized by the scavenger receptors (SR) expressed on the macrophage surface. These receptors, including type I and II class A SR [5], CD36 [6], the Fc receptor γ RII-B2 for IgG [7], SR-BI [8] and CD68 [9, 10], can internalize high levels of modified LDL because they are not negatively controlled when the intracellular concentration of cholesterol increases [11].

CLASSIFICATION OF ATHEROSCLEROTIC LESIONS

Herbert Stary [2, 3] has categorized the morphological spectrum of lesions found in the first three decades of life into five stages or types:

Isolated Macrophage Foam Cells (Type-1 Lesion)

Increased lipid deposition in the intima occur in some infants and consist of macrophages overloaded with lipid droplet inclusions (macrophage foam cells). Type-I lesions are macroscopically invisible in Sudan-IV stained coronary arteries. Macrophages without inclusions are increased over the number normally present. Macrophage foam cells usually occur as isolated cells and sometimes as small clusters of cells. Macrophage foam cells and macrophages are more numerous in eccentric thickening than in coronary locations with only diffuse thickening. While macrophages remained in the immediately subendothelial region, macrophage foam cells are frequent in the deep part of the glycosaminoglycan (GAG)-rich intima layer. Macrophage foam cells occur in 45% of infants in the first eight months of life. Over the next four years of life the number of cases with macrophage foam cells is only 17%. Macrophage foam cells reappeared in greater number at puberty as a component of the type-II lesion. Lipid-laden smooth muscle cells are not a component of type-I lesions and the phenotypic range of intimal smooth muscle cells does not differ from normal.

Fatty Streak - Superficial or Submerged (Type II Lesion)

The lesion classified microscopically as a fatty streak appears at puberty. It is present in 43% of children 10-14 years old. Fatty streaks are composed of multiple layers of cells overloaded with lipid droplet inclusions. Cells with inclusions are intimal smooth muscle cells and macrophages. Intimal smooth muscle cells assume a synyhetic phenotype with more rough endoplasmic reticulum (RER). The number of macrophages and macrophage foam cells is greater than it is in infants with foam cells. Sometimes smooth muscle cells with inclusions predominate and at other times macrophage foam cells predominate. Occasionally, the proportions of the two cell types that contain droplets are about equal.

Superficial fatty streak: Because the GAG layer of the intima is narrow in diffuse thickening and at the shoulders of eccentric thickening, lipid-laden cells, particularly the more lipid-laden macrophage foam cells, occupy the full thickness of the GAG layer. Therefore, since the lipid-laden cells are located superficially they are macroscopically visible as fatty streaks.

Submerged fatty streak: The GAG layer is wide at the thick center of eccentric thickening, the lipid-laden macrophages aggregate in its deepest part. Thus, these foam cells occupying the center of eccentric thickening are concealed some distance below the endothelial surface, and macroscopically less obvious, even after staining with Sudan-IV.

Preatheroma (Type-III Lesion)

This is the lesion that is transitional or intermediate between the fatty streak and the atheroma (type-IV). In young people only the fatty streaks located in eccentric thickening are disposed to this type of transformation. In addition to the components that make up fatty streaks, such intimal lesions contain many separate pool-like aggregates of extracellular lipid particles. The pools of lipid particles are below the foam cell layer, that is, they are in the musculoelastic intima layer of eccentric thickening. By increasing the volume of the intercellular space, they disrupt the coherence of structural smooth muscle cells in the affected regions, and somewhat dispersed them. This is associated with a phenotypic change in these normally myofilament-rich cells. The cells are now thin, long, very rich in RER and encased in gigantic basement membranes. The thickness of the basement membranes is in no relation to the diameter of the cells, sometimes exceeding this several fold. Myofilaments often are entirely absent.

Atheroma (Type-IV Lesion)
The feature that distinguishes atheroma is a single and massive expanse of extracellular lipid particles (the lipid core) that is grossly visible. Lipid cores occupy the musculoelastic layer of eccentric thickening. Thus, the location is the same as that of the separate pools of extracellular lipid particles found in other cases, and classified as preatheroma. The lipid core separates the musculoelastic intima layer into a luminal (inner) and a medial (outer) part, and it greatly thickens the intima. However, a simultaneous mild dilatation of the arterial wall forestalls a significant loss of arterial lumen. Lipid cores are the result of progression and coalescence of smaller, separate pools of extracellular lipid particles. The lipid particles that constitute cores are identical in their fine structure to the more or less degraded lipid droplets within the cytoplasm of macrophage foam cells and smooth muscle cells. Macrophage foam cells do not occur within lipid cores. Instead, macrophage foam cells, some dead and decomposing, border the core along its inner periphery, that is, along the aspect that is towards the lumen. The preexisting structural intimal smooth muscle cells that are trapped within the region of the lipid core are widely separated and scattered. They are thinner and more elongated still than those trapped within the multiple smaller lipid pools of preatheroma. Their basement membranes are equally massive.

Fibroatheroma (Type-V Lesion)
The feature that distinguishes fibroatheroma is a more or less thick "cap" of smooth muscle cells embedded in a matrix of collagen. This cap forms the uppermost part of the lesion. They are located just above the foam cell layer and above the lipid core. The smooth muscle cells of caps are stratified in parallel layers, and longitudinally arranged around the artery circumference. The fine structure of cap smooth muscle cells vary somewhat. Their cytoplasm is completely packed with cisternae of RER, and basement membranes are not massive. However, in some regions of the caps RER-rich smooth muscle cells resemble those in the lipid cores in their basement membrane gigantism. Lipid droplet inclusions in cap smooth muscle cells are small and often infrequent or absent.

LIPID INFILTRATION

The view that lipid infiltration in the arterial wall is the *primum movens* of atherosclerosis and precedes the inflammatory reaction in the intima was first proposed by Anitschkov and Chalatov [12] at the beginning of the century. In 1954 Irvine Page brilliantly developed this concept as follows [13]:

The filtration concept is based on the view that atherogenesis is due to the tissue reaction to substances filtered from plasma as lipoprotein by arterial pressure, and deposited in the intima as foreign lipid. Most of the filtered materials pass on harmlessly to be picked up by the adventitial capillaries or the lymph. But some may stay behind, whether the vessel fails to function properly as a filter or because the size, shape and charge of the lipoproteins is such as to allow them to stick. Changes in the arrangement, amount and chemical nature of subendothelial ground substance conceivably may initiate a focal change in filter function. The reaction which ensues depends on the nature of the lipid deposited and the responsiveness of the tissue to it.

This view remains of great modernity and contains all elements that constitute our most recent approach of atherogenesis: lipids (low density lipoprotein, LDL) deposited in the arterial wall, role of mechanical factors (arterial pressure) as risk factor, modification of lipids (LDL oxidation), and tissue reaction (inflammatory response).

Smith [14] and Hoff [15] were the first to show that substantial amounts of apolipoprotein (apo) B are located in human atherosclerotic lesions, which indicates that LDL accumulated in the plaque are derived from plasma. LDL preferentially accumulate in the lesion-susceptible areas as a result of intimal trapping instead of endothelial injury [16]. Due to their large size plasma LDL are forced by pressure-driven convection into the arterial wall that behaves as a porous material, whereas smaller macromolecules, such as albumin, can move freely across the media and avoid accumulation in the intima [17, 18], and larger ones do not cross the endothelial barrier [19].

In cholesterol-fed rabbits, intimal lipoprotein accumulation precedes macrophage recruitment [16] suggesting that intimal LDL accumulation is the first event leading to atherosclerosis. But direct evidence in humans was lacking until Napoli et al. showed that 1) fetal aortas from hypercholesterolemic mothers contain much more fatty streaks than aortas from normocholesterolemic mothers, and 2) LDL and oxidized LDL are frequently found in very early lesions in the absence of monocytes/macrophages, whereas the opposite is very rare [20].

IMMUNOCOMPETENT CELLS AND ATHEROSCLEROSIS

The central role of the macrophage as a target cell for the accumulation of cholesterol, which has been confirmed by immunohistochemical studies, emphasizes the importance of inflammatory processes in the pathogenesis of atherosclerosis [21]. The finding of T-lymphocytes and major histocompatibility complex (MHC) class-II HLA-DR molecules in the human plaque [4] was unexpected, suggesting that dysimmune, or even autoimmune mechanisms might be directly involved in atherogenesis [22].

Monocytes-macrophages

Macrophages are the predominant cells in the human atherosclerotic plaque as well as in animal models of atherosclerosis induced by a cholesterol-enriched diet. Macrophages are derived from circulating monocytes that adhere to the endothelium and then migrate to the subendothelial space. These macrophages are involved in the formation of the plaque, as evidenced by the decreased atherosclerosis in apoE -/- mice deficient in macrophage-stimulating factor (*op/op* mice), which have a decreased blood monocyte differential count [23-25]. It is thus likely that macrophages enter the intima following primary LDL accumulation to scavenge cholesterol overload and that they activate endothelial cells through cytokine release, leading to increase endothelial permeability to LDL, hence triggering a vicious circle.

The importance of scavenger receptor SR-A1 to lesion development was demonstrated by experiments in apoE -/- mice that do not express SR-A1. These mice show markedly reduced lesions compared to apoE-/- [26].

Macrophages synthesize a number of proinflammatory cytokines and vascular growth factors. The immunohistochemical analyses of human carotid atheroma lesions shows the presence of platelet-derived growth factor B in macrophages [27], and increased expression of its β-type receptor on the adjacent smooth muscle cells [28]. This supports the hypothesis that the proliferation of smooth muscle cells is induced by their cooperation with macrophages. Two types of macrophages have been described in human atherosclerotic pulmonary arteries [29]: (i) a population of neointimal non-foam macrophages associated with the expression of fibronectin and type-1 procollagen, and (ii) an underlying population of foam macrophages with modified secretory phenotypes that do not express the products of the extracellular matrix. Moreover, the macrophages isolated from atheromatous tissue removed by carotid endarterectomy show evidence of activation: they express more tumor necrosis factor α (TNFα) than circulating monocytes normally do [30].

One important finding is that macrophages in the plaque can multiply in situ in the vessel wall [31]. Monocyte-colony stimulating factor (M-CSF), a factor of differentiation and proliferation of stem cells into monocytes, is locally produced by the endothelial and smooth muscle cells from human atheromatous plaques [32]. It may be involved in macrophage survival and proliferation. M-CSF increases the expression of the scavenger receptor and the secretion of apo E by macrophages. The clearance of lipoproteins of the plaque and the reverse transport of cholesterol may therefore be more easily performed. Moreover, the expression of M-CSF in the plaque may insure the survival of macrophages, by limiting or delaying the onset of apoptosis and the appearance of necrotic foci.

We now know that macrophages, as well as smooth muscle cells, can also die from apoptosis in the plaque [33-38]. Therefore the macrophage content in atherosclerotic lesions results from the net balance between proliferation and death. Mechanisms and potential roles of apoptosis in atherosclerosis are presented above in the chapter on *Apoptosis in the Vascular System*.

T-lymphocytes and atheroma

T-lymphocytes have been found in human atheromatous lesions by various laboratories. T-cell subsets are heterogeneous: some are CD4+, and fewer are CD8+. CD4+ lymphocytes are polyclonal, mainly expressing α/β T-cell receptors and some γ/δ T-cell receptors [39]. Natural killer cells are rarely present in atheroma lesions. Quantitative immunocytochemical analysis of coronary plaques obtained by in-vivo atherectomy shows that lymphocytes and monocytes account for over 20% cells, and that CD4+ T-cells predominate over suppressor CD8+ T-cells and B-lymphocytes. Variable interleukin-2 receptor subtype expression occurs in mononuclear leukocytes infiltrating chronic human atheroma [40].

T-lymphocytes are important for both the etiology and the evolution of the disease. They show evidence of an immune response and can modulate the secretory and contractile function of vessels. Most T-lymphocytes are memory cells expressing the CD45 RO [41], indicating that they have rather chronic activation phenotypes. The smooth muscle cells that are adjacent to lymphocytes often express major histocompatibility complex class-II HLA-DR molecules, probably induced by lymphocyte-derived γ-interferon (IFNγ).

Potential antigens

The presence of lymphocytes in atheromatous lesions can be interpreted in two ways: either they are resting T-lymphocytes, infiltrated and sequestered in the lesions, or they are activated and involved in the pathogenesis of atherosclerosis. Some arguments support the second hypothesis. First, T-cells infiltrating the plaque express the β-type IL-2 receptor [40]; lymphocytes express MHC class-II molecules, IFNγ, IL-2 and its receptors, and very late antigen-1 integrin. However the antigen(s) involved in lymphocyte activation have not been identified. Stemme et al. [42] established T-cell clones from human atherosclerotic plaques using polyclonal mitogens as stimuli and exposed the clones to potential antigens in the presence of autologous monocytes as antigen-presenting cells. Some clones proliferated and secreted IFNγ in response to oxidized LDL whereas no significant response to oxLDL was detected in CD4+ T-cell clones derived from the peripheral blood of the same individuals. This proliferative T cell response required the presence of antigen presenting cells (autologous mononuclear cells) and was restricted to MHC class II HLA-DR. The epitotes generated upon oxidation of LDL and specifically recognized by T-cell clones from the plaque have not yet been identified, but these findings suggest that a T-cell-dependent, autoimmune response to oxLDL occurs in the atherosclerotic plaque.

Heat shock proteins and atheroma

Heat shock proteins are a family of proteins with highly phylogenically conserved sequences, from microorganisms to mammalians. The early increase in HSP cell expression, induced by varied stresses (infection, high temperature, free radicals, or mechanical stress) provides cells with protection from microenvironmental insults. The functions of HSP are to stabilize and protect newly synthesized proteins during folding.

Atheromatous lesions can be induced in normocholesterolemic rabbits by immunizing them with HSP65, a HSP which is expressed at high levels in human atheromatous lesions [43]. In other words, an immune or auto-immune response to HSP65 can initiate atherosclerosis lesions in rabbits. Moreover, HSP70 is markedly present at the centre of atheromatous lesions [44], which is consistent with the presence of foam cells and smooth muscle cells and surrounding necrotic and lipid accumulation foci. It seems that HSP70 expression and distribution are correlated to the thickness of the plaque. Human atheromatous lesions also contain HSP60 [39].

An epidemiological study shows a significant increase in serum anti-HSP65 antibodies in subjects aged over 60 with carotid atherosclerosis, compared to a group of subjects without carotid lesions [43]. The presence of anti-HSP65 antibodies is a strong and independent risk factor for atherosclerosis, statistically as strong as high cholesterol levels and hypertension. Tissue damage associated with inflammatory and autoimmune diseases is thought to induce high-level expression and extracellular leakage of HSP, which would induce an autoimmune response. Considering this hypothesis, the HSP autoimmune process appears as secondary, and may be involved in the evolution of the atheromatous disease. A second hypothesis is that, taking into account the high similarities between bacterial and human HSP, the antibody response initially directed against bacterial HSP will cross-react with human HSP. In that case, the infectious processes due to some microorganisms could be considered as etiological factors of human atherosclerosis [45]. Human atherosclerotic lesions show immunoreactive chlamydial HSP60 in 47% of the cases and human HSP60 in 89% of the cases [46]. Moreover, chlamydial HSP60 colocalizes with human HSP60 within plaque macrophages [46]. Whatever the pathological signification of the presence of anti-HSP antibodies (whether primary or secondary phenomenon), they could be considered as markers of progressive atherosclerosis.

Role of T lymphocytes in atherosclerosis

To determine the precise role of the cellular and humoral immune systems in atherogenesis, the immune system has been manipulated in different ways. Yet the results remain controversial. Pharmacological studies using cyclosporin A in hyperlipidemic C57BL/6 mice [47] or cholesterol-fed rabbits [48, 49] have shown both increased [47, 48] and decreased [49] atherosclerosis. Also, a marked reduction in atherosclerotic lesions was reported in hyperlipidemic C57BL/6 euthymic mice with CD4 T-cell ablated with monoclonal antibodies and in T-cell deficient C57BL/6 nude (*nu/nu*) mice [50]. Others presented evidence that the classical immune system including T lymphocytes is not essential for the development of atherosclerotic lesions. Mice with severe combined immunodeficiency (SCID), *nu/nu* mice, and class II MHC deficient mice were found to show similar lesion development as control mice when fed an atherogenic diet [51]. In this study, only class I MHC deficient mice demonstrated a 3-fold increase in lesion area. A common criticism to these studies is that they used C57BL/6 mice fed an atherogenic diet, which developed small lesions limited to the aortic root and containing no or few T cells. Therefore, the effect of total lymphocyte deficiency on atherogenesis was investigated by crossing apoE-deficient mice, which develop complex atherosclerotic lesions, with mice deficient in recombinase activating gene (*Rag*)-1 or 2 required for normal B and T lymphocyte development [52, 53]. These mice have no T lymphocytes and a markedly reduced number of MHC class II-positive macrophages in atherosclerotic lesions. Despite these differences, *Rag-1* or *Rag-2* deficient mice develop atherosclerosis to a similar extent as their immunocompetent littermates, indicating that the absence of autoantibodies and T lymphocytes does not influence the development of early atherosclerotic lesions.

However controversial the results obtained in mice regarding the role of the specific immune system, they are all confined to the early stage of the atherogenic process. The role of T lymphocytes in lesion formation and in plaque (un)stability might be totally different. In particular, T cell-derived IFNγ inhibits collagen production by activated smooth muscle cells [54], which can be viewed as highly deleterious in terms of plaque stability.

CYTOKINES IN ATHEROMA

Peptide mediators, known as cytokines, coordinate the interactions within the immune system. Cytokines form a family comprising several classes: interleukins (IL), tumor necrosis factors (α and β), interferons, transforming growth factors (TGF), colony stimulating factors (CSF) and chemotactic factors (also known as chemokines), such as IL-8 and monocyte chemotactic protein (MCP)-1 and 2.

Cytokines are produced by the cells of the immune system, although by some other cells as well, including vascular (endothelial and smooth muscle) cells. Vascular cells are therefore possibly involved in the local inflammatory response (table 1). Cytokines act as activators for vascular cells as they do for the cells of the immune system, hence the idea of an "activation" of vascular cells, which includes morphological, functional or antigenic modifications. Activating cytokines are mainly proinflammatory cytokines, TNFα, IL-1, IL-8, IL-12, oncostatin M (OSM) and IFNγ. Bacterial endotoxins (lipopolysaccharides) are also potent agents of activation.

Proinflammatory cytokines in the atherosclerotic plaque

A large number of cytokines have been found in the human [55, 58] and rabbit [59] atherosclerotic plaque. TNFα is present in human atheromatous lesions. It can be immunohistochemically found in 88% of atheromatous vascular tissues whereas it is absent from normal vascular tissue [58]. It is mainly located in the smooth muscle cells of the intima (both cytoplasm and membrane are TNFα-positive), but can also be found in the foam cells of the plaque and in endothelial cells. In early atheromatous lesions, smooth muscle cells are the first to express TNFα. It is noteworthy that human aorta smooth muscle cells incubated with LDL produce TNFα. Another proinflammatory cytokine IL-1β is expressed in the foam cells of plaques in hypercholesterolemic rabbits. Furthermore, it has been demonstrated by in situ hybridization that the gene encoding MCP-1 is expressed in large amounts in atherosclerosis, especially in the macrophages and smooth muscle cells of the plaque [60]. These cytokines may therefore be actively involved in recruiting monocytes onto the atheromatous site. M-CSF is expressed as well by the endothelial and smooth muscle cells of the plaque. OSM, a potent inflammatory cytokine [61], is present in large amounts in the human atherosclerotic plaque [Mallat, unpublished data], and might activate smooth muscle cells which possess the specific OSM receptor type II [62]. Finally, IL-8 [63] and IL-12 [64] have also been found in the human atherosclerotic plaque.

The pro-inflammatory cytokines alone or in conjunction contribute to the local inflammatory response and may have great impact on plaque formation and progression [22, 65]. Indeed, pro-inflammatory cytokines have the potential to induce excessive extracellular matrix degradation and cell death promoting plaque instability [66].

The implication of proinflammatory cytokines in the atherosclerotic process has been recently explored using transgenic mice. For example, the role of IFNγ was investigated in apoE -/- mice crossed with IFNγ receptor deficient mice [67]. Compared to the apoE-/- mice, the compound knock-out mice exhibited a substantial reduction in atherosclerotic lesion size and a decrease in lesion cellularity, with a marked increase in lesion collagen content. Evaluation of the plasma lipoproteins showed that the compound knockout mice had a marked increase in potentially atheroprotective phospholipid/apoA-IV rich particles as well. These observations suggest that IFNγ promotes and modifies atherosclerosis through both local effects in the arterial wall and systemic effects on plasma lipoproteins.

Others have attempted to evaluate the role of TNFα in atherosclerosis by studying lesion development in C57BL/6 mice which lack TNF receptor type 1 (TNFR1, or p55) or type 2 (TNFR2, or p75), fed an atherogenic diet. p55 receptor mediates most of the biological effects of soluble TNF, whereas p75 receptor is an important mediator of cell-cell TNF-mediated function. Surprisingly, p55 receptor -/- mice show increased size of lesions, but contain fewer cell nuclei, while p75 receptor -/- mice have lesions similar to +/+ mice [68]. Mice lacking both p55 and p75 receptors also have larger lesions than +/+ mice, and lesions are rich in macrophages. The larger lesions seen in p55 -/- could be accounted for by larger, more lipid-laden macrophages. Indeed, TNF is known to downregulate scavenger receptor activity in macrophages [69, 70], and macrophages from p55 -/- mice show increased scavenger receptor activity [68]. Therefore the primary role of the p55 receptor in fatty streak development is likely to be to downregulate scavenger receptor activity, protecting the wall from exaggerated foam cell formation.

In complementary studies, female apoE -/- mice chronically treated with TNF binding protein to prevent interaction of TNF with its receptors were found to have decreased atherosclerotic lesions [71]. Furthermore, male and female apoE -/- mice treated with IL-1 receptor antagonist (IL-1ra) to block the effects of IL-1 also showed decreased lesion development, underlying the potentially deleterious effects of proinflammatory cytokines in atherogenesis.

CD40 ligand (CD40L) is a transmembrane protein structurally related to the cytokine TNF that is present on CD4+ T cells. CD40 is found on B cells, monocytes, macrophages and endothelial and vascular smooth muscle cells. Interaction of CD40 with CD40L results in humoral and cell-mediated immune responses, but recent findings indicate that it also activates typical inflammatory responses including induction of proinflammatory cytokines, MMPs, adhesion molecules and tissue factor. Cells in human atherosclerotic plaques, including endothelial cells, smooth muscle cells and macrophages, express CD40 and CD40L [72]. The involvement of CD40 signalling in atherogenesis has been demonstrated by treating LDL receptor deficient mice fed an atherogenic diet with an antibody directed against mouse CD40L [73]. This antibody markedly reduces atherosclerotic lesion size as well as macrophage and T lymphocyte content.

Secondary mediators of inflammation and atheroma

Inducible NO-synthase in atheromatous lesions

NO is a key molecule in intercellular signalling, and it plays a major role in cardiovascular physiology and physiopathology (see chapter on Endothelial Function and Dysfunction). Modifications in NO synthesis might be involved in atherosclerosis. For example, esterified cholesterol enrichment of smooth muscle cells in culture upregulates cytokine- and lipopolysaccharide-inducible NO synthase (iNOS) activity [74]. Moreover, in atheromatous aortas from hypercholesterolemic rabbits, Verbeuren et al [75] show an increased endothelium-independent contractility in response to an NO-synthesis inhibitor, suggesting that iNOS is expressed in the walls of atheromatous aortas by smooth muscle cells and may play a functional role. iNOS expression in human atherosclerotic plaque has also been confirmed by immunohistochemical techniques [76, 77].

We found a strong association between iNOS expression and apoptosis (TUNEL labeling) in the plaque, which could be due to the reported apoptotic effects of inflammatory NO [78, 79]. In presence of superoxide anion, NO might lead to local peroxynitrite formation [80, 81] which is also a potent inducer of cell death [82].

Prostaglandin pathway

The esterified cholesterol enrichment of smooth muscle cells in culture can modulate the transcription of many genes. Oxidized LDL, in addition to inducing NO, upregulate the expression of inducible cyclooxygenase (COX-2) mRNA [74]. COX-2 is expressed in the human atherosclerotic plaque [Mallat, unpublished data].

Antiinflammatory cytokines in the atherosclerotic plaque

We have seen that a variety of pro-inflammatory cytokines, including TNFα, IL-1b, IL-6, IL-8, IL-12 and IFNγ are expressed in the human atherosclerotic plaque. However, the inflammatory response is known to be balanced by anti-inflammatory cytokines, including IL-4, IL-10 and IL-13 [83]. Among the anti-inflammatory cytokines, IL-10 is produced by Th2 cells as well as by macrophages [83] and has potent deactivating properties on these cells [84]. IL-4 is not expressed in the atherosclerotic plaque, and IL-13 can barely be found in lymphocytes [Mallat, unpublished data]. IL-10 mRNA expression is found in non complicated atherosclerotic plaques [64], and IL-10 mRNA and protein are highly expressed in human advanced atherosclerotic plaques [77]. IL-10 is detected in the extracellular matrix, and it is expressed mainly by macrophages. Interestingly, IL-10 also localizes to the cytoplasm of a number of smooth muscle cells, suggesting that these cells are producing IL-10. However, IL-10 mRNA or IL-10 protein cannot be detected in human aortic smooth muscle cells in culture, whether unstimulated or stimulated by minimally modified LDL, oxidized LDL or a mixture of TNFα, IL-1β and IFNγ (Mallat, unpublished data). Therefore, very specific environmental conditions or cell phenotype might be required for smooth muscle cells to produce IL-10.

Expression of IL-10 in human atherosclerotic plaques may have several potential effects. Because IL-10 is a potent anti-inflammatory cytokine with deactivating properties in macrophages [83-85], it is likely that its expression by plaque macrophages would limit the inflammatory response and promote plaque healing. Indeed, endogenous production of IL-10 by human monocytes in response to LDL stimulation inhibits IL-12 production [64], indicating a cross-regulatory action of IL-10 that may counterbalance the pro-inflammatory response. IL-10 expression in advanced human atheroma is associated with a 3-fold decrease in the probability of iNOS expression [77], suggesting a local anti-inflammatory role for IL-10 in human atherosclerosis.

IL-10 has been reported to have anti-apoptotic properties in cultured macrophages [86] and in T lymphocytes [87]. In agreement with these findings, IL-10 expression in human atherosclerotic plaques is associated with low levels of apoptosis (TUNEL labeling) [77]. Therefore, the counterregulatory effect induced by anti-inflammatory cytokines, especially IL-10, on the inflammatory response may protect from excessive cell damage and death in the plaque.

VASCULAR EFFECTS OF CYTOKINES

Vascular cells are preferential targets for cytokines (see the chapter on Vessels and Inflammation). The anticoagulant and antiadhesive properties of endothelial cells are strongly modified under the effect of cytokines. The phenotypes of activated smooth muscle cells will be rather synthetic and proliferative, as opposed to resting cells whose phenotypes are contractile.

Modification of the antithrombotic properties of the endothelium

The antithrombotic properties of endothelial cells are deeply altered by IL-1 and by TNFα. They can increase the tissue procoagulant activity, and suppress the anticoagulant activity mediated by the thrombomodulin-protein C system, by decreasing the expression of thrombomodulin [88]. These cytokines also modify the fibrinolytic properties of endothelial cells: they decrease the production of tissue plasminogen activator (tPA) and they increase the production of an inhibitor of the tissue plasminogen activator (PAI-1). IL-1 and TNFα, produced in situ in the plaque may therefore be involved in the thrombotic complications associated with atherosclerosis.

Modulation of the expression of leukocyte adhesion molecules

IL-1 and TNF are cytokines that stimulate the membrane expression of leukocyte adhesion molecules (ICAM-1, ICAM-2, E-selectin and VCAM-1) by endothelial cells. These molecules interact with specific ligands expressed by neutrophils, lymphocytes and circulating monocytes. In early inflammation, the rapid induction of E-selectin, which is maximal after about four to six hours, promotes the adhesion of neutrophils. The accompanying secretion of IL-8 (a chemotactic factor for neutrophils) by the activated endothelium amplifies the primary polymorphonuclear response. Then, the ICAM-1 expression, which is maximal after 24 hours, promotes the adhesion of monocytes to the endothelium, whereas VCAM-1, which has a late, rather long-lasting expression, binds monocytes by the means of VLA-4 integrin. These monocytes are attracted by MCP-1, secreted by endothelial and smooth muscle cells in response to IL-1 and TNFα. It is noteworthy that neutrophils do not seem to be involved in any stage of atherosclerosis.

VCAM-1 plays an important role in atherogenesis. It selectively promotes the adhesion of mononuclear cells to the vascular endothelium that constitutively expresses VLA-4 ligand. It may also promote their fixation on the intima. In hypercholesterolemic rabbits, endothelial cells express VCAM-1 very early on the preferential foci for the formation of the lesion, even before intimal foam cells appear [89]. Furthermore, VCAM-1 is present in human coronary atherosclerotic lesions, and it is expressed by endothelial cells as well as by the intimal neovessels of the plaques [90]. VCAM-1 is found in intimal smooth muscle cells, in association with the inflammatory infiltrate and with the expression of class-II HLA molecules. Although the functional significance of the pluricellular VCAM-1 expression during the development of atherosclerosis is not well understood, these findings suggest that the cells involved in the atheromatous plaque are in a state of activation.

Increase in the synthetic activity of vascular cells

Proinflammatory cytokines stimulate the production of various vasoactive agents by endothelial and smooth muscle cells, such as platelet activating factor (PAF) and prostaglandins (PGE_2, PGI_2), which have vasodilatory actions and characterize the inflammatory response. The marked increase in prostaglandin production in vascular cells under the influence of cytokines is due to activated COX-2, the inducible isoform of cyclooxygenase [91]. IL-1 and TNFα also stimulate NO synthesis, by activating iNOS in endothelial and smooth muscle cells in culture, but they suppress the activity of endothelial eNOS [92]. Finally, cytokines have been shown to stimulate the production of growth factors such as PDGF in endothelial and smooth muscle cells in culture.

Trophic and morphologic effects of cytokines on vascular cells

IL-1 and TNF increase the permeability of the endothelium by reorganizing the structure of the endothelial cell monolayer, and may also promote leukocyte diapedesis after a long period of exposure. These cytokines have trophic effects on vascular cells, and they also have an antiproliferative effect on endothelial cells, whereas IL-1, by inducing PDGF-AA, has a mitogenic effect on smooth muscle cells [93]. IL-1 also increases the production of basic FGF and the number of FGF receptors, thus showing evidence that the stimulation of smooth muscle cell growth may be autocrine and/or paracrine. TNF is also a potent activator of angiogenesis; it is necessary to tissue lesion repairs and might be responsible for the neovascularisation of the atheromatous plaque.

Regulation of extracellular matrix proteins

The production by smooth muscle cells of type I and III collagen, which contribute to the formation of the fibrous cap favoring the stability of the plaque [54], is slightly increased by IL-1 and TNF, whereas TGFβ is a potent inducer of collagen synthesis. In contrast, IFNγ inhibits collagen synthesis [54].

Proinflammatory cytokines also induce the expression by smooth muscle cells of matrix metalloproteinases (MMPs) which are able to degrade extracellular matrix proteins, including elastin and collagen. Cytokines stimulate the activity of MMP-2 (72 kD gelatinase) which is expressed constitutively by smooth muscle cells, and induce the expression of MMP-9 (92 kD gelatinase) and MMP-3 (stromelysin) which degrade proteoglycans and elastin [94]. The activity of MMPs is negatively regulated by endogenous tissue inhibitors of metalloproteinases (TIMPs), including TIMP-1, -2 and -3 which are constitutively expressed by smooth muscle cells [95]. TIMP-1 and -2 are unaffected by IL-1 or TNF [94], whereas PDGF and TGFβ can induce TIMP-3 [95]. Therefore in a given plaque area where IL-1, TNF and IFNγ are expressed, one might expect IL-1/TNF to induce MMP expression and IFNγ to decrease matrix protein expression by smooth muscle cells, without a parallel increased TIMP production, leading to matrix degradation and plaque fragilization [66].

LIPIDS, OXIDATION AND INFLAMMATORY REACTION

Experimental studies have shown the relationship between hypercholesterolemia and an inflammatory reaction in vascular tissues. The biological mechanisms responsible for this association, however, remain to be clarified. Incubating human monocytes in the presence of cholesterol induces an increased class-II HLA molecule expression which is responsible for the enhancement of their antigen-presenting functions [96]. Many studies were performed in order to assess the effect of oxidized LDL as inflammatory agents. Oxidized LDL have a chemoattracting activity on monocytes, they promote their differentiation into macrophages, but inhibit their mobility. Incubating human blood mononuclear cells (monocytes, T-lymphocytes and a few B-lymphocytes) with oxidized LDL results in T-lymphocyte activation, which is assessed by an increased expression of IL-2 receptors and HLA-DR antigens on T-cells [97]. In vivo administration of minimally modified LDL in mice causes rapid induction of circulating M-CSF and of pro-inflammatory genes encoding JE (the murine equivalent of MCP-1), as well as of other inflammatory proteins in tissues. This response is similar to that induced by an atherogenic diet; it supports the hypothesis that lipid-derived oxidation products are responsible for the inflammatory reaction [98]. This response, however, might be genetically controlled. Indeed, it is not found in C3H mice which do not develop any atheromatous lesions when fed with a hypercholesterolemia-inducing diet [99].

Oxidized LDL may thus cause autoimmune reactions, as suggested by the presence of specific circulating antibodies to these molecules in man. Data concerning the effects of oxidized LDL on macrophage activation, however, are conflicting. Oxidized LDL do not induce IL-1β expression in murine peritoneal macrophages. They even inhibit the induction of IL-1, TNFα and IL-6 by lipopolysaccharides [100]. It cannot be ruled out, however, that the profile of cytokines produced by activated macrophages depends on their phenotypes. TNF released by macrophage-derived foam cells is enhanced compared to that released by circulating monocytes [30]. On the contrary, IL-1 production by both these cells is similar, suggesting local and selective augmentation of cytokine production in the atherosclerotic plaque.

Significantly higher levels of autoantibodies to oxidized LDL are found in patients with progressive carotid atherosclerosis rather than in age-matched control groups [101]. Again, the serum level of antibodies to oxidized LDL (malondialdehyde antibodies) is an independent predictor of the evolution of atherosclerosis.

VASCULAR REMODELING IN ATHEROSCLEROSIS

Compensatory vascular enlargement

At the initial phase of atherosclerosis, the plaque develops in the absence of reduction of the lumen area. This has been found in coronary arteries [102-105], in carotid arteries [106], as well as in femoral arteries [107]. Glagov [102] showed for the first time using histologic sections of human coronary arteries fixed under normal physiological pressure, that the area delimited by the internal elastic lamina increases with the size of the atheromatous plaque, in such a way that the lumen area remains constant as long as the size of the plaque occupies less than 40% of the total surface area of the arterial wall. This adaptive vascular process in atherosclerosis is known as "remodeling" or "vascular enlargement". The histologic data first provided by Glagov [102] have been confirmed in vivo by use of intravascular ultrasound imaging techniques [103, 104].

A retrospective histological study by Clarkson et al. [105] performed on hearts from 416 primates fed an atherogenic diet, and from 100 men or women, confirm the existence of vascular enlargement remodeling in atherosclerosis. The mean size of the vascular lumen is not reduced in subjects with large atherosclerotic plaques. The tendency is toward a parallel increase in the size of the lumen with the increase in the size of the plaque. However, there is no correlation between lumen caliber and classical risk factors, including hyperlipidemia, hypertension, and smoking, whereas plaque size is positively correlated with arterial pressure and cholesterol plasma level. Interestingly, the Clarkson study reveals that following subclassification of patients into symptomatic and asymptomatic, the lumen area does not increase with the size of the coronary lesion, whereas it does in subjects without coronary heart disease. Therefore, a defect in vascular remodeling or adaptive vascular enlargement of atheromatous coronary arteries is likely to be a major determinant of clinical coronary events associated with atherosclerosis.

MMPs and TIMPs in atheroma : Relation to plaque stability

The rupture of atherosclerotic plaques is the main cause of myocardial infarction [108, 109]. The stability of the plaque depends on its composition, especially on the structural integrity of the collagen-rich fibrous cap [110, 111]. Cells able to produce extracellular matrix proteins in the plaque are essentially smooth muscle cells, even though macrophages can produce extracellular proteoglycans, including chondroitin sulfate, heparan sulfate and dermatan sulfate [112]. The deposition of collagen in the fibrous cap allows it to support high tensile stress and prevents plaque rupture. It is now well accepted that smooth muscle cell proliferation in the plaque, thanks to its ability to produce extracellular matrix proteins, is an adaptive protective mechanism which ultimately contributes to plaque stabilization [113]. However, within the plaque macrophages and smooth muscle cells are able to produce MMPs which degrade matrix proteins. MMP-1, MMP-3 and MMP-9 are expressed in atherosclerotic plaques and are localized to the fibrous cap, to lesion shoulders and to the base of the lipid core [114-116], while MMP-7 (matrilysin) is predominantly expressed in macrophages overlying the lipid core [117]. The activity of MMPs is counterregulated by endogenous inhibitors, known as TIMPs. TIMP-1, -2, and -3 have similar localization in the plaque as MMPs [95], indicating that TIMPs contribute to increase the stability of the plaque by limiting matrix degradation by MMPs.

Polarity of MMP secretion

The mechanisms responsible for vascular remodeling in atherosclerosis are largely unknown. It is likely that the atrophy of the native media underlying the lipid necrotic core, which result from macrophage protease activity [118], creates a zone of low resistance to tensile strength where the intimal mass is pushed outwards by the arterial pressure, leading to the deformation of the internal elastic lamina to ensure an unchanged coronary lumen despite the presence of the plaque. However, an excessive proliferative response can cause obstruction of the vascular lumen. When the intimal mass exceeds 40% of the total wall surface area, artery dilatation is no longer sufficient to compensate for the plaque development, and the vascular lumen is then reduced [102].

CONCLUSION

A network of cytokines secreted by leukocytes and vascular cells regulate two phenomena characteristic of atherosclerosis: the inflammatory reaction and the proliferation of smooth muscle cells in the intima. The effects of cytokines on vascular cells are multiple and very complex. The inflammatory reaction, however, must not be considered only as promoting the development of the atherosclerotic plaque. Indeed, macrophages act as scavengers and eliminate the cholesterol overload of the intima when LDL have accumulated. Older studies have shown that macrophages can move from the blood into the vessel wall and *vice-versa* [44]. Macrophages appear to be efficient at the very beginning of the atherosclerotic process. Stary [2] showed, on autopsy specimens, that macrophage foam cells in coronary arteries occur in 45% of infants in the first eight months of life, but only in 17% over the next four years. Not all fatty streaks develop into atherosclerotic plaques and some of them will regress through the action of macrophages. However, when phagocytosis is overrun, the intracellular metabolism of cholesterol in macrophages is reduced, cholesterol ester accumulates, and macrophages are immobilized in the intima as foam cells. Cytokines can promote the formation of fibromuscular lesions by promoting smooth muscle cell proliferation and synthesis of extracellular matrix proteins. This process might be beneficial if it is limited. It may stabilize and strengthen the atheromatous plaque and increase its abilities to resist the stresses to which it is subjected. But cytokines can also trigger and/or amplify the activation of endothelial cells allowing increased LDL uptake and inflammatory cell penetration in the intima, favoring the development of early lesions. At a later stage, they can stimulate the production of MMPs which degrade the extracellular matrix and weaken the plaque, eventually leading to rupture and thrombosis.

REFERENCES

1. Jonasson L, Holin J, Skalli 0, Bondjers G, Hansson GK.. Regional accumulations of T cells, macrophages and smooth muscle cells in the human atherosclerotic plaque. Arteriosclerosis. 1986;6:131-138.

2. Stary HC, Chandler AB, Glagov S, Guyton JR, Insull W Jr, Rosenfeld ME, Schaffer SA, Schwartz CJ, Wagner WD, Wissler RW. A definition of initial, fatty streak, and intermediate lesions of atherosclerosis. A report from the Committee on Vascular Lesions of the Council on Arteriosclerosis, American Heart Association. Circulation. 1994;89:2462-2478.

.3. Stary HC, Chandler AB, Dinsmore RE, Fuster V, Glagov S, Insull W Jr, Rosenfeld ME, Schwartz CJ, Wagner WD, Wissler RW. A definition of advanced types of atherosclerotic lesions and a histological classification of atherosclerosis. A report from the Committee on Vascular Lesions of the Council on Arteriosclerosis, American Heart Association. Circulation. 1995; 92:1355-1374.

4. Hansson GK, Holin J, Jonasson L. Detection of activated T lymphocytes in the human atherosclerotic plaque. Am J Pathol. 1989;135:169-175.

5. Resnick D, Chatterton JE, Schwartz K, Slayter H, Krieger M J. Structures of class A macrophage scavenger receptors. Electron microscopic study of flexible, multidomain, fibrous proteins and determination of the disulfide bond pattern of the scavenger receptor cysteine-rich domain. J Biol Chem. 1996;271:26924-26930.

6. Endemann G, Stanton LW, Madden KS, Bryant CM, White RT, Protter AA. CD36 is a receptor for oxidized low density lipoprotein. J Biol Chem. 1993;268:11811-11816.

7. Stanton LW, White RT, Bryant CM, Protter AA, Endemann G. A macrophage Fc receptor for IgG is also a receptor for oxidized low density lipoprotein. J Biol Chem. 1992;267:22446-22451.

8. Acton S, Rigotti A, Landschulz KT, Xu S, Hobbs HH, Krieger M. Identification of scavenger receptor SR-BI as a high density lipoprotein receptor. Science. 1996;271:518-520.

9. Ramprasad MP, Fischer W, Witztum JL, Sambrano GR, Quehenberger O, Steinberg D. The 94- to 97-kDa mouse macrophage membrane protein that recognizes oxidized low density lipoprotein and phosphatidylserine-rich liposomes is identical to macrosialin, the mouse homologue of human CD68. Proc Natl Acad Sci U S A. 1995;92:9580-9584.

10. Ramprasad MP, Terpstra V, Kondratenko N, Quehenberger O, Steinberg D. Cell surface expression of mouse macrosialin and human CD68 and their role as macrophage receptors for oxidized low density lipoprotein. Proc Natl Acad Sci U S A. 1996;93:14833 14838.

11. Steinberg D, Parthasarathy S, Carew TE, Khoo JC, Witztuin JL. Beyond cholesterol. Modifications of low-density lipoprotein that increase its atherogenicity. N Engl J Med. 1989;320:915-924.

12. Anitschkow N, Chalatow S. Über experimentelle cholesterin-steatose und ihre bedeutung für die entstechung einiger pathologishen prozesses. Zentrablatt für Allgemeine Pathologie und Pathologishe Anatomie, vol 24(1), 1913, p 1-9. Translated in Arteriosclerosis. 1983;3:178-182, entitled: On experimental cholesterin steatosis and its significance in the origin of some pathological processes.

13. Page IH. Atherosclerosis. An introduction. Circulation. 1954;10:1-27.

14. Smith EB, Slater R. The chemical and immunological assay of low density lipoproteins extracted from human thoracic aortic intima. Atherosclerosis. 1970;11:417-438.

15. Hoff HF, Heideman CL, Gaubatz JW, Titus JL, Gotto AM. Quantification of apo-B in human aortic fatty streaks. Atherosclerosis. 1978;30:263-272.

16. Schwenke DC, Carew TE. Initiation of atherosclerotic lesions in cholesterol-fed rabbits. I. Focal increases in arterial LDL concentrations precede development of fatty streaks lesions. Arteriosclerosis. 1989;9:895-907.

17. Curmi PA, Juan L, Tedgui A. Effect of transmural pressure on LDL and albumin transport and distribution across the intact arterial wall. Circ Res. 1990;66:1692-1702.

18. Meyer G, Merval R, Tedgui A. Effects of pressure-induced wall stretching and convection on low density lipoprotein and albumin uptake in the rabbit aorta. Circ Res. 1996;79: 532-540.

19. Nordestgaard BG, Stender S, Kjeldsen K. Reduced atherogenesis in cholesterol-fed diabetic rabbits. Giant lipoproteins do not enter the arterial wall. Arteriosclerosis. 1988;8:421-428.

20. Napoli C, D'Armiento FP, Mancini FP, Postiglione A, Witztum JL, Palumbo G, Palinski W. Fatty streak formation occurs in human fetal aortas and is greatly enhanced by maternal hypercholesterolemia. Intimal accumulation of low density lipoprotein and its oxidation precede monocyte recruitment into early atherosclerotic lesions. J Clin Invest. 1997;100:2680-2690.

21. Ross R. 1993. The pathogenesis of atherosclerosis: a perspective for the 1990s. Nature. 362:801-809.

22. Libby P, Hansson GK. Biology of disease. Involvement of the immune system in human atherogenesis; current knowledge and unanswered questions. Lab Invest. 1991;64:5-15.

23. Smith JD, Trogan E, Ginsberg M, Grigaux C, Tian J, Miyata M. Decreased atherosclerosis in mice deficient in both macrophage colony-stimulating factor (op) and apolipoprotein E. Proc Natl Acad Sci U S A. 1995;92:8264-8268.

24. Qiao JH, Tripathi J, Mishra NK, Cai Y, Tripathi S, Wang XP, Imes S, Fishbein MC, Clinton SK, Libby P, Lusis AJ, Rajavashisth TB. Role of macrophage colony-stimulating factor in atherosclerosis: studies of osteopetrotic mice. Am J Pathol. 1997;150:1687-1699.

25. Rajavashisth T, Qiao JH, Tripathi S, Tripathi J, Mishra N, Hua M, Wang XP, Loussararian A, Clinton S, Libby P, Lusis A. Heterozygous osteopetrotic (op) mutation reduces atherosclerosis in LDL receptor-deficient mice. J Clin Invest. 1998;101:2702-2710.

26. Suzuki H, Kurihara Y, Takeya M, Kamada N, Kataoka M, Jishage K, Ueda O, Sakaguchi H, Higashi T, Suzuki T, Takashima Y, Kawabe Y, Cynshi O, Wada Y, Honda M, et al. A role for macrophage scavenger receptors in atherosclerosis and susceptibility to infection. Nature. 1997;386:292-296.

27. Ross R, Masuda J, Paines EW, Gown AM, Katuda S, Sasahara M, Malden LT, Masuko H, Sato H. Localization of PDGF-B protein in macrophages in all phases of atherogenesis. Science. 1990;248:1009-1012.

28. Wilcox JN, Smith KM, Williams LT, Schwartz SM, Gordon D. Platelet derived growth factor mRNA detection in human atherosclerotic plaques by in situ hybridization. J Clin Invest. 1988; 82:1134-1143.

29. Liptay MJ, Parks WC, Mecham RP, Roby J, Kaiser LR, Cooper JD, Botney MD. Neointimal macrophages colocalize with extracellular matrix gene expression in human atherosclerotic pulmonary arteries. J Clin Invest. 1993;91:588-594.

30. Tipping PG, Hancock WW. Production of tumor necrosis factor and interleukin by macrophages from human atheromatous plaques. Am J Pathol. 1993;42:1121-1728.

31. Rosenfeld ME, Ross R. Macrophage and smooth muscle cell proliferation in atherosclerotic lesions of WHHL and comparably hypercholesterolemic fat-fed rabbits. Arteriosclerosis. 1990;10:680-687.

32. Clinton SK, Underwood R, Hayes L, Sherman ML, Kufe DW, Libby P. Macrophage colony-stimulating factor gene expression in vascular cells and in experimental and human atherosclerosis. Am J Pathol. 1992;140:301-316.

33. Geng Y-J, Libby P. Evidence for apoptosis in advanced human atheroma. Colocalization with interleukin-1β-converting enzyme. Am J Pathol. 1995;147:251-266.

34. Isner JM, Kearney M, Bortman S, Passeri J. Apoptosis in human atherosclerosis and restenosis. Circulation. 1995;91:2703-2711.

35. Han DKM, Haudenschild CC, Hong MK, Tinkle BT, Leon MB, Liau G. Evidence for apoptosis in human atherogenesis and in a rat vascular injury model. Am J Pathol. 1995;147:267-277.

36. Björkerud S, Björkerud B. Apoptosis is abundant in human atherosclerotic lesions, especially in inflammatory cells (macrophages and T cells), and may contribute to the accumulation of gruel and plaque instability. Am J Pathol. 1996;149:367-380.

37. Mallat Z, Ohan J, Lesèche G, Tedgui A. Colocalization of CPP-32 with apoptotic cells in human atherosclerotic plaques. Circulation. 1997;96:424-428.

38. Kockx MM, De Meyer GR, Muhring J, Jacob W, Bult H, Herman AG. Apoptosis and related proteins in different stages of human atherosclerotic plaques. Circulation. 1998;97:2307-2315.

39. Kleindiest, Xu Q, Willeit J, Waldenberger FR, Weimann, Wick G. Immunology of atherosclerosis. Demonstration of Heat Shock Protein 6 expression and T lymphocytes bearing α/β or γ/δ receptor in human atherosclerotic lesions. Am J Pathol. 1993;142:1927-1937.

40. Miller DD, Craig FE, Dressler FA, Aguirre FV, Farrar MA, Breland CM, Donohue TJ, Kern MJ, Bach RG. Immunohistochemical characterization of immune cell composition and cytokine receptor expression in human coronary atherectomy tissue. Coron Artery Dis. 1995;6:965-972.

41. Stemme S, Holm J, Hansson GK. T lymphoctes in human atherosclerotic plaques are memory cells expressing CD45RO and the integrin VLA-1. Arterioscler Thromb. 1992;12:206-211.

42. Stemme S, Faber B, Holm J, Wiklund O, Witzum JL, Hansson GK. T lymphocytes from human atherosclerotic plaques recognize oxidized low density lipoprotein. Proc Nat Acad Sci USA. 1995;92:3893-3897.

43. Xu Q, Willeit J, Marosi M, Kleindienst R, Oberhollenzer F, Kiechl S, Stulnig T, Luef G, Wick G. Association of serum antibodies to heat-shock protein 65 with carotid atherosclerosis. Lancet. 1993;341:255-259.

44. Berberian PA, Myers W, Tytell M, Challa V, Bond MG. Immununohistochemical localisation of heat shiock protein-70 in normal-appearing and atherosclerotic specimens of human arteries. Am J Pathol. 1990; 36:7 1-80.

45. Hansson GK. Immunological markers of atherosclerosis. Lancet, 1993; 341:278.

46. Kol A, Sukhova GK, Lichtman AH, Libby P. Chlamydial heat shock protein 60 localizes in human atheroma and regulates macrophage tumor necrosis factor-α and matrix metalloproteinase expression. Circulation. 1998;98:300-307.

47. Emeson EE, Shen ML. Accelerated atherosclerosis in hyperlipidemic C57BL/6 mice treated with cyclosporin A. Am J Pathol. 1993;142:1906-1915.

48. Roselaar SE, Schonfeld G, Daugherty A. Enhanced development of atherosclerosis in cholesterol-fed rabbits by suppression of cell-mediated immunity. J Clin Invest. 1995;96:1389-1394.

49. Drew AF, Tipping PG. Cyclosporine treatment reduces early atherosclerosis in the cholesterol-fed rabbit. Atherosclerosis. 1995;116:181-189.

50. Emeson EE, Shen ML, Bell CG, Qureshi A. Inhibition of atherosclerosis in CD4 T-cell-ablated and nude (nu/nu) C57BL/6 hyperlipidemic mice. Am J Pathol. 1996;149:675-685.

51. Fyfe A, Qiao JH, Lusis AJ. Immune-deficient mice develop typical atherosclerosis fatty streaks when fed an atherogenic diet. J Clin Invest. 1994;94:2516-2520.

52. Daugherty A, Pure E, Delfel-Butteiger D, Chen S, Leferovich J, Roselaar SE, Rader DJ. The effects of total lymphocyte deficiency on the extent of atherosclerosis in apolipoprotein E-/- mice. J Clin Invest. 1997;100:1575-1580.

53. Dansky HM, Charlton SA, Harper MM, Smith JD. T and B lymphocytes play a minor role in atherosclerotic plaque formation in the apolipoprotein E-deficient mouse. Proc Natl Acad Sci U S A. 1997;94:4642-4646.

54. Amento EP, Ehsani N, Palmer H, Libby P. Cytokines positively and negatively regulate interstitial collagen gene expression in human smooth-muscle cells. Arteriosclerosis. 1991;11:1223-1230.

55. Clinton SK, Libby P. Cytokines and growth factors in atherogenesis. Arch Pathol Lab Med. 1992;116:1292-1300.

56. Kishikawa H, Shimokama T, Watanabe T. Localization of T lymphocytes and macrophages expressing IL-1, IL-2 receptor, IL-6 and TNF in human aortic intima. Role of cell-mediated immunity in human atherogenesis. Virchows Arch A Pathol Anat Histopathol. 1993;423:433-442.

57. Seino Y, Ikeda U, Ikeda M, Hasegawa T, Misawa Y, Yamamoto K, Kano S, Shimada K. Interleukin 6 gene transcripts are expressed in human atherosclerotic lesions. Cytokine. 1994;6:87-91.

58. Barath P, Fishleni MC, Cao J, Berenson J, Helfant RH, Forrester JS. Detection and localisation of tumor necrosis factor in human atheroma. Am J Cardiol. 1990; 65:297-302.

59. Clinton SK, Fleet JC, Loppnow H, Salomon RN, Clark BD, Cannon JG, Shaw AR, Dinarello CA, Libby P. Interleukin-1 gene expression in rabbit vascular tissue in vivo. Am J Pathol 1991;138:1005-1014.

60. Nelken NA, Coughlin SR, Gordon D, Wilcox JN. Monocyte chemoattractant protein-1 in human atheromatous plaques. J Clin Invest. 1991;88:1121-1127.

61. Modur V, Feldhaus MJ, Weyrich AS, Jicha DL, Prescott SM, Zimmerman GA, McIntyre TM. Oncostatin M is a proinflammatory mediator. In vivo effects correlate with endothelial cell expression of inflammatory cytokines and adhesion molecules. J Clin Invest 1997;100:158-168.

62. Bernard C, Merval R, Lebret M, Delerive P, Dusanter-Fourt I, Lehoux S, Créminon C, Staels B, Maclouf J, Tedgui A. Activation of human vascular smooth muscle cells by oncostatin M. Circ Res. 1998;(in press)

63. Moreau M, Brocheriou I, Petit L, Ninio E, Chapman MJ, Rouis M. Interleukin-8 mediates downregulation of TIMP-1 expression in cholesterol-loaded human macrophages: relevance to the stability of the atherosclerotic plaque. Circulation. 1998;(in press).

64. Uyemura K, Demer LL, Castle SC, Jullien D, Berliner JA, Gately MK, Warrier RR, Pham N, Fogelman AM, Modlin RL. Cross-regulatory roles of interleukin (IL)-12 and IL-10 in atherosclerosis. J Clin Invest. 1996;97:2130-2138.

65. Libby P, Galis ZS. Cytokines regulate genes involved in atherogenesis. Ann N Y Acad Sci. 1995;748:158-168.

66. Libby P, Sukhova G, Lee RT, Galis ZS. Cytokines regulate vascular functions related to stability of the atherosclerotic plaque. J Cardiovasc Pharmacol. 1995;25(Suppl 2):S9-S12.

67. Gupta S, Pablo AM, Jiang Xc, Wang N, Tall AR, Schindler C. IFN-γ potentiates atherosclerosis in ApoE knock-out mice. J Clin Invest. 1997;99:2752-2761.

68. LeBoeuf RC, Schreyer SA. The role of tumor necrosis factor-α receptors in atherosclerosis. Trends Cardiovasc Med. 1998;8:131-138.

69. van Lenten BJ, Fogelman AM. Lipopolysaccharide-induced inhibition of scavenger receptor expression in human monocyte-macrophages is mediated through tumor necrosis factor-α. J Immunol. 1992;148:112-116.

70. Hsu HY, Nicholson AC, Hajjar DP. Inhibition of macrophage scavenger receptor activity by tumor necrosis factor-α is transcriptionally and post-transcriptionally regulated. J Biol Chem. 1996;271:7767-7773.

71. Elhage R, Maret A, Pieraggi MT, Thiers JC, Arnal JF, Bayard F. Differential effects of interleukin-1 receptor antagonist and tumor necrosis factor binding protein on fatty-streak formation in apolipoprotein E-deficient mice. Circulation. 1998;97:242-244.

72. Mach F, Schonbeck U, Sukhova GK, Bourcier T, Bonnefoy JY, Pober JS, Libby P. Functional CD40 ligand is expressed on human vascular endothelial cells, smooth muscle cells, and macrophages: implications for CD40-CD40 ligand signaling in atherosclerosis. Proc Natl Acad Sci U S A. 1997;94:1931-1936.

73. Mach F, Schonbeck U, Sukhova GK, Atkinson E, Libby P. Reduction of atherosclerosis in mice by inhibition of CD40 signalling. Nature. 1998;394:200-203.

74. Pomerantz KB, Hajjar DP, Levi R, Gross SS. Cholesterol enrichment of arterial smooth muscle cells upregulates cytokine-induced nitric oxide synthesis. Biochem Biophys Res Comm. 1993;191:103-109.

75. Verbeuren TJ, Bonhomme E, Laubie M, Simonet S. Evidence for induction of nonendothelial NO synthase in aortas of cholesterol-fed rabbits. J Cardiovasc Pharmacol. 1993;21:841-845.

76. Buttery LD, Springall DR, Chester AH, Evans TJ, Standfield EN, Parums DV, Yacoub MH, Polak JM. Inducible nitric oxide synthase is present within human atherosclerotic lesions and promotes the formation and activity of peroxynitrite. Lab Invest. 1996;75:77-85.

77. Mallat Z, Heymes C, Ohan J, Lesèche G, Tedgui A. Expression of Interleukin-10 in advanced human atherosclerotic plaques. Relation to inducible nitric oxide synthase and cell death. Arterioscler Thromb Vasc Biol. 1998;(in press)

78. Geng YJ, Wu Q, Muszynski M, Hansson GK, Libby P. Apoptosis of vascular smooth muscle cells induced by in vitro stimulation with interferon-γ, tumor necrosis factor-α, and interleukin-1 β. Arterioscler Thromb Vasc Biol. 1996;16:19-27.

79. Albina JE, Cui SJ, Mateo RB, Reichner JS. Nitric-oxide-mediated apoptosis in murine peritoneal macrophages. J Immunol. 1993, 150:5080-5085.

80. Kockx MM, Muhring J, Knaapen MW, de Meyer GR. RNA synthesis and splicing interferes with DNA in situ end labeling techniques used to detect apoptosis. Am J Pathol. 1998;152:885-888

81. Luoma JS, Stralin P, Marklund SL, Hiltunen TP, Sarkioja T, Yla-Herttuala S. Expression of extracellular SOD and iNOS in macrophages and smooth muscle cells in human and rabbit atherosclerotic lesions: colocalization with epitopes characteristic of oxidized LDL and peroxynitrite-modified proteins. Arterioscler Thromb Vasc Biol. 1998;18:157-167.

82. Szabo C. DNA strand breakage and activation of poly-ADP ribosyltransferase: a cytotoxic pathway triggered by peroxynitrite. Free Radic Biol Med. 1996;21:855-869.

83. de Vries JE. Immunosuppressive and anti-inflammatory properties of interleukin 10. Ann Med. 1995;27:537-541.

84. de Waal Malefyt R, Abrams RJ, Bennett B, Figdor CG, de Vries JE. Interleukin-10 (IL-10) inhibits cytokine synthesis by human monocytes - an autoregulatory role of IL-10 produced by monocytes. J Exp Med. 1991;174:1209-1220.

85. Gérard C, Bruyns C, Marchant A, Abramowicz D, Vandebeele P, Delvaux A, Fiers W, Goldman M, Velu T. Interleukin-10 reduces the release of tumor necrosis factor and prevents lethality in experimental endotoxemia. J Exp Med. 1993;177:547-550.

86. Arai T, Hiromatsu K, Nishimura H, Kimura Y, Kobayashi N, Ishida H, Nimura Y, Yoshikai Y. Endogenous interleukin 10 prevents apoptosis in macrophages during salmonella infection. Biochem Biophys Res Commun. 1995;213:600-607.

87. Cohen SB, Crawley JB, Kahan MC, Feldmann M, Foxwell BM. Interleukin-10 rescues T cells from apoptotic cell death: association with an upregulation of Bcl-2. Immunology. 1997;92:1-5.

88. Pober JS, Cotran RS. The role of endothelial cells in inflammation. Transplantation. 1990;50:537-544.

89. Li H, Cybulsky MI, Gimbrone Jr MA, Libby P. An atherogenic diet rapidly induces VCAM-1, a cytokine-regulatable mononuclear leukocyte adhesion molecule, in rabbit aortic endothelium. Arterioscler Thromb. 1993;13:197-204.

90. O'Brien KD, Allen MD, McDonald TO, Chait A, Harlan JM, Fishbein D, McCarty J, Ferguson M, Hudkins K, Benjamin CD, Lobb R, Alpers CE. Vascular cell adhesion molecule-1 is expressed in human coronary atherosclerotic plaques. Implications for the mode of progression of advanced coronary atherosclerosis. J Clin Invest. 1993;92:945-951.

91. Hla T, Neilson K. Human cyclooxygenase-2 cDNA. Proc Natl Acad Sci USA, 1992;89:7384-7388.

92. Yoshizumi M, Perrella MA, Burnett JC Jr, Lee ME. Tumor necrosis factor downregulates an endothelial nitric oxide synthase mRNA by shortening its half-life. Circ Res. 1993;73:205-209.

93. Raines E, Dower S, Ross R.. Interleukin-1 mitogenic activity for fibroblast and smooth muscle cells is due to PDGF-AA. Science. 1989;243:393-396.

94. Galis Z, Muszynski M, Sukhova G, Simon-Morrissey E, Unemori EN, Lark MW, Amento E, P Libby. Cytokine-stimulated human vascular smooth cells synthesize a complement of enzymes required for extracellular matrix digestion. Circ Res. 1994;75:181-189.

95. Fabunmi RP, Sukhova GK, Sugiyama S, Libby P. Expression of Tissue Inhibitor of Matalloproteinases-3 in human atheroma and regulation in lesion-associated cells. Circ Res. 1998;83:270-278.

96. Hughes DA, Townsend PJ, Haslam PL. Enhancement of the antigen-presenting function of monocytes by cholesterol: possible relevance to inflammatory mechanisms in extrinsic allergic alveolitis and atherosclerosis. Clin Exp Immunol. 1992;87:279-286.

97 Frostegard J, Wu R, Giscombe R, Holm G, Lefvert AK, Nilsson J. Induction of T-cell activation by oxidized low density lipoprotein. Arterioscler Thromb. 1992;12:461-467.

98 Liao F, Berliner JA, Mehrabian M, Navab M, Demer LL, Lusis AJ, Fogelman AM. Minimally modified low density lipoprotein is biologically active in vivo in mice. J Clin Invest. 1991;87:2253-2257.

99. Liao F, Andalabi A, deBeer FC, Fogelman AM, Lusis AJ. Genetic control of inflammatory gene induction and NF-kB-like transcription factor activation in response to an etherogenic diet in mice. J Clin Invest. 1993;91:2572-2579.

100. Fong LG, Fong TAT, Copper AD. Inhibition of lipopolysaccharide-induced interleukin-1β mRNA expression in mouse macrophages by oxidized low density lipoprotein. J Lipid Res. 1991;32:1899-1910.

101. Salonen JT, Yla-Herttuala S, Yamamoto R, Butler S, Korpela H, Salonen R, Nyyssonen K, Palinski W, Witztum JL. Autoantibodies against oxydised LDL and progression of carotid atherosclerosis. Lancet. 1992;339:883-887.

102. Glagov S, Weisenberg E, Zarins CK, Stankunavicius R, Kolettis GJ. Compensatory enlargement of human atherosclerotic coronary arteries. N Engl J Med. 1987;316:1371-1375.

103. McPherson DD, Sirna SJ, Hiratza LF, Thorpe LJ, Armstrong ML, Marcus ML, Kerber RE. Coronary arterial remodeling studied by high-frequency epicardial echocardiography: an early compensatory mechanism in patients with obstructive coronary atherosclerosis. J Am Coll Cardiol. 1991;17:79-86.

104. McPherson DD, Johson MR, Alvarez NM, Rewcastle NB, Collins SM, Armstrong ML, Kieso RA, Thorpe LJ, Marcus ML, Kerber RE. Variable morphology of coronary atherosclerosis: characterization of atherosclerotic plaque and residual arterial lumen size and shape by epicardial echocardiography. J Am Coll Cardiol. 1992;19:593-599.

105. Clarkson TB, Prichard RW, Morgan TM, Petrick GS, Klein KP. Remodeling of coronary arteries in human and nonhuman primates. JAMA. 1994;271:289-294.

106. Steinke W, Els T, Hennerici M. Compensatory carotid artery dilatation in early atherosclerosis. Circulation. 1994;89:2578 2581.

107. Losordo DW, Rosenfield K, Kaufman J, Pieczek A, Isner JM. Focal compensatory enlargement of human arteries in response to progressive atherosclerosis. In vivo documentation using intravascular ultrasound. Circulation. 1994;89:2570-2577.

108. Davies MJ, Thomas AC. Plaque fissuring: the cause of acute myocardial infarction, sudden ischemic death, and cescendo angina. Br Heart J. 1985;53:363-373.

109. Fuster V, Badimon I, Badimon J, Chesebro J. The pathogenesis of coronary artery disease and the acute coronary syndromes. N Engl J Med. 1992;326:242-250.

110. Richardson PD, Davies MJ, Born GV. Influence of plaque configuration and stress distribution on fissuring of coronary atherosclerotic plaques. Lancet. 1989;2:941-944.

111. Lee RT, Libby P. The unstable atheroma. Arterioscler Thromb Vasc Biol. 1997;17:1859-1867.

112. Kaplan M, Aviram M. Oxidized LDL binding to a macrophage-secreted extracellular matrix. Biochem Biophys Res Commun. 1997;237:271-276.

113. Weissberg PL, Clesham GJ, Bennett MR. Is vascular smooth muscle cell proliferation beneficial? Lancet. 1996;347:305-307.

114. Henney AM, Wakeley PR, Davies MJ, Foster K, Hembry R, Murphy G, Humphries S. Localization of stromelysin gene expression in atherosclerotic plaques by in situ hybridization. Proc Natl Acad Sci U S A 1991;88:8154-8158.

115. Galis ZS, Sukhova GK, Lark MW, Libby P. Increased expression of matrix metalloproteinases and matrix degrading activity in vulnerable regions of human atherosclerotic plaques. J Clin Invest. 1994;94:2493-2503.

116. Nikkari ST, Geary RL, Hatsukami T, Ferguson M, Forough R, Alpers CE, Clowes AW. Expression of collagen, interstitial collagenase, and tissue inhibitor of metalloproteinases-1 in restenosis after carotid endarterectomy. Am J Pathol. 1996;148:777-783.

117. Halpert I, Sires UI, Roby JD, Potter-Perigo S, Wight TN, Shapiro SD, Welgus HG, Wickline SA, Parks WC. Matrilysin is expressed by lipid-laden macrophages at sites of potential rupture in atherosclerotic lesions and localizes to areas of versican deposition, a proteoglycan substrate for the enzyme. Proc Natl Acad Sci U S A. 1996;93:9748-9753.

118. van der Wal AC, Becker AE, Das PK. Medial thinning and atherosclerosis-evidence for involvement of a local inflammatory effect. Atherosclerosis. 1993;103:55-64.

119. Joris I, Zand T, Nunnari JJ, Krolikowsk FJ, Majno G. Studies on the pathogenesis of atherosclerosis. 1. Adhesion and emigration of mononuclear cells in the aorta of hypercholesterolemic rats. Am J Pathol. 1983;113:341-358.

13. ARTERIAL GENE TRANSFER

Laurent J. Feldman, Ph. Gabriel Steg
Cardiology Department and U460 INSERM
Faculté Xavier Bichat, Paris, France

Atherosclerosis and its complications represent the first cause of mortality and morbidity in the western world. Major advances have been made in the treatment and prevention of symptomatic atherosclerosis. Improved understanding of the pathophysiology of atherosclerosis and its complications [1], as well as spectacular advances in the molecular biology of the vascular wall [2] may open new perspectives for treatment based upon local delivery of genetic material designed to modify the atherosclerotic plaque at the molecular level [3]. Transfer of a functional gene into arterial wall cells, termed arterial gene therapy, may be used to replace or palliate a defective gene, or to express a protein with a therapeutic effect [4].

Effective arterial gene therapy requires techniques to introduce (transfect) a foreign gene (transgene) into the cells of the arterial wall. These techniques rely on transfer vectors which facilitate cellular penetration and intra-cellular trafficking of the transgene, as well as local delivery systems, either catheter-based or surgical, to deliver the vector to the vicinity of the target cells.

VECTORS

Under certain conditions, it is feasible to introduce foreign DNA into the nucleus of eukaryotic cells. This has been used to obtain transient or stable expression of several genes in cell lines. To achieve expression of foreign DNA, however, the transferred gene should enter the cell, escape degradation by lysosomal enzymes, cross the nuclear membrane, escape degradation by endonucleases and, eventually, be expressed. Each of these steps represents a potential limitation to the efficacy of gene transfer, which make spontaneous transfer and expression of foreign DNA into eukaryotic cells a rare phenomenon. Indeed, several investigators have used "naked" plasmidic DNA to transfect the arterial wall. They all agreed, however, that, although the technique is simple and safe, it results in very low transfection efficiency [5]. Transfer vectors are therefore required to increase the efficiency of the process [6]. These may be viruses, non-viral vectors or mixed systems combining viral and non-viral elements.

Viral vectors

Viral-based vectors are viral particles which retain their ability to enter target cells and transfer in these cells foreign genes, but have been engineered to incorporate the transgene in their genome and loose their replicative activity. Some of them (retroviruses, lentiviruses and possibly adeno-associated viruses) are the only methods ensuring stable integration of transferred DNA into the chromosomal DNA of the target cell. In order to transform viruses into safe vectors, genomic sequences which are required for viral replication have been deleted. Although, retroviruses have been the first viral vectors used to perform arterial gene transfer in live animals, their efficiency is extremely low [7]. In fact, the only viral vectors which have unequivocally improved transfection efficiency are adenoviral vectors.

First generation recombinant adenoviral vectors currently used (for a review, see [8]) are obtained through homologous recombination between the genome of a serotype 5 or 2 adenovirus and a shuttle plasmid in which the transgene has been inserted. The vectors are made replication-defective by deletion of the E1 region, which encodes a transactivating factor required for viral DNA replication.

Adenoviral vectors allow to circumvent some of the problems encountered with retroviruses: 1) they can infect quiescent as well as replicating cells; 2) high titer stocks can be produced easily, generally ranging from 10^{11} to 10^{12} plaque forming units (pfu)/mL; 3) they accommodate large DNA inserts; and 4) after infection, the adenovirus genome remains episomal, therefore avoiding the risk of insertional mutagenesis.

Nevertheless, several drawbacks of adenoviral vectors have been identified. First, current adenovectors are associated with only transient, 2- to 4-week, transgene expression [9]. Second, at a high concentration, replication-competent adenoviral particles may be observed. Third, it has been demonstrated that first-generation adenovectors evoke a strong cellular immune response, targeted at viral proteins as well as certain transgene products [10, 11], resulting in destruction of these cells by cytotoxic T lymphocytes. In addition, infection of the target cells by adenoviral vectors leads to an acute inflammatory reaction, characterized by a neutrophil- [12] and macrophage-rich [13] cellular infiltrate. Finally, it has been observed that local delivery of adenoviral vectors in normal arteries upregulates the expression of vascular cell adhesion molecules and may even trigger mild, although significant, neointima formation [14].

The main consequence of the inflammatory response directed against recombinant adenoviral vectors is a rapid extinction of transgene expression (within a few weeks). Repeated injections of adenoviral vectors might not result in prolonged transgene expression due to a humoral immune response targeted at both viral proteins and transgene products [15]. Interestingly, the duration of transgene expression is substantially prolonged when transfer is achieved in newborn animals [16], probably because these are not yet immunocompetent and can therefore tolerate viral proteins, or in animals in which cellular immune response has been pharmacologically depressed [17]. Immunosuppression, however, cannot be used routinely in clinical practice. Thus, new adenoviral vectors, termed second and third-generation vectors, are currently developed [18-20]. These vectors are modified to prevent residual expression of adenoviral proteins, thereby mitigating adenoviral protein-specific cellular immune response. However, it must be borne in mind that, in certain indications, such as prevention of restenosis, transient gene expression may be sufficient and may even confer a relative safety to adenoviral vector-based strategies for gene therapy, since potential deleterious effects related to transgene expression would be limited to a few weeks.

Other viral vectors, such as the AAV (adeno-associated viruses) and herpes viruses are being developed for gene transfer but have not been tested in models of arterial transfection.

Non viral vectors

The limitations and risks associated with viral vectors are powerful stimulants for the development of non viral vectors.

Cationic liposomes

Cationic liposomes are positively-charged artificial lipid vesicles, which incorporate negatively-charged plasmid DNA. In contrast with viral vectors, there is no size constraint for the transgene and preparation of the vectors is easy. DNA-liposomes complexes contain an excess of positive charges, which facilitates fusion between liposomes and the negatively-charged cell membranes [21]. Once in the cytoplasm, most of the DNA-liposome complexes are degraded by lysosomal enzymes, and approximately 1% of the DNA which originally penetrated the cell enter the nucleus where it remains extra-chromosomal. This explains why transgene expression is transient when liposomes are used as vectors. The efficiency of cationic liposomes for gene transfer is superior to that of other non-viral methods and varies with the preparation used, the DNA/liposome concentration ratio, the type and the proliferative status of the transfected cells (transfer appears increased in proliferating cells). While in theory, in vitro transfection efficiencies can reach up to 90 % of target cells, experience with in vivo arterial gene transfer suggests that efficiency remains disturbingly low [22].

Conjugated vectors

Conjugated vectors (for a review, see [23]) represent a heterogeneous class of vectors in which the transgene is conjugated to a polycation - e.g., polylysine - which is chemically bound to a proteic ligand. Polylysine forms a complex with negatively-charged DNA via electrostatic interaction, and condenses the DNA into a macromolecule-like structure. Binding of the ligand to its specific membrane receptor mediates vector internalization into the cell. Although, several studies have established that selective in vitro gene transfer can be achieved using this method, its efficacy in an in vivo setting has been disappointing.

Conjugated vectors associated with the Hemagglutinating Virus of Japan (HVJ)

These vectors integrate liposomes, DNA and UV-inactivated HVJ particles. It is assumed that when one HVJ particle gains access to the cell, via a specific receptor-mediated pathway, one or several liposome/DNA complexes are internalized in the cell and a certain amount of internalized DNA reaches the nucleus where it is expressed. The absence of toxicity of these vectors and their efficiency (~10-fold that of cationic liposomes) for vascular smooth muscle cell transfection, make them an attractive tool for arterial gene transfer [24]. These vectors have been used for a host of cardiovascular applications, including the study of the effect of autocrine-paracrine vasoactive modulators - e.g., the renin-angiotensin system - on vascular smooth muscle cells in vitro, as well as the inhibition of intimal hyperplasia using nitric oxide synthase cDNA [25], antisense oligonucleotides directed towards cell-cycle positive regulators [26], or transcription factor "decoys" [27].

GENE DELIVERY TECHNIQUES

Percutaneous local delivery systems

Most potential targets of arterial gene therapy would require local expression of the transgene at a specific arterial site. This can be achieved either by delivering the gene vector in the vicinity of the target site - i.e., local delivery - or by incorporating in the vector design a ligand which will drive transgenic expression at that site, even if the vector is injected systemically. Advances in local gene delivery to the arterial wall have largely benefited from progress made in the technology of angioplasty balloon catheters. The simplest method for local arterial gene transfer is the "dwell method," which requires one to isolate surgically between two temporary ligatures an arterial segment, withdraw blood, inject into the isolated segment a solution containing the vector, then withdraw this solution after a variable incubation time and reestablish blood flow. Dwelling allows to perform arterial gene transfer in a well controlled manner. The invasive nature of the method, however, is a major drawback for clinical application. For this reason, various local gene delivery systems have been developed (for a review, see [28]).

Table 1: Currently available local delivery catheters.

device	mechanism	target cells	released for clinical use
double-balloon (USCI)	diffusion	endothelium	no
Dispatch (Scimed)	diffusion	endothelium	yes
hydrogel balloon (Boston Sc.)	diffusion + pressure	SMC	yes
coated stents (Johnson & Johnson)	diffusion + apposition	SMC	yes
porous balloon (USCI)	high pressure	SMC, adventitia	no
microporous balloon (Cordis)	low pressure	SMC?	no
channelled balloon (Boston Sc.)	low pressure	SMC	yes
Transport (Scimed)	low pressure	SMC?	yes
InfusaSleeve (Localmed)	low pressure	SMC?	yes
iontophoretic balloon (CorTrak)	electric gradient	SMC?, adventitia?	no
needle balloon	puncture	SMC? adventitia?	no
nipple balloon (Interventional Ther.)	puncture	SMC? adventitia?	no

SMC, smooth muscle cell

Catheters (Table 1)

The "ideal" local delivery catheter should incorporate the following features to perform optimal arterial gene transfer. First and foremost it should achieve very efficient gene transfer. Second, if it is intended to be used for prevention of restenosis, balloon angioplasty and gene transfer should be performed simultaneously during a one-step straightforward procedure using the same catheter. Third, the local delivery device should be safe, and, in particular, should not induce excessive injury to the arterial wall, in addition to the injury associated with the angioplasty per se. Fourth, gene delivery catheters should incorporate a perfusion design to limit myocardial ischemia during gene incubation. Finally, site-specificity is a key issue for arterial gene therapy. Local delivery catheters should be designed in order to minimize gene leakage in the bloodstream, which may limit local efficacy and result in systemic toxicity, especially when viral vectors are required.

Schematically, local delivery through catheters involves 3 basic, device-related mechanisms: passive diffusion, pressure facilitation, and mechanical facilitation.

Passive diffusion

The prototypic double balloon catheter incorporates two latex balloons which, when inflated into the target arterial segment, delineate a "transfection chamber" into which the transfer vector can be instilled via an infusion port. This catheter was the first to be used for catheter-based arterial gene transfer [7]. It has, however, several shortcomings. Passive diffusion of the gene vector requires protracted incubation times, which may generate tissue ischemia. There is a risk of vector diffusion into the systemic circulation via side branches (such side branches emerge every 2-4 mm in the epicardial coronary arteries). Finally, inflation of the two latex balloons may generate additional arterial trauma both upstream and downstream to the target site.

New catheters represent "improved" designs of the double-balloon catheter. The Dispatch™ catheter is a sophisticated catheter allowing simultaneous distal perfusion and isolation of multiple infusion chambers between the catheter and the vessel wall. The main advantage of this device is that even protracted incubations do not induce significant tissue ischemia. This system has been successfully used to achieve substantial gene delivery into the endothelium and superficial medial layers of both normal and atherosclerotic rabbit arteries [29].

The hydrogel-coated balloon catheter [9] is a conventional angioplasty balloon which is coated with a hydrophilic polymer which swells like a sponge in presence of a solution containing a drug or a gene vector. Upon inflation of the angioplasty balloon in the target vessel, the polymer expresses the adsorbed materials toward the arterial wall. This balloon was initially designed to cross high-grade complex lesions. It turned out to be a very efficient local delivery system which has been released for clinical use in Europe and United-States. This catheter is currently used for local delivery of the VEGF cDNA in the peripheral arteries of patients suffering end-stage arteriopathy [30]. Its main shortcoming is that during exposure of the catheter in the bloodstream, most of the hydrogel content is "washed" off the balloon. However, retention into the polymer may be enhanced by using a protective sheath.

Pressure facilitation

The Wolinsky catheter incorporates an angioplasty balloon with 25 μm diameter pores. During balloon angioplasty, inflation pressure drives the solution filling the balloon into the arterial wall [31]. However, at high inflation pressures typically required for optimal angioplasty (5 to 10 atm), porous balloons generate high velocity jets, which result in arterial perforation as well as reactive intimal •hyperplasia, and carries a risk of gene dissemination. Therefore, improved porous balloon catheters - such as the microporous balloon catheter [32] and the channeled balloon catheter [33] - have been designed to overcome these limitations, principally to dissociate balloon inflation pressure from vector instillation pressure.

Mechanical facilitation

The mechanism used to facilitate local delivery may be either an electrical field, in the case of the iontophoretic catheter, or a physical injury to the vessel wall in the case of the Infiltrator catheter. These systems deserve to be tested in experimental models of arterial gene transfer.

Intracoronary stents

Intracoronary metallic stents are exponentially used in clinical practice to reduce the rate of restenosis following interventional procedures. In addition, stents may be used as vehicles for local delivery of drugs or gene vectors into the arterial wall (for a review see [34]). One of the approaches would be to coat the stent struts with endothelial cells, previously transfected with a therapeutic gene, the product of which may be released locally to exert therapeutic effects [35]. However, clinical applicability of this technique remains limited due to the requirement of previous isolation and in vitro transfection of endothelial cells - a long and costly procedure - as well as the poor adhesion of these cells to the stent struts following expansion under flow condition. Another strategy is to use polymer-coated stents, or even fully-biodegradable polymeric stents, which may act as a reservoir for a gene vector

Polymers

Biodegradable polymers could be used for arterial gene transfer in the form of intracoronary stents, micro- or nano-particles injected locally via a catheter [36] , gels coating the endoluminal aspect of the artery [37], or periadventitial wrapping. The latter method has been used to transfer plasmidic DNA [38] as well as antisense oligonucleotides [39] into the rat carotid artery, in order to prevent intimal thickening following arterial injury. In addition, we recently demonstrated that co-delivery of adenoviral vectors together with the block co-polymer poloxamer 407 facilitates arterial transfection and allows for shorter incubation times [13].

Myocardial delivery

Pioneered by Wolff and colleagues [40], intramuscular gene transfer represents an alternative method to target the myocardium. Direct injection of plasmid DNA into the myocardium, although feasible, has been associated with short-lived (2 to 4 weeks) expression of the transgene in only a small number of cells, as well as with potentially deleterious myocardial fibrosis and inflammation. When intravenous injection of recombinant replication defective adenoviral vectors is used, only minor myocardial uptake is observed [16]. In contrast, direct injection of such vectors in the heart resulted in efficient transfection [41], albeit limited by the previously described shortcomings associated with adenoviral vectors and by the need for direct access to the myocardium. Alternatively, at least three studies have reported efficient transfection of the myocardium after intracoronary catheter delivery of adenoviral vectors [42-44]. Pressure conditions during delivery as well as pretreatment of the vessel with various vasoactive agents may impact heavily on the feasibility of myocardial gene transfer via the intracoronary route [44]. Other methods for myocardial gene transfer include percutaneous coronary artery transfection, percutaneous intracavitary injection, and pericardial apposition.

FEASIBILITY OF ARTERIAL GENE TRANSFER

Two strategies can be used for gene transfer: indirect gene transfer involves in vitro transfection of vascular cells which are then implanted back into the vasculature. Conversely, direct gene transfer is the direct introduction of a foreign gene into the arterial wall in vivo.

Indirect gene transfer

In their seminal experiments, Nabel et al. used recombinant retroviral vectors to transfer the ß-galactosidase reporter gene into porcine endothelial cells in vitro [45]. Transfected endothelial cells could be easily recognized and selected for subsequent introduction into the iliac artery in vivo using a double-balloon catheter. Expression of the transgene was found in these arteries several weeks after transfer. Indirect gene transfer allows for in vitro selection of successfully transduced cells prior to arterial delivery. However, it is only suitable for autologous transfection. Moreover, it requires cell isolation and culture prior to transfer, which makes it a cumbersome and costly technique for routine clinical application.

Direct gene transfer

In contrast, direct gene transfer is a one-step method in which the transgene is directly delivered into the target arterial site. Nabel and her colleagues were the first to achieve direct arterial gene transfer into porcine iliac arteries using replication-defective retroviral vectors and cationic liposomes expressing the ß-galactosidase reporter gene [7]. ß-galactosidase activity was observed in all the transfected animals, up to 21 weeks in those animals transfected using retroviral vectors and 6 weeks when liposomes were used. Other studies using retroviral vectors [46], cationic liposomes [22] or plasmid DNA [5], have confirmed these results. However, in all theses studies, transfection efficiency was far below 0.1 %.

Low transfection efficiency is a major limitation of these methods, in particular when the transgene encodes a protein which remains intracellular, a frequent feature for many of the candidate genes used in attempts to prevent restenosis. Conversely, when the transgene encodes a secreted protein, such as secreted growth factor, a substantial biological effect may be observed even when a small number of target cells express the transgene.

Replication-defective adenoviral vectors have been used for gene transfer into the arterial wall, using surgical [47] or percutaneous [9] transfer techniques. When such a vector expressing a nuclear-targeted ß-galactosidase gene (Figure 1) is delivered in contact with the endothelial layer without previous injury, transfer is strictly localized to the endothelium (Figure 2A) [9, 48]. Transfection efficiency is related to the concentration of the adenovirus stocks used [48]. Under certain conditions, however, transfection efficiency approximates 100% [9, 47].

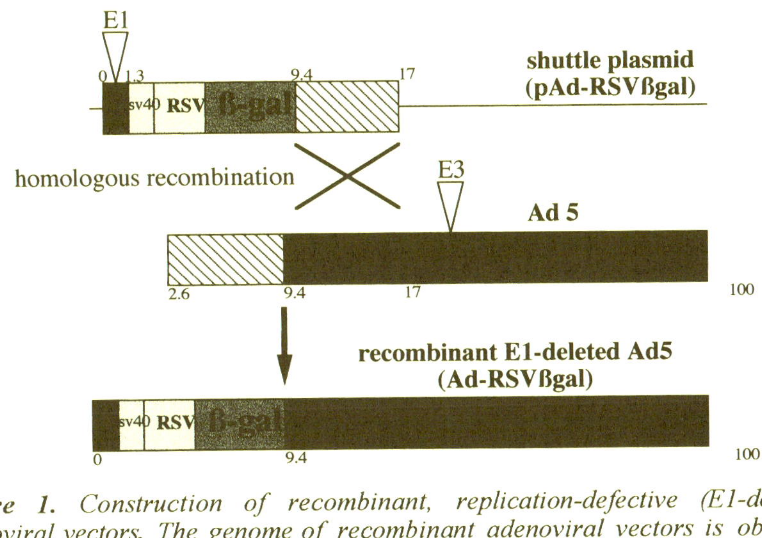

Figure 1. *Construction of recombinant, replication-defective (E1-deleted) adenoviral vectors. The genome of recombinant adenoviral vectors is obtained through homologous recombination between (1) a shuttle plasmid (pAd-RSVβgal) carrying the left end (map units 0 to 17) of the Ad5 genome, except E1 which is replaced by a nuclear localization signal (SV40), the Rous Sarcoma Virus long terminal repeat (LTR) and the β-galactosidase reporter gene, and (2) the rightward part of the Ad5 genome (map units 26 to 100), in which the E3 region has been deleted.*

When adenovirus-mediated gene transfer is attempted following abrasion of the endothelial layer, transgene expression is found mostly in smooth muscle cells located in the superficial layers of the media (Figure 2B) [9, 33]. Reported transfection efficiencies, expressed as the percentage of transfected medial cells, range from 2 to 70%. Adenoviral vectors are thus, by far, the most effective vectors for medial transfection. Transfection efficiency in the media depends upon various factors, including virus preparation (concentration of viral stocks, promoter sequences), delivery technique (dwell > catheter-based), catheter-type (hydrogel balloon > double balloon), duration of incubation (long > short), ligation of side-branches, as well as co- or pre-treatment with adjunctive agents. The endothelium and internal elastic lamina are the main barriers to penetration of the adenoviral vectors in the media of non atherosclerotic vessels [9, 49, 50], whereas the neointima is a relatively resistant layer to adenovirus penetration in severely atherosclerotic arteries [33]. Recently, Low transfection efficiency in atherosclerotic vessels must be considered when potential clinical applications relate to atheromatous arteries. Methods to circumvent low transfer efficiency include the use of therapeutic genes encoding for secreted proteins, which may therefore affect not only transduced but also neighboring untransduced cells. Alternatively, transfection efficiency can be enhanced by using adjunctive agents such as elastase, to permeate the internal elastic lamina [50], or polymeric compounds [13].

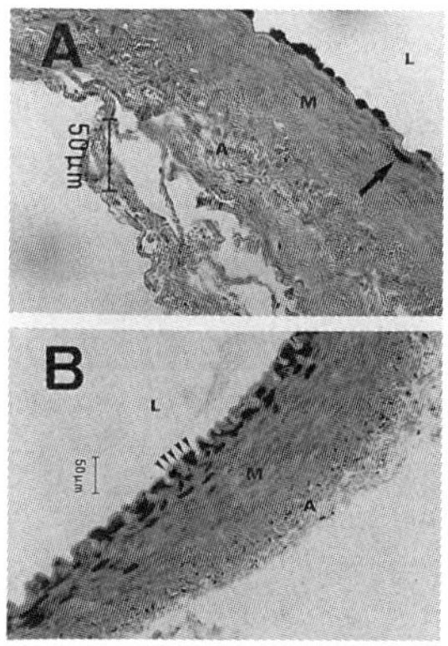

Figure 2. *Percutaneous adenovirus-mediated arterial gene transfer.(A): Endothelium-specific arterial gene transfer, (B): Smooth-muscle cell-specific arterial gene transfer. L, lumen; M, media; A, adventitia*

LIMITATIONS

The risk of systemic dissemination of viral particles is high in the case of amphotropic adenoviral vectors. Previous studies have shown that the risk is low when adenoviral vectors are introduced into a surgically isolated arterial segment [13, 47] . When percutaneous delivery is attempted, however, the risk of viral dissemination becomes substantial, most of extra-arterial transfection being located in the liver [9, 33] .The risk depends, at least in part, upon the design of the delivery catheter [9]. Other factors may facilitate viral spreading such as the degree of arterial trauma induced by the gene delivery device, the use of non-specific viral promoters, high viral titers, as well as the presence of developed vasa vasora in the arterial wall, a typical feature of the atherosclerotic plaque.

Another issue is that in most of currently available transfection systems transgene expression cannot be regulated by physiological or exogenous signals, which may represent an important limitation for clinical application of gene therapy to some diseases in which precise control over the level of protein production is required to achieve a therapeutic effect and/or prevent toxicity. Several approaches have been tested in transgenic animals and in models of somatic gene transfer to regulate transgene expression. For example, Bohl et al. have recently designed a system in which two genes packaged into distinct retroviral vectors are transferred in primary myoblasts in vitro: an erythropoietin cDNA driven by a tetO-CMV promoter and a reverse transactivator (rtTA). When doxycycline is added, rtTA binds the tetO-CMV promoter and activates transcription of the erythropoietin cDNA [51]. To date, none of these systems have been tested in models of cardiovascular gene transfer. However, tight regulation of transgene expression may not be as crucial in cardiology as in inheritable metabolic diseases.

CONCLUSIONS AND PERSPECTIVES OF CLINICAL APPLICATION

In vivo transfection of foreign genes in the cardiovascular system has broad applications. Retroviral vectors have been used to study cardiovascular development and to learn more on the role of growth factors in the pathophysiology of atherosclerosis [52]. Clinical use of transfection techniques to treat cardiovascular disease represents a more challenging issue [3]. In vivo application of gene therapy requires a unique combination of appropriate vector, delivery system, target cell and therapeutic gene which is likely to be specifically tailored to each application. Therefore there is no "best" vector or delivery system. Current indications for arterial gene therapy should fulfil the following requirements: 1) it should be a disease without proven efficient conventional therapy (due to either ineffectiveness or major adverse effects); 2) the arterial lesions to be treated should be amenable to local delivery techniques; 3) pathophysiology should be clear enough for identification of candidate therapeutic genes.

There are currently two major potential indications for arterial gene therapy: prevention of restenosis after angioplasty of coronary arteries, and therapeutic angiogenesis to treat chronic ischemia, either in the myocardium or in the limbs (for reviews, see [3, 53, 54]). Other potential indications are prevention of degeneration of aortocoronary saphenous vein bypass grafts [55] and atheromatous plaque stabilization [56], but nearly all vascular diseases are theoretically amenable to gene therapy [2, 57], as well as a host of non vascular diseases in which arterial access to the diseased organs is available for local genetic treatment.

Genetic prevention of restenosis best exemplifies the importance of gene delivery techniques in achieving therapeutic success. Indeed, it has long been considered that restenosis - i.e., the recurrence of luminal narrowing in the months following angioplasty - results almost exclusively from intimal hyperplasia, a proliferating process involving medial and intimal layers of vascular smooth muscle cells [58]. Therefore, most of current strategies aimed at preventing restenosis by gene therapy consist in local transluminal delivery of antiproliferative genes in the abluminal arterial smooth muscle cells in order to inhibit intimal hyperplasia [3, 59]. However, recent advances in the understanding of the pathophysiology of restenosis indicate that, both in experimental models [60] and in humans [61], chronic constrictive remodeling (or the absence of compensatory vessel enlargement) plays a major role in late luminal loss and that lumen diameter is strongly correlated to the magnitude of constriction but not to intimal thickness. The mechanism of constrictive remodeling, although still unclear, may involve biological changes in the adventitia including inflammation, myocellular proliferation and migration, and fibrosis [62]. Should constrictive remodeling become a target for antiproliferative gene therapy, maybe periadventitial gene delivery will be more appropriate than current intraluminal delivery techniques. Adventitial transfection, however, is likely to induce unacceptable arterial trauma which may outweigh the potential benefit of the therapy. Alternatively, endovascular stents are both efficient and safe to prevent arterial constriction. It has been convincingly demonstrated that in-stent restenosis, which occurs in roughly 25% of the patients, is almost exclusively composed of intimal hyperplasia [63]. Based on these recent data, the concept of integrated strategies to prevent restenosis arose, in which anti-remodeling stents are combined with antiproliferative genes to achieve optimal prevention of restenosis [64]. If integrated strategies are to be applied to the prevention of restenosis, we should consider developing in-stent gene delivery techniques as a priority [65].

The first clinical use of gene therapy to cure cardiovascular diseases in humans, however, involved the delivery of genes encoding angiogenic growth factors in patients suffering from end-stage peripheral artery disease [66]. In this strategy, termed therapeutic angiogenesis, the physiological process of angiogenesis in response to ischemia [67], is accelerated by local delivery of a plasmid expressing a secreted form of the angiogenic growth factor VEGF. Whether the best route for gene transfer in the ischemic skeletal muscle is direct intramuscular injection or percutaneous, transarterial delivery still remains debated [68]. It is also unknown whether the promising results reported in peripheral artery disease can be extended to the far more challenging issue of coronary artery disease [54]. Positive results of therapeutic angiogenesis have been obtained in animal models of myocardial ischemia using gene vectors expressing VEGF as well as genes from the fibroblast growth factor (FGF) family. In some of these studies, successful gene transfer was associated with increased collateral vessels and improved left ventricular function [43]. Surprisingly, even the low transfection efficiency typically achieved using single administration of "naked" DNA seems to be sufficient to bring about a localized and prolonged increase of VEGF or FGF in the ischemic myocardium. These observations paved the way for currently ongoing phase I and II clinical trials of myocardial angiogenesis [54]. The results of these trials will be available within 2 to 3 years. Until then, it is critical to remember that clinical applications of gene transfer techniques to the field of cardiovascular diseases is still in its infancy.

REFERENCES

1. Ross R. The pathogenesis of atherosclerosis: a perspective for the 1990s. *Nature*. 1993;362:801-809.

2. Gibbons GH, Dzau VJ. Molecular therapies for vascular diseases. *Science*. 1996;272:689-693.

3. Nabel EG. Gene therapy for cardiovascular disease. *Circulation*. 1995;91:541-548.

4. Isner JM, Feldman LJ. Gene therapy for arterial disease. *Lancet*. 1994;344:1653-4.

5. Riessen R, Rahimizadeh H, Takeshita S, Gal D, Barry JJ, Isner JM. Successful vascular gene transfer using a hydrogel coated balloon angioplasty catheter. *Hum Gene Ther*. 1993;4:749-758.

6. Mulligan RC. The basic science of gene therapy. *Science*. 1993;260:926-932.

7. Nabel EG, Plautz G, Nabel GJ. Site-specific gene expression in vivo by direct gene transfer into the arterial wall. *Science*. 1990;249:1285-1288.

8. Berkner KL. Expression of heterologous sequences in adenoviral vectors. *Curr. Top. Microbiol. Immunol*. 1992;158:39-66.

9. Steg PG, Feldman LJ, Scoazec JY, Tahlil O, Barry JJ, Boulechfar S, Ragot T, Isner JM, Perricaudet M. Arterial gene transfer to rabbit endothelial and smooth muscle cells using percutaneous delivery of an adenoviral vector. *Circulation*. 1994;90:1648-56.

10. Tripathy SK, Black HB, Goldwasser E, Leiden JM. Immune responses to transgene-encoded proteins limit the stability of gene expression after injection of replication-defective adenovirus vectors. *Nature Med*. 1996;2:545-550.

11. Yang Y, Nunes FA, Berencsi K, Furth EE, Gönczöl E, Wilson JM. Cellular immunity to viral antigens limits E1-deleted adenoviruses for gene therapy. *Proc Natl Acad Sci USA*. 1994;91:4407-4411.

12. Schulick AH, Newman KD, Virmani R, Dichek DA. In vivo gene transfer into injured carotid arteries. Optimization and evaluation of acute toxicity. *Circulation*. 1995;91:2407-2414.

13. Feldman LJ, Pastore CJ, Aubailly N, Kearney M, Chen D, Perricaudet M, Steg PG, Isner JM. Improved efficiency of arterial gene transfer by use of poloxamer 407 as a vehicle for adenoviral vectors. *Gene Ther*. 1997;4:189-98.

14. Newman KD, Dunn PF, Owens JW, Schulick AH, Virmani R, Sukhova G, Libby P, Dichek DA. Adenovirus-mediated gene transfer into normal rabbit arteries results in prolonged vascular cell activation, inflammation, and neointimal hyperplasia. *J Clin Invest*. 1995;96:2955-2965.

15. Van Ginkel F, Liu C, Simecka J, Dong J, Greenway T, Frizzel R, Kiyono H, McGhee J, Pascual D. Intratracheal gene delivery with adenoviral vector induces elevated systemic IgG and mucosal IgA antibodies to adenovirus and ß-galactosidase. *Hum Gene Ther*. 1995;6:895-903.

16. Stratford-Perricaudet LD, Makeh I, Perricaudet M, Briand P. Widespread long-term gene transfer to mouse skeletal muscles and heart. *J Clin Invest*. 1992;90:626-630.

17. Vilquin J-T, Guérette B, Kinoshita I, Roy B, Goulet M, Gravel C, Roy R, Tremblay JP. FK506 immunosuppression to control the immune reactions triggered by first-generation adenovirus-mediated gene transfer. *Hum Gene Ther*. 1995;6:1391-1401.

18. Engelhardt JF, Ye X, Doranz B, Wilson JM. Ablation of E2A in recombinant adenoviruses improves transgene persistence and decreases inflammatory response in mouse liver. *Proc Natl Acad Sci USA*. 1994;91:6196-6200.

19. Yeh P, Dedieu J-F, Orsini C, Vigne E, Denefle P, Perricaudet P. Efficient dual transcomplementation of adenovirus E1 and E4 regions from a 293-derived cell line expressing a minimal E4 functional unit. *J Virol*. 1996;70:559-565.

20. Wang Q, Finer MH. Second-generation adenovirus vectors. *Nature Med*. 1996;2:714-716.

21. Felgner PL, Ringold GM. Cationic liposome-mediated transfection. *Nature*. 1989;337:387-388.

22. Leclerc G, Gal D, Takeshita S, Nikol S, Weir L, Isner JM. Percutaneous arterial gene transfer in a rabbit model. Efficiency in normal and balloon-dilated atherosclerotic arteries. *J Clin Invest*. 1992;90:936-944.

23. Michael SI, Curiel DT. Strategies to achieve targeted gene delivery via the receptor-mediated endocytosis pathway. *Gene Therapy*. 1994;1:223-232.

24. Morishita R, Gibbons G, Kaneda Y, Ogihara T, Dzau V. Novel and effective gene transfer technique for study of vascular renin angiotensin system. *J Clin Invest*. 1993;91:2580-2585.

25. von der Leyen HE, Gibbons GH, Morishita R, Lewis NP, Zhang L, Nakajima M, Kaneda Y, Cooke JP, Dzau VJ. Gene therapy inhibiting neointimal vascular lesion: in vivo transfer of endothelial cell nitric oxide synthase gene. *Proc Natl Acad Sci USA*. 1995;92:1137-1141.

26. Morishita R, Gibbons GH, Ellison KE, Nakajima M, von der Leynen H, Zhang L, Kaneda Y, Ogihara T, Dzau VJ. Intimal hyperplasia after vascular injury is inhibited by antisense cdk 2 kinase oligonucleotides. *J Clin Invest*. 1994;93:1458-1464.

27. Morishita R, Gibbons GH, Horiuchi M, Ellison KE, Nakajima M, Zhang L, Kaneda Y, Ogihara T, Dzau VJ. A gene therapy strategy using a transcription factor decoy of the E2F binding site inhibits smooth muscle proliferation in vivo. *Proc Natl Acad Sci USA*. 1995;92:5855-5859.

28. Riessen R, Isner JM. Prospects for site-specific delivery of pharmacologic and molecular therapies. *J Am Coll Cardiol*. 1994;23:1234-1244.

29. Tahlil O, Brami M, Feldman LJ, Branellec D, Steg PG. The Dispatch catheter as a delivery tool for arterial gene transfer. *Cardiovasc Res*. 1997;33:181-7.

30. Isner JM, Pieczek A, Schainfield R, Blair R, Haley L, Asahara T, Rosenfield K, Razvi S, Walsh K, Symes JF. Clinical evidence of angiogenesis after arterial gene transfer of phVEGF165 in patient with ischaemic limb. *Lancet*. 1996;348:370-374.

31. Wolinsky H, Thung SN. Use of a perforated balloon catheter to deliver concentrated heparin into the wall of the normal canine artery. *J Am Coll Cardiol*. 1990;15:475-481.

32. Lambert CR, Leone JE, Rowlands SM. Local drug delivery catheters: functional comparisson of porous and microporous designs. *Cor Art Dis*. 1993;4:469-475.

33. Feldman LJ, Steg PG, Zheng LP, Chen D, Kearney M, McGarr SE, Barry JJ, Dedieu JF, Perricaudet M, Isner JM. Low-efficiency of percutaneous adenovirus-mediated arterial gene transfer in the atherosclerotic rabbit. *J Clin Invest*. 1995;95:2662-71.

34. Lincoff AM, Topol EJ, Ellis SG. Local drug delivery for the prevention of restenosis. Fact, fancy, and future. *Circulation*. 1994;90:2070-2084.

35. Dichek DA, Neville RF, Zwiebel JA, Freeman SM, Leon MB, Anderson WF. Seeding of intravascular stents with genetically engineered endothelial cells. *Circulation*. 1989;80:1347-1353.

36. Dev V, Zeng H, Forrester JS, Eigler N, Tian Y, Hickey AJ, Fishbein MC, Kupfer J, Litvak F. Microspheres for drug delivery to the arterial wall: a study of kinetics, toxicity, and effects of corticosteroid loaded microspheres. *J Am Coll Cardiol*. 1994;23:19A. Abstract.

37. Slepian MJ, Sawhney A, Pathak CP, Khosravi F, Roth L, Massia SP, Hubbell JA. In situ photo-polymerized thin hydrogel barriers applied following arterial injury reduce intimal thickening. *J Am Coll Cardiol*. 1994;23:473A. Abstract.

38. Indolfi C, Avvedimento EV, Rapacciuolo A, Di Lorenzo E, Esposito G, Stabile E, Mele E, Giuliano P, Condorelli GL, Chiariello M. Inhibition of cellular *ras* prevents smooth muscle cell proliferation after vascular injury *in vivo*. *Nature Med*. 1995;1:541-545.

39. Simons M, Edelman ER, DeKeyser J-L, Langer R, Rosenberg RD. Antisense c-myb oligonucleotides inhibit intimal arterial smooth muscle cell accumulation in vivo. *Nature*. 1992;359:67-70.

40. Wolff JA, Malone RW, Williams P, Chong W, Acsadi G, Jani A, Felgner PL. Direct gene transfer into mouse muscle in vivo. *Science.* 1990;247:1465-1468.

41. French BA, Mazur W, Geske RS, Bolli R. Direct in vivo gene transfer into porcine myocardium using replication-deficient adenoviral vectors. *Circulation.* 1994;90:2414-2424.

42. Barr E, Carroll J, Kalynych AM, Tripathy SK, Kozarsky K, Wilson JM. Efficient catheter-mediated gene transfer into the heart using replication-defective adenovirus. *Gene Ther.* 1994;1:51-58.

43. Giordano FJ, Ping P, McKirnan MD, Nozaki S, DeMaria AN, Dillman WH, Mathieu-Costello O, Hammond HK. Intracoronary gene transfer of fibroblast growth factor-5 increases blood flow and contractile function in an ischemic region of the heart. *Nature Med.* 1996;2:534-539.

44. Logeart D, Hatem SN, Inamo J, Haddada H, Perricaudet M, Mercadier JJ. Increased efficiency of gene transfer to cardiac myocytes with recombinant adenovirus by altering the endothelial barrier permeability. *Circulation.* 1996;94(suppl I):I-590. Abstract.

45. Nabel EJ, Plautz G, Boyce DM, Stanley JC, Nabel GJ. Recombinant gene expression in vivo within endothelial cells of the arterial wall. *Science.* 1989;244:1342-1344.

46. Flugelman MY, Jaklitsch MT, Newman KD, Casscells W, Brathauer GL, Dichek DA. Low level in vivo gene transfer into the arterial wall through a perforated balloon catheter. *Circulation.* 1992;85:1110-1117.

47. Lemarchand P, Jones M, Yamada I, Crystal RG. In vivo gene transfer and expression in normal uninjured blood vessels using replication-deficient recombinant adenovirus vectors. *Circ Res.* 1993;72:1132-1138.

48. Schulick AH, Dong G, Newman KD, Virmani R, Dichek DA. Endothelium-specific in vivo gene transfer. *Circ Res.* 1995;77:475-485.

49. Rome JJ, Shayani V, Flugelman MY, Newman KD, Farb A, Virmani R, Dichek DA. Anatomic barriers influence the distribution of in vivo gene transfer into the arterial wall. Modeling with microscopic tracer particles and verification with a recombinant adenoviral vector. *Arterioscler Thromb.* 1994;14:148-161.

50. Maillard L, Ziol M, Tahlil O, Le Feuvre C, Feldman LJ, Branellec D, Bruneval P, Steg PG. Pre-treatment with elastase improves the efficiency of percutaneous adenovirus-mediated gene ransfer to the arterial media. *Gene Ther.* 1998. In press.

51. Bohl D, Naffakh N, Heard J-M. Long-term control of erythropoietin secretion by doxyxycline in mice transplanted with engineered primary myoblasts. *Nature Med.* 1997;3:299-305.

52. Nabel EJ, Yang Z, Liptay S, San H, Gordon D, Haudenschild CC, Nabel GJ. Recombinant platelet-derived growth factor B gene expression in porcine arteries induces intimal hyperplasia in vivo. *J. Clin. Invest.* 1993;91:1822-1829.

53. Isner JM, Walsh K, Symes J, Pieczek A, Takeshita S, Lowry J, Rossow S, Rosenfield K, Weir L, Brogi E, Schainfeld R. Arterial gene therapy for therapeutic angiogenesis in patients with peripheral artery disease. *Circulation.* 1995;91:2687-2691.

54. Lee JS, Feldman AM. Gene therapy for therapeutic myocardial angiogenesis: a promising synthesis of two emerging technologies. *Nature Med.* 1998;4:739-742.

55. Chen S-J, Wilson JM, Muller DWM. Adenovirus-mediated gene transfer of soluble vascular cell adhesion molecule to porcine interposition vein grafts. *Circulation.* 1994;89:1922-1928.

56. Feldman LJ, Isner JM. Gene therapy for the vulnerable plaque. *J Am Coll Cardiol.* 1995;26:826-35.

57. Lafont A, Guérot C, Lemarchand P. Prospects for gene therapy in cardiovascular disease. *Eur Heart J.* 1996;17:1312-1317.

58. Clowes AW, Reidy MA, Clowes MM. Mechanisms of stenosis after arterial injury. *Lab Invest.* 1983;49:208-215.

59. Feldman LJ, Tahlil O, Steg PG. Perspectives of arterial gene therapy for the prevention of restenosis. *Cardiovasc Res.* 1996;32:194-207.

60. Lafont A, Guzman LA, Whitlow PL, Goormastic M, Cornhill JF, Chisolm GM. Restenosis after experimental angioplasty. Intimal, medial, and adventitial changes associated with constrictive remodeling. *Circ Res.* 1995;76:996-1002.

61. Mintz GS, Popma JJ, Pichard AD, Kent KM, Satler LF, Wong SC, Hong MK, Kovach JA, Leon MB. Arterial remodeling after coronary angioplasty. A serial intravascular ultrasound study. *Circulation.* 1996;94:35-43.

62. Shi Y, Pieniek M, Fard A, O'Brien J, Mannion JD, Zalewski A. Adventitial remodeling after coronary arterial injury. *Circulation.* 1996;93:340-348.

63. Hoffmann R, S MG, Dussaillant GR, Popma JJ, Pichard AD, Satler LF, Kent KM, Griffin J, Leon MB. Patterns and mechanisms of in-stent restenosis. A serial intravascular ultrasound study. *Circulation.* 1996;94:1247-1254.

64. Lafont A, Guerot C, Lemarchand P. Which gene for which restenosis? *Lancet.* 1995;346:1442-1443.

65. Van Belle E, Maillard L, Rivard A, Fabre JE, Couffinhal T, Kearney M, Branellec D, Feldman LF, Walsh K, Isner JM. Effects of poloxamer 407 on transfection time and percutaneous adenovirus-mediated gene transfer in native and stented vessels. *Hum Gene Ther.* 1998;9:1013-1024.

66. Baumgartner I, Pieczek AM, Blair R, Manor O, Walsh K, Isner JM. Evidence of therapeutic angiogenesis in patients with critical limb ischemia after intramuscular phVEGF165 gene transfer. *Circulation.* 1997;96:I-32. Abstract.

67. Schaper W, Ito WD. Molecular mechanisms of coronary collateral vessel growth. *Circ Res.* 1996;79:911-919.

68. Tsurumi Y, Takeshita S, Chen D, Kearney M, Rossow ST, Passeri J, Horowitz JR, Symes JF, Isner JM. Direct intramuscular gene transfer of naked DNA encoding vascular endothelial growth factor augments collateral development and tissue perfusion. *Circulation.* 1996;94:3281-3290.

—A—

—B—

—C—

—D—

—E—

—F—

—G—

—H—

—I—

—L—

—M—

—N—

—O—

—P—

—R—

—S—

—T—

—V—